Synthesis Lectures on Human Language Technologies

Series Editor

Graeme Hirst, Department of Computer Science, University of Toronto, Toronto, Canada

The series publishes topics relating to natural language processing, computational linguistics, information retrieval, and spoken language understanding. Emphasis is on important new techniques, on new applications, and on topics that combine two or more HLT subfields.

Zekun Wang · Ge Zhang · Chenghua Lin ·
Jie Fu

Editors

Interactive Natural Language Processing

Language Model as Agent

Editors
Zekun Wang
Beihang University
Beijing, China

Ge Zhang
University of Waterloo
Waterloo, Canada

Chenghua Lin
The University of Manchester
Manchester, UK

Jie Fu
Hong Kong University of Science
and Technology
Clear Water Bay, Hong Kong

ISSN 1947-4040   ISSN 1947-4059 (electronic)
Synthesis Lectures on Human Language Technologies
ISBN 978-3-032-06263-5   ISBN 978-3-032-06264-2 (eBook)
https://doi.org/10.1007/978-3-032-06264-2

This Springer imprint is published by the registered company Springer Nature Switzerland AG
The registered company address is: Gewerbestrasse 11, 6330 Cham, Switzerland

If disposing of this product, please recycle the paper.

Preface

The purpose of writing this book is to introduce a new topic within the field of Natural Language Processing (NLP): Interactive NLP (iNLP). Although the term might suggest that the book focuses on the intersection of NLP and human-computer interaction, our definition of "interaction" is broader. We approach the subject from an NLP-centric perspective, examining the paradigm shifts in NLP following the advent of large language models (LLMs).

NLP is dedicated to studying how computers can process various forms of language. Early research in NLP primarily focused on rule-based methods, such as symbolic parsing and pattern matching. Since the early 2010s, with the maturation of machine learning and deep learning technologies, NLP has undergone significant paradigm shifts. This era saw the rise of techniques likeWord2Vec and GloVe, which focused on solving problems like word representation and semantic similarity. Subsequently, the emergence of Transformer-based pre-trained language models (PLMs) marked a pivotal moment in NLP's evolution. These models can be broadly categorized into two types: encoder-only models and sequence-to-sequence (seq2seq) models. Encoder-only models, represented by BERT, excel at understanding and encoding input text, making them highly effective for tasks like text classification, sentiment analysis, and information retrieval. In contrast, seq2seq models, such as T5, GPT, and BART, are designed to generate texts based on the understanding of the inputs, enabling them to perform tasks like machine translation, summarization, and question answering with remarkable flexibility.

The rise of these Transformer-based architectures not only expanded the range of tasks NLP could handle but also highlighted the potential of leveraging unified architectures for both understanding and generation. Building on these foundations, the field witnessed the emergence of large language models (LLMs), which primarily adopt decoder-only architectures. These models, exemplified by GPT-3, ChatGPT, GPT-4, and GPT-4V, focus on autoregressive generation, combining deep contextual understanding with the ability to generate fluent and coherent text. This paradigm shift has redefined NLP by enabling models to tackle more complex, human-level challenges, including mathematical reasoning, code generation, role-playing, embodied intelligence, and decision-making, marking a new era in natural language processing.

This book, situated in the era of LLMs, aims to provide a comprehensive review, summarizing the key technologies of the past and inspiring the next phase of NLP development beyond LLMs. This is one of the core motivations behind the writing of this book.

So, what will be the next paradigm for NLP after LLMs?

In fact, as early as 2020, Yoshua Bengio et al. outlined the stages of NLP development in his paper *"Experience Grounds Language."* He divided the evolution of NLP into five "world scopes": World Scope I (Corpus), World Scope II (Internet), World Scope III (Perception), World Scope IV (Embodiment), and World Scope V (Social).With the release of multimodal LLMs (MLLMs) like GPT-4V and Gemini, we believe that we are nearing the end of World Scope III: Perception, and are gradually moving towards the era of Embodiment and Social. If we view embodiment as the interaction between an intelligent agent and its environment, we can unify these two world scopes under the concept of Interactive NLP. This notion is one of the inspirations behind this book.

To systematically introduce the theme of Interactive NLP, we have conducted a comprehensive and complete breakdown. Centered around LLMs, the book first explains the basic definition of Interactive NLP by introducing the entities that can interact with LLMs and the language interfaces involved. We then discuss how to build LLMs and MLLMs, optimize them, and interact with external entities with these models, covering topics ranging from prompting and agents to data synthesis, supervised fine-tuning (SFT), and reinforcement learning, etc. Finally, we explore the evaluation methods in the scope of Interactive NLP, its application scenarios, ethical considerations, and future directions.

This book is primarily aimed at students, researchers, and practitioners in the field of LLMs, helping them understand the technological chain beyond LLMs, seek potential research inspiration, or find guides for expanding the functionality of LLMs. The book surveys nearly 1000 references, primarily covering developments up to July 2024, with some new content extending to June 2025. Therefore, it offers a relatively comprehensive overview of cutting-edge developments and technologies related to LLMs. However, it should be noted that the book's comprehensiveness lies in the breadth of the topics and technologies it covers, not in the specificity of its operational guidance. Thus, readers looking to learn practical coding guidances may need to consult other practice-focused books, as this book only provides high-level technical suggestions.

Finally, we would like to express our gratitude to Pengfei Liu for his constructive comments on this work. We also extend our thanks to Haoran Zhang, Yang Liu, Wenzhen Miao, and Iman Yeckehzaare for their insightful discussions during the early stages of this book. Additionally, we appreciate the efforts of Ziwei Zhu, Ziqiao Ma, Yichi Zhang, Renliang Sun, Xingran Chen, and Chenghao Xiao for their careful proofreading of the manuscript. On the publication side, we are deeply grateful to Graeme Hirst, the editor of *Synthesis Lectures on Human Language Technologies*, for his invaluable guidance throughout this project. We also wish to thank Susanne Filler, Executive Editor at Springer Nature, as well as Harini Ganapathi, Ashok Kumar, Boopalan Renu, and Lara Glueck, the production editors, for their timely and comprehensive assistance whenever required.

Beijing, China   Zekun Wang
June 2025

Contents

Contributors

Wenhu Chen Cheriton School of Computer Science, University of Waterloo, Waterloo, Canada

Xiuying Chen King Abdullah University of Science and Technology, Thuwal, Saudi Arabia

Jie Fu Hong Kong University of Science and Technology, Hong Kong, China; Beijing, China

Shaochun Hao Xi'an Jiaotong University, Hangzhou, China

Yizhi Li University of Manchester, Manchester, UK

Chenghua Lin Department of Computer Science, University of Manchester, Manchester, UK

Dayiheng Liu Sichuan University, Chengdu, China

Qi Liu City University of Hong Kong, G2328, Yeung Kin Man Academic Building, Kowloon Tong, Hong Kong, China

Ruibo Liu Dartmouth College, Department of Computer Science, Class of 1982 Engineering & Computer Science Center, Hanover, USA

Ning Shi University of Alberta, Edmonton, AB, Canada

Mong Yuan Sim University of Adelaide, Adelaide, SA, Australia

Zekun Wang Beihang University, Haidian District, Beijing, China

Guangzheng Xiong Haidian District, Beijing, China; Hong Kong University of Science and Technology, Hong Kong, China

Ke Xu Beihang University, Beijing, China

Kexin Yang Chengdu, China

Zhenzhu Yang Haidian District, Beijing, China

Ge Zhang University of Michigan, Haidian District, Beijing, China

Wangchunshu Zhou AI Waves, Hangzhou, Zhejiang, China

Qingqing Zhu Peking University, Beijing, China

In this part, we introduce the basics of Interactive Natural Language Processing (iNLP), covering its background, the current state of development, key definitions, the interactive entities, and the language interfaces that facilitate this interaction.

Interactive NLP is a paradigm that considers Large Language Models (LLMs) as agents capable of observing and interacting with external entities, and iteratively adjusting their behavior based on external feedback. The basic concept of iNLP involves identifying the external entities that an LLM agent can interact with and understanding how these interactions are facilitated and enhanced through various language interfaces.

The typical interactive entities in iNLP include humans, knowledge bases, models, tools, and environments. The language interfaces that enable interaction encompass natural language, formal language, representation language, edits, and shared memories.

We will provide a detailed discussion on how LLMs utilize these various interfaces and interact with different entities. Through these examples, we aim to elucidate the fundamental concepts and propositions of Interactive NLP.

Introduction

1

Zekun Wang and Chenghua Lin

Natural Language Processing (NLP) has witnessed a remarkable revolution in recent years, thanks to the development of generative pre-trained language models (PLMs) such as BART [1], T5 [2], GPT-3 [3], PaLM [4], to name a few. These models can generate coherent and semantically meaningful text, making them useful for various NLP tasks such as machine translation [5], summarization [6, 7], and question answering [2, 3, 8]. However, these models also have clear limitations such as misalignment with human needs [9, 10], lack of interpretability [11, 12], hallucinations [12–14], imprecise mathematical operations [15, 16], inadequate experience grounding [17], and limited ability for complex reasoning [18, 19], among others [20].

To address these limitations, a new paradigm of natural language processing has emerged: **interactive natural language processing (iNLP)** [17, 21]. There have been a variety of definitions for *"interactive"* in the NLP and Machine Learning literature, where the term typically refers to the involvement of humans in the process. For example, [22] define Interactive Machine Learning (iML) as *"an active machine learning technique in which models are designed and implemented with human-in-the-loop manner."* Faltings et al. [23] view Interactive Text Generation as *"a task that allows training generation models interactively without the costs of involving real users, by using user simulators that provide edits that guide the model towards a given target text."* Wang et al. [24] describe Human-in-the-loop (HITL) as *"where model developers continuously integrate human feedback into different*

Z. Wang (✉)
Beihang University, Beijing, China
e-mail: zenmoore@buaa.edu.cn

C. Lin
Department of Computer Science, University of Manchester, Manchester, UK
e-mail: chenghua.lin@manchester.ac.uk

© The Author(s), under exclusive license to Springer Nature Switzerland AG 2026
Z. Wang et al. (eds.), *Interactive Natural Language Processing*, Synthesis Lectures
on Human Language Technologies, https://doi.org/10.1007/978-3-032-06264-2_1

steps of the model deployment workflow." The popularity of ChatGPT[1] also demonstrated the impressive capabilities of human-LM interaction via reinforcement learning from human feedback (RLHF). Although humans are the most common type of objects for interacting with language models, recent research has revealed other important object types for interaction, which include Knowledge Bases (KBs) [25, 26], Models/Tools [16, 18, 27–30], and Environments [31–36]. Therefore, in our survey, we first define interactive natural language processing which accounts for a broader scope of objects that can interact with language models:

Interactive Natural Language Processing (iNLP) considers language models as agents capable of observing, acting, and receiving feedback in a loop with external objects such as humans, knowledge bases, tools, models, and environments.[2]

Specifically, through interaction, a language model (LM) can leverage external resources to improve its performance and address its limitations mentioned in the first paragraph. For example, interacting with humans aligns language models better with human needs and human values (e.g., helpfulness, harmlessness, honesty) [37, 38] and interacting with KBs can help language models alleviate hallucinations [14]. Likewise, interacting with models or tools can improve the abilities of LMs such as reasoning, faithfulness, and exactitude of mathematical operations [15, 16]. And finally, interacting with environments can enhance the grounded reasoning capability of LMs [39] and promote the applications of LMs in embodied tasks [32, 40].

Furthermore, interaction may hold the potential to unlock future milestones in language processing, which can be considered the holy grail of artificial intelligence [17]. In 2020, [17] have examined the future direction of natural language processing and proposed five levels of world scope to audit progress in NLP: "(1) Corpus; (2) Internet; (3) Perception (multimodal NLP); (4) Embodiment; (5) Social." Notably, the recent release of GPT-4 [12] and PaLM-2 [41], which are large multimodal language models, has brought significant advancements to the third level "Perception". Embodied AI and Social Embodied AI fundamentally posit that a more comprehensive language representation can be learned through the establishment of an interactive loop involving language model agents, environments, and humans [17, 21, 42–46]. This perspective highlights the need for the NLP community to shift its attention towards the fourth and fifth levels ("Embodiment" and "Social Interaction") to propel the field forward. In addition to models, humans, and environments, tools and knowledge bases that facilitate connections between language models and the external world also play a significant role in enabling (social) embodiment [17, 30, 47, 48]. The future achievement of social embodiment of language models may lead to significant phenomena, including artificial self-awareness [36, 49] and the emergence of a language model society [50, 51].

---

[1] https://openai.com/blog/chatgpt.

[2] **Observation** involves all kinds of inputs to language models. **Action** involves all kinds of outputs of language models such as text generation [37], requesting for external objects [15, 28], text editing [23], etc. **Feedback** involves feedback messages passed from external objects to language models such as scoring from humans [37].

Therefore, interactive NLP (iNLP) is beneficial for both NLP researchers and practition-ers, since it has the potential to address limitations such as hallucination [14] and align-ment [9], while also aligning with the ultimate goals of AI [17, 30, 36]. Notably, with the recent release of ChatGPT and GPT-4 [12], which have overwhelmed the NLP community and are considered the spark of artificial general intelligence (AGI) by some researchers due to their remarkable universal capabilities [36], the NLP community is now experiencing a shift in focus towards posing new challenges in the field. This transition has prompted numerous surveys and position papers that aim to propose novel research directions, with many of them addressing the theme of interaction. For example, [16] survey the strategies that PLMs employ cascading mechanisms for reasoning [27] and utilize tools for taking action. But [16] lacks an in-depth discussion on interactivity, and focuses solely on tool use and reasoning, while overlooking other topics such as interaction with knowledge bases, and simulation of social behavior. Yang et al. [32] investigate the cross-disciplinary research field of foundation models and decision making, with a particular emphasis on exploring the interactions of language models with humans, tools, agents, and environments. But they primarily focus on decision-making settings and reinforcement learning formalisms, without providing a comprehensive discussion on interacting with knowledge bases or the interaction methodology from the perspective of NLP techniques, such as chain-of-thought prompting [52]. Bubeck et al. [36] discuss the interactions of language models with the world based on tool-use and embodiment, as well as their interactions with humans based on Theory of Mind (ToM) and self-explanation. But they primarily focus on evaluating the abilities of large language models (LLMs) and lack a comprehensive discussion of the interaction methodology employed in the studies. Other surveys and works [30, 35, 46, 53] have also contributed valuable insights to the theme of interaction. However, they are also specific to certain aspects and do not offer a unified and systematic review that covers the entire spectrum of interactive NLP.

Clearly, the field of interactive NLP has undergone significant development in the past few years, with the emergence of new forms of interactive objects that go beyond the stan-dard Human-in-the-loop approach. These new forms of objects encompass knowledge bases, models/tools, and environments. While the aforementioned works provide some coverage of interactions involving models/tools and environments, there is a notable absence of dis-cussions regarding interactions with language models using knowledge bases (KB). Fur-thermore, there is a lack of a comprehensive review of methodologies in the context of interactive NLP. Hence, the main goals of our survey are:

1. **Unified Definition and Formulation**: to provide a unified definition and formulation of interactive NLP, establishing it as a new paradigm of NLP.
2. **Comprehensive Classification**: to provide a comprehensive breakdown of iNLP along dimensions such as interactive objects, interaction interfaces, and interaction methods, enabling a systematic understanding of its different aspects and components.

3. **Further Discussion**: to survey the evaluation methodologies used in iNLP, examine its diverse applications, and discuss the ethical and safety issues as well as the future directions in this field.

We believe that conducting such a survey is highly timely, and our paper aims to fill the gaps of aforementioned surveys by serving as an entry point for researchers who are interested in pursuing research in this important and fast-evolving area but may not yet be familiar with it. As illustrated in Fig. 1.1, we will start with an in-depth discussion about interactive objects (Chap. 2), followed by an overview of interaction interfaces by which the language models communicate with the external objects (Chap. 3). We then organize a variety of interaction methods by which the language models fan in and out interaction messages (Part II). This is followed by a discussion about evaluation in the context of iNLP

(a) Interacting with Humans.   (b) Interacting with Knowledge Bases.

(c) Interacting with Models and Tools[a].   (d) Interacting with Environments.

Fig. 1.1 The paradigm of interactive natural language processing (Self-interaction is also included)

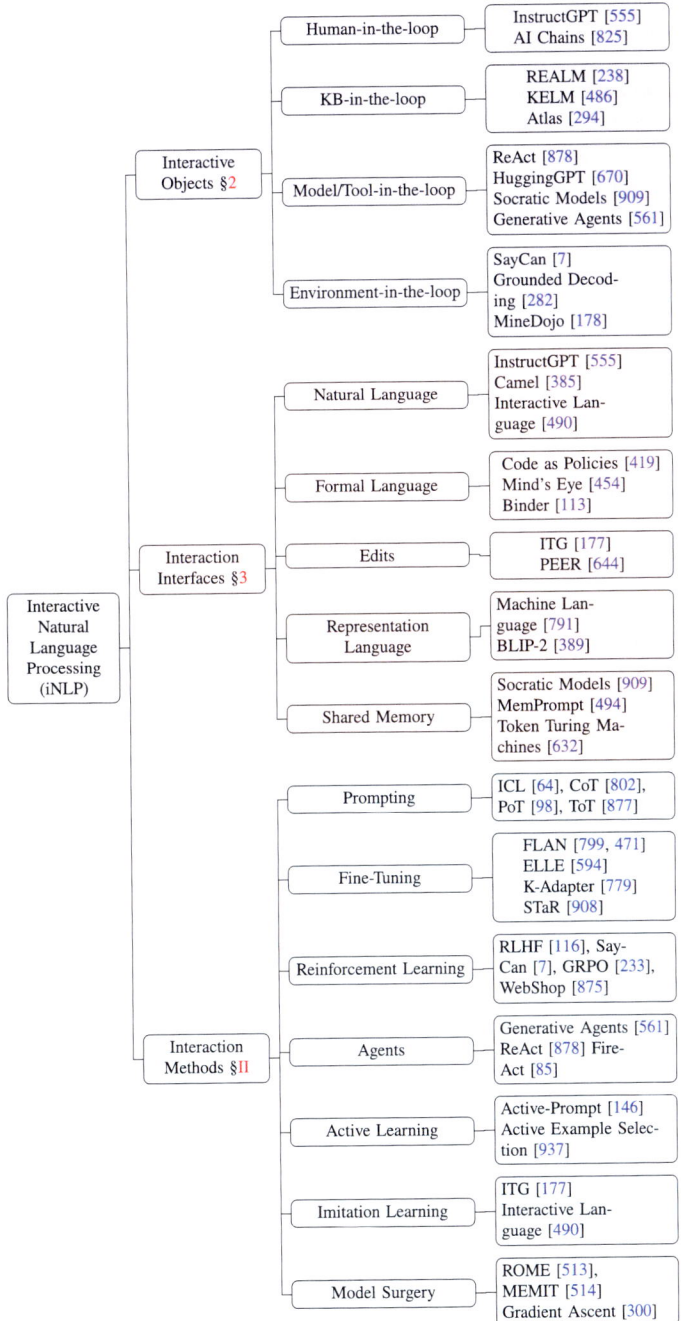

Fig. 1.2 Taxonomy of interactive NLP (iNLP)

(Chap. 10). Finally, we will examine the current applications of iNLP (Chap. 11), discuss ethical and safety issues (Chap. 12), and suggest future directions and challenges (Chap. 13). Taxonomy 1.2 gives a bird's-eye view of our book.

References

1. Lewis, M., Liu, Y., Goyal, N., Ghazvininejad, M., Mohamed, A., Levy, O., Stoyanov, V., Zettle-moyer, L.: Bart: Denoising sequence-to-sequence pre-training for natural language generation, translation, and comprehension (2019)
2. Raffel, C., Shazeer, N., Roberts, A., Lee, K., Narang, S., Matena, M., Zhou, Y., Li, W., Liu, P.J.: Exploring the limits of transfer learning with a unified text-to-text transformer. Journal of Machine Learning Research **21**(140), 1–67 (2020). http://jmlr.org/papers/v21/20-074.html
3. Brown, T., Mann, B., Ryder, N., Subbiah, M., Kaplan, J.D., Dhariwal, P., Neelakantan, A., Shyam, P., Sastry, G., Askell, A., et al.: Language models are few-shot learners. Advances in neural information processing systems **33**, 1877–1901 (2020)
4. Chowdhery, A., Narang, S., Devlin, J., Bosma, M., Mishra, G., Roberts, A., Barham, P., Chung, H.W., Sutton, C., Gehrmann, S., et al.: Palm: Scaling language modeling with pathways. arXiv preprint arXiv:2204.02311 (2022)
5. Liu, Y., Gu, J., Goyal, N., Li, X., Edunov, S., Ghazvininejad, M., Lewis, M., Zettlemoyer, L.: Multilingual denoising pre-training for neural machine translation. arXiv preprint arXiv:2001.08210 (2020)
6. Liu, Y.: Fine-tune bert for extractive summarization. ARXIV (2019)
7. Liu, Y., Lapata, M.: Text summarization with pretrained encoders. Conference On Empirical Methods In Natural Language Processing (2019). https://doi.org/10.18653/v1/D19-1387
8. Radford, A., Wu, J., Child, R., Luan, D., Amodei, D., Sutskever, I.: Language models are unsupervised multitask learners (2019)
9. Wolf, Y., Wies, N., Levine, Y., Shashua, A.: Fundamental limitations of alignment in large language models. arXiv preprint arXiv:2304.11082 (2023)
10. Kenton, Z., Everitt, T., Weidinger, L., Gabriel, I., Mikulik, V., Irving, G.: Alignment of language agents. arXiv preprint arXiv:2103.14659 (2021)
11. Wu, T.S., Terry, M., Cai, C.J.: Ai chains: Transparent and controllable human-ai interaction by chaining large language model prompts. International Conference On Human Factors In Computing Systems (2021). https://doi.org/10.1145/3491102.3517582
12. OpenAI: GPT-4 technical report. PREPRINT (2023)
13. Welleck, S., Kulikov, I., Roller, S., Dinan, E., Cho, K., Weston, J.: Neural text generation with unlikelihood training (2019)
14. Ji, Z., Lee, N., Frieske, R., Yu, T., Su, D., Xu, Y., Ishii, E., Bang, Y., Madotto, A., Fung, P.: Survey of hallucination in natural language generation. CoRR **abs/2202.03629** (2022). https://arxiv.org/abs/2202.03629
15. Schick, T., Dwivedi-Yu, J., Dessì, R., Raileanu, R., Lomeli, M., Zettlemoyer, L., Canceddda, N., Scialom, T.: Toolformer: Language models can teach themselves to use tools. CoRR **abs/2302.04761** (2023). https://doi.org/10.48550/arXiv.2302.04761. https://doi.org/10.48550/arXiv.2302.04761
16. Mialon, G., Dessì, R., Lomeli, M., Nalmpantis, C., Pasunuru, R., Raileanu, R., Roziére, B., Schick, T., Dwivedi-Yu, J., Celikyilmaz, A., Grave, E., LeCun, Y., Scialom, T.: Augmented language models: a survey. arXiv preprint arXiv: Arxiv-2302.07842 (2023)

17. Bisk, Y., Holtzman, A., Thomason, J., Andreas, J., Bengio, Y., Chai, J., Lapata, M., Lazaridou, A., May, J., Nisnevich, A., Pinto, N., Turian, J.: Experience grounds language. arXiv preprint arXiv:2004.10151 (2020)

18. Qiao, S., Ou, Y., Zhang, N., Chen, X., Yao, Y., Deng, S., Tan, C., Huang, F., Chen, H.: Reasoning with language model prompting: A survey. arXiv preprint arXiv:2212.09597 (2022)

19. Huang, J., Chang, K.: Towards reasoning in large language models: A survey. ARXIV.ORG (2022). https://doi.org/10.48550/arXiv.2212.10403

20. Borji, A.: A categorical archive of ChatGPT failures. arxiv.org (2023). https://doi.org/10.48550/arXiv.2302.03494

21. Bolotta, S., Dumas, G.: Social neuro ai: Social interaction as the "dark matter" of ai. Frontiers in Computer Science **4** (2022). https://doi.org/10.3389/fcomp.2022.846440. https://www.frontiersin.org/articles/10.3389/fcomp.2022.846440

22. Wondimu, N.A., Buche, C., Visser, U.: Interactive machine learning: A state of the art review. arXiv preprint arXiv:2207.06196 (2022)

23. Faltings, F., Galley, M., Peng, B., Brantley, K., Cai, W., Zhang, Y., Gao, J., Dolan, B.: Interactive text generation. arXiv preprint arXiv:2303.00908 (2023)

24. Wang, Z.J., Choi, D., Xu, S., Yang, D.: Putting humans in the natural language processing loop: A survey. arXiv preprint arXiv:2103.04044 (2021)

25. Li, H., Su, Y., Cai, D., Wang, Y., Liu, L.: A survey on retrieval-augmented text generation. arXiv preprint arXiv:2202.01110 (2022)

26. Hu, L., Liu, Z., Zhao, Z., Hou, L., Nie, L., Li, J.: A survey of knowledge-enhanced pre-trained language models. arXiv preprint arXiv:2211.05994 (2022)

27. Dohan, D., Xu, W., Lewkowycz, A., Austin, J., Bieber, D., Lopes, R.G., Wu, Y., 0Michalewski, H., Saurous, R.A., Sohl-dickstein, J., Murphy, K., Sutton, C.: Language model cascades. arXiv preprint arXiv:2207.10342 (2022)

28. Yao, S., Zhao, J., Yu, D., Du, N., Shafran, I., Narasimhan, K., Cao, Y.: React: Synergizing reasoning and acting in language models. arXiv preprint arXiv:2210.03629 (2022)

29. Shen, Y., Song, K., Tan, X., Li, D., Lu, W., Zhuang, Y.: Hugginggpt: Solving ai tasks with ChatGPT and its friends in huggingface. arXiv preprint arXiv:2303.17580 (2023)

30. Qin, Y., Hu, S., Lin, Y., Chen, W., Ding, N., Cui, G., Zeng, Z., Huang, Y., Xiao, C., Han, C., Fung, Y., Su, Y., Wang, H., Qian, C., Tian, R., Zhu, K., Liang, S., Shen, X., Xu, B., Zhang, Z., Ye, Y., Li, B., Tang, Z., Yi, J., Zhu, Y., Dai, Z., Yan, L., Cong, X., Lu, Y.T., Zhao, W., Huang, Y., Yan, J.H., Han, X., Sun, X., Li, D., Phang, J., Yang, C., Wu, T., Ji, H., Liu, Z., Sun, M.: Tool learning with foundation models. ARXIV.ORG (2023). https://doi.org/10.48550/arXiv.2304.08354

31. Li, S., Puig, X., Paxton, C., Du, Y., Wang, C., Fan, L., Chen, T., Huang, D.A., Akyürek, E., Anandkumar, A., et al.: Pre-trained language models for interactive decision-making. Advances in Neural Information Processing Systems **35**, 31199–31212 (2022)

32. Yang, S., Nachum, O., Du, Y., Wei, J., Abbeel, P., Schuurmans, D.: Foundation models for decision making: Problems, methods, and opportunities. arXiv preprint arXiv:2303.04129 (2023)

33. Ahn, M., Brohan, A., Brown, N., Chebotar, Y., Cortes, O., David, B., Finn, C., Gopalakrishnan, K., Hausman, K., Herzog, A., et al.: Do as i can, not as i say: Grounding language in robotic affordances. arXiv preprint arXiv:2204.01691 (2022)

34. Huang, W., Xia, F., Xiao, T., Chan, H., Liang, J., Florence, P., Zeng, A., Tompson, J., Mordatch, I., Chebotar, Y., et al.: Inner monologue: Embodied reasoning through planning with language models. arXiv preprint arXiv:2207.05608 (2022)

35. Vemprala, S., Bonatti, R., Bucker, A., Kapoor, A.: ChatGPT for robotics: Design principles and model abilities. Tech. Rep. MSR-TR-2023-8, Microsoft (2023). https://www.microsoft.com/en-us/research/publication/chatgpt-for-robotics-design-principles-and-model-abilities/

36. Bubeck, S., Chandrasekaran, V., Eldan, R., Gehrke, J., Horvitz, E., Kamar, E., Lee, P., Lee, Y.T., Li, Y., Lundberg, S., Nori, H., Palangi, H., Ribeiro, M.T., Zhang, Y.: Sparks of artificial general intelligence: Early experiments with GPT-4 (2023). https://www.microsoft.com/en-us/research/publication/sparks-of-artificial-general-intelligence-early-experiments-with-gpt-4/

37. Ouyang, L., Wu, J., Jiang, X., Almeida, D., Wainwright, C.L., Mishkin, P., Zhang, C., Agarwal, S., Slama, K., Ray, A., et al.: Training language models to follow instructions with human feedback. arXiv preprint arXiv:2203.02155 (2022)

38. Bai, Y., Jones, A., Ndousse, K., Askell, A., Chen, A., DasSarma, N., Drain, D., Fort, S., Ganguli, D., Henighan, T., Joseph, N., Kadavath, S., Kernion, J., Conerly, T., El-Showk, S., Elhage, N., Hatfield-Dodds, Z., Hernandez, D., Hume, T., Johnston, S., Kravec, S., Lovitt, L., Nanda, N., Olsson, C., Amodei, D., Brown, T., Clark, J., McCandlish, S., Olah, C., Mann, B., Kaplan, J.: Training a helpful and harmless assistant with reinforcement learning from human feedback. arXiv preprint arXiv:2204.05862 (2022)

39. Liu, R., Wei, J., Gu, S.S., Wu, T.Y., Vosoughi, S., Cui, C., Zhou, D., Dai, A.M.: Mind's eye: Grounded language model reasoning through simulation. arXiv preprint arXiv:2210.05359 (2022)

40. Zeng, A., Attarian, M., Ichter, B., Choromanski, K., Wong, A., Welker, S., Tombari, F., Purohit, A., Ryoo, M., Sindhwani, V., Lee, J., Vanhoucke, V., Florence, P.: Socratic models: Composing zero-shot multimodal reasoning with language. arXiv preprint arXiv:2204.00598 (2022)

41. Google: Palm 2 (2023). https://ai.google/discover/palm2

42. Bandura, A.: Social learning theory. Prentice-Hall. https://books.google.co.jp/books?id=mjpbjgEACAAJ

43. Tamari, R., Shani, C., Hope, T., Petruck, M.R.L., Abend, O., Shahaf, D.: Language (re)modelling: Towards embodied language understanding. In: Proceedings of the 58th Annual Meeting of the Association for Computational Linguistics, pp. 6268–6281. Association for Computational Linguistics, Online (2020). https://doi.org/10.18653/v1/2020.acl-main.559. https://aclanthology.org/2020.acl-main.559

44. Lake, B., Ullman, T., Tenenbaum, J., Gershman, S.: Building machines that learn and think like people. Behavioral And Brain Sciences (2016). https://doi.org/10.1017/S0140525X16001837

45. Driess, D., Xia, F., Sajjadi, M.S.M., Lynch, C., Chowdhery, A., Ichter, B., Wahid, A., Tompson, J., Vuong, Q., Yu, T., Huang, W., Chebotar, Y., Sermanet, P., Duckworth, D., Levine, S., Vanhoucke, V., Hausman, K., Toussaint, M., Greff, K., Zeng, A., Mordatch, I., Florence, P.: Palm-e: An embodied multimodal language model. In: arXiv preprint arXiv:2303.03378 (2023)

46. Yuan, L., Zhu, S.C.: Communicative learning: A unified learning formalism. Engineering (2023). https://doi.org/10.1016/j.eng.2022.10.017. https://www.sciencedirect.com/science/article/pii/S2095809923001339

47. Xie, T., Wu, C.H., Shi, P., Zhong, R., Scholak, T., Yasunaga, M., Wu, C.S., Zhong, M., Yin, P., Wang, S.I., et al.: Unifiedskg: Unifying and multi-tasking structured knowledge grounding with text-to-text language models. arXiv preprint arXiv:2201.05966 (2022)

48. Weser, V., Proffitt, D.R.: Tool embodiment: The tool's output must match the user's input. Frontiers in Human Neuroscience 12 (2019). https://www.frontiersin.org/articles/10.3389/fnhum.2018.00537. Original Research

49. Kosinski, M.: Theory of mind may have spontaneously emerged in large language models. arXiv preprint arXiv:2302.02083 (2023)

50. Park, J.S., O'Brien, J.C., Cai, C.J., Morris, M.R., Liang, P., Bernstein, M.S.: Generative agents: Interactive simulacra of human behavior. arXiv preprint arXiv:2304.03442 (2023)

51. Li, G., Hammoud, H.A.A.K., Itani, H., Khizbullin, D., Ghanem, B.: Camel: Communicative agents for "mind" exploration of large scale language model society. arXiv preprint arXiv:2303.17760 (2023)

52. Wei, J., Wang, X., Schuurmans, D., Bosma, M., Chi, E.H., Le, Q., Zhou, D.: Chain of thought prompting elicits reasoning in large language models. CoRR **abs/2201.11903** (2022). https://arxiv.org/abs/2201.11903

53. Lee, M., Srivastava, M., Hardy, A., Thickstun, J., Durmus, E., Paranjape, A., Gerard-Ursin, I., Li, X.L., Ladhak, F., Rong, F., Wang, R.E., Kwon, M., Park, J.S., Cao, H., Lee, T., Bommasani, R., Bernstein, M.S., Liang, P.: Evaluating human-language model interaction. CoRR **abs/2212.09746** (2022). https://doi.org/10.48550/arXiv.2212.09746. https://doi.org/10.48550/arXiv.2212.09746

Interaction Objects

Zekun Wang and Wenhu Chen

2.1 Human-in-the-Loop

Human-in-the-loop NLP represents a paradigm that emphasizes information exchange between humans and language models [1]. This approach seeks to more effectively address users' needs and uphold human values, a concept known as Human-LM Alignment [2–5]. In contrast, earlier research on text generation primarily concentrated on the input and output of samples, overlooking aspects such as human preferences, experiences, personalization, diverse requirements, and the actual text generation process [6]. In recent years, as pre-trained language models (PLMs) and large language models (LLMs) have matured, optimizing human-model interactions has emerged as a prevalent concern within the community. Incorporating human prompts, feedback, or configurations during the model training or inference stages, using either real or simulated users, proves to be an effective strategy for enhancing the Human-LM alignment [4, 7, 8] (Fig. 2.1).

Subsequently, we divide human-in-the-loop NLP into three types according to the schemes of user interaction, along with an additional section that delves into the simulation of human behaviors and preferences for these types, in order to enable scalable deployment of human-in-the-loop systems. These categories are:

Z. Wang (✉)
Beihang University, Beijing, China
e-mail: zenmoore@buaa.edu.cn

W. Chen
Cheriton School of Computer Science, University of Waterloo, Waterloo, Canada
e-mail: wenhuchen@uwaterloo.ca

© The Author(s), under exclusive license to Springer Nature Switzerland AG 2026
Z. Wang et al. (eds.), *Interactive Natural Language Processing*, Synthesis Lectures
on Human Language Technologies, https://doi.org/10.1007/978-3-032-06264-2_2

Fig. 2.1 Human-in-the-loop

1. Communicating with Human Prompts: users can interact with the model consecutively in a conversation.
2. Learning from Human Feedback: users can provide feedback to update the parameters of LMs.
3. Regulating via Human Configuration: users can configure the settings of LMs.
4. Learning from Human Simulation: simulations of users are employed for the three afore-mentioned types, ensuring practical implementation and scalability.

Communicating with Human Prompts. This is the most general form of Human-LM interaction, which allows a language model to interact with a human in a conversational manner. The main purpose of this interaction scheme is to maintain real-time and continuous interaction, with typical application scenarios including dialogue systems, real-time translation, and multiple rounds of question answering [9, 10]. This interactive process of alternating iterations allows the output of the model to realign gradually to meet user requirements.

Generally, this interaction scheme does not update the model's parameters during the interaction, instead requiring users to continuously input or update prompts to elicit more meaningful responses from the language model. As a result, conversation can be inflexible and labor-intensive due to the need for prompt engineering or dialogue engineering. To address these limitations, editing-based methods have been proposed by [7, 11–14] to encourage the language model to modify existing output (c.f., Sect. 3.3). Additionally, context-based methods have been developed that enhance model output by adding examples or instructions to the input context, such as few-shot prompting or in-context learning [15].

However, since these approaches do not involve adapting language models to accommodate human users, numerous trial edits or prompts may be required to achieve the desired outcome, resulting in lengthier dialogue rounds. As such, this interaction scheme can be inefficient and may lead to a suboptimal user experience.

Learning from Human Feedback. In contrast to "Communicating with Human Prompts", this interaction scheme provides feedback on the model's outputs, such as scoring, ranking, and offering suggestions, for model optimization. This feedback is therefore used to adjust the model's parameters, rather than simply acting as prompts for language models to respond. The primary objective of this interaction is to better adapt LMs for user needs and human values [2].

For instance, [16, 17] employ active learning to provide human feedback. By labeling a few examples based on model predictions, they update the model parameters to improve its understanding of human needs. More recently, [18] enhance a language model through continuous learning from user feedback and dialogue history. InstructGPT [4] initially trains GPT-3 using supervised instruction tuning and subsequently fine-tunes it via reinforcement learning from human feedback (RLHF), where the reward model is trained on annotated human preference data. This reward model, in turn, serves as a user simulator which can provide feedback for model's predictions. Ramamurthy et al. [19] demonstrate that RLHF is more data- and parameter-efficient than supervised methods when a learned reward model provides signals for an RL method, not to mention that preference data is easier to collect than ground-truth data. References [1, 20–23] provide comprehensive surveys on the topic of "learning from feedback" and alignment. We refer the readers to these surveys and Sect. 6.4 for more information.

Regulating via Human Configuration. The two interaction schemes previously discussed involve engagement with simulated or real humans through prompts or feedback. Regulation through human configuration, on the other hand, relies on users to customize and configure the language model system according to their needs. This customization can include adjustments to the system's structure, hyperparameters, decoding strategy, and more. Although it may not be the most flexible method, it is one of the simplest ways to facilitate interaction between the user and the system.

For example, [8] predefine a set of LLM primitive operations, such as "ideation", "split points", "compose points", etc.; each operation being controlled by a specific prompt template. Users can customize the usage and chaining schemes of different operations to meet a set of given requirements. Similarly, PromptChainer [24] is an interactive interface designed to facilitate data transformation between different steps of a chain. It also offers debugging capabilities at various levels of granularity, enabling users to create their own LM chains. Users can also configure some hyperparameters to control the performance of LLMs. This includes, but is not limited to, temperature (which controls the stochasticity of the output), the maximum number of tokens to generate, and "top-p" controlling diversity via nucleus sampling [25].[1] Vemprala et al. [26] have proposed the concept of *"user-on-the-loop"*, implying

---

[1] https://platform.openai.com/playground.

that users can configure the LM-robot interaction with human instructions, ensuring that the process and results of the interaction are centered around the user's needs.

Learning from Human Simulation. In many cases, training or deploying language models with real users is impractical, prompting the development of various user simulators to emulate user behavior and preferences. For instance, [4] initially rank generated responses with real annotators based on their preferences and then train a reward model–initialized from GPT-3 [15]–on this preference data to serve as a user preference simulator. Kim et al. [27] propose a method to simulate human preference by utilizing a transformer model that captures important events and temporal dependencies within segments of human decision trajectories. Additionally, this approach relies on a weighted sum of non-Markovian rewards. Faltings et al. [7] simulate user editing suggestions through BertScore-based [28] token-wise similarity scores and dynamic programming to compute an alignment between a draft and a target. Lynch et al. [9] collect numerous language-annotated trajectories, with the policy trained using behavioral cloning on the dataset. These collected trajectories can also be viewed as a user simulator.

The design of a user simulator is critical for the successful training and evaluation of language models. For example, to accurately replicate the behavior and preferences of real users when developing a generic dialogue system, it is vital to collect a diverse and extensive range of user data for training the simulator. This allows it to encompass the full spectrum of user preferences and behaviors. Moreover, when developing language models for rapidly changing application scenarios, it is essential to continually update and refine the simulator to adapt to shifts in user demographics and their evolving preferences.

2.2   KB-in-the-Loop

KB-in-the-loop NLP has two main approaches: one focuses on utilizing external knowledge sources to augment language models during inference time [29–40], while the other aims to employ external knowledge to enhance language model training, resulting in better language representations [41–49]. Interacting with KB during training can help improve the model's representation to incorporate more factual knowledge. In contrast, interacting with KB during inference can assist the language model in generating more accurate, contextually relevant, and informed responses by dynamically leveraging external knowledge sources based on the specific input or query at hand (Fig. 2.2).

In the following sections, we will discuss knowledge sources and knowledge retrieval. As for knowledge integration, we refer the readers to Sect. 9.4 for more details.

Knowledge Sources. Knowledge sources are normally categorized into the following types:

(1) Corpus Knowledge: Typically, corpus knowledge is stored in an offline collection from a specific corpus, which the language model accesses to enhance its generation capabilities. Common examples of corpus knowledge include the Wikipedia Corpus [50], WikiData

Fig. 2.2 KB-in-the-loop

Corpus [51], Freebase Corpus [52], PubMed Corpus,[2] and CommonCrawl Corpus,[3] among others. Most previous research has focused on corpus knowledge due to its controllability and efficiency. Retrieval-Augmented Language Models [30, 31, 33, 35, 37] have been proposed to develop language models capable of utilizing external knowledge bases for more grounded generation [49, 53]. To further improve interpretability, subsequent studies [39, 40, 54] have suggested using extracted Question-Answer pairs as the corpus for more fine-grained knowledge triple grounding. Recently, there has been growing interest in incorporating citations to enhance grounding in language models, as demonstrated by GopherCite [34]. Another line of work, including KELM [41], ERNIE [44, 46, 55], and others [45, 47], primarily employs recognized entities as the foundation for integrating knowledge graph information into neural representations.

(2) Internet Knowledge: One challenge associated with corpus knowledge is its limited coverage and the need for specialized retrieval training. A potential solution involves offloading the retrieval process to search engines and adapting them to find the desired content. The Internet-augmented language model [56] was first introduced, to answer open-domain questions by grounding responses in search results from the Internet. This approach has since been demonstrated to effectively answer time-sensitive questions [57]. The Internet has also been employed for post-hoc attribution [58]. WebGPT [36] proposes powering language models with a web browser, which searches the web before generating knowledgeable or factual text. MineDojo [59] equips a video-language model with Internet-scale knowledge to tackle

---

[2] https://pubmed.ncbi.nlm.nih.gov/.

[3] https://commoncrawl.org/.

diverse tasks within a *Minecraft* environment. ToolFormer [60] similarly integrates a search engine into the tool-use adaptation of language models. ReAct [61] suggests leveraging the Internet to augment reasoning capabilities in black-box large language models.

While corpus knowledge and internet knowledge are both valuable resources that language models can utilize to enhance their capabilities, they inherently differ in terms of controllability and coverage. Corpus knowledge is pre-collected and stored offline in a controlled setting, making it easy to access and integrate into a language model. However, it is limited by the information within the corpus and may not be up-to-date or comprehensive. In contrast, internet knowledge offers a vast and diverse pool of constantly updated information, providing more comprehensive coverage. However, controlling and curating internet knowledge is challenging, as the information obtained from the internet may be more noisy, or even more misleading. Additionally, it is worth noting that there are other miscellaneous types of knowledge sources, such as visual knowledge [62], rule-based knowledge [63–66], implicit knowledge [67], database knowledge [68], and documentation knowledge [69]. These can be categorized into either corpus knowledge or internet knowledge, depending on their nature.

Knowledge Retrieval. Enhancing language models with knowledge requires careful consideration of knowledge quality. Knowledge quality is primarily affected by issues such as knowledge missing and knowledge noise [70]. Knowledge missing can be mitigated by changing or extending the knowledge source to provide more comprehensive information. To tackle knowledge noise, an intuitive approach is to filter out the noisy information. References [42, 70] propose addressing this issue by using a visibility matrix that functions on the attention scores between the knowledge and input. This helps in better integration of high-quality knowledge into the language model. Despite the success of these methods, improving knowledge retrieval remains the most critical aspect of addressing these challenges. This is because improving knowledge retrieval directly impacts the precision and recall of knowledge that is selected and integrated into the language model, leading to better overall performance. There are overall three methods for knowledge retrieval:

(1) **Sparse Retrieval**: In this approach, knowledge is retrieved based on lexical matches between words or phrases in the input text and a knowledge source or the similarity between sparse representations. For example, ToolFormer [60] employs BM25 [71] as a metric to retrieve knowledge from Wikipedia. DrQA [72] retrieves documents using TF-IDF vectors. RepoCoder [73] incorporates the Jaccard index [74] as one of its retrieval metrics. Moreover, researchers explore on utilizing the sparse representations from pre-trained language model compound with the lexical matching methods [75–77].

(2) **Dense Retrieval**: Dense retrieval approach retrieves knowledge based on the meaning of the input text rather than merely matching exact words or phrases. The meaning is typically encoded by a learned retriever. A dual encoder or cross encoder can be used as the retriever. For example, REALM [30] employs a latent knowledge retriever that is trained in an unsupervised manner to extract relevant information and context from a vast corpus during both the training and inference stages. Retro [35] retrieves chunks from an external

knowledge base using a dual encoder and integrates the retrieved chunks into language models through cross attention. Cai et al. [78] jointly train a translation memory retriever and neural machine translation model. RepoCoder [73] also employs an embedding model to compute the cosine similarity between input and knowledge. Atlas [33] retrieves knowledge with Contriever [79], a dense dual encoder-based retriever trained via contrastive learning. Izacard and Grave [80] and RePlug [81] propose distilling knowledge from a reader to a retriever model, which requires very few annotated training data.

(3) **Generative Retrieval**: Instead of retrieving knowledge through matching, a generative retriever directly produces the document id or content as knowledge. As such, the generative retriever, typically in the form of a language model, can be considered a type of knowledge base, which is also known as implicit knowledge [67, 82, 83]. For example, DSI [84] encodes numerous documents with their ids into the language model's parameters. During inference, the model generates the id of the most relevant document. Sun et al. [85] propose augmenting language models with recitations, which are relevant knowledgeable content generated by language models. Yu et al. [86] prompt a large language model to generate diverse contextual documents based on a given question and then read the generated documents to produce a final answer, where the in-context demonstrations for the LLM prompting are sampled from a clustered document pool. It is worth noting that knowledge distillation may also fall within this category. For example, [87] allows large language models to serve as teachers, distilling their reasoning skills into smaller language models. The knowledgeable large language model can be viewed as a generative retriever-like knowledge base for the smaller language models.

(4) **Reinforcement Learning**: Knowledge retrieval can also be formulated as a reinforcement learning problem. For example, WebGPT [36] learns to retrieve and select documents via behavior cloning (BC) and reinforcement learning from human feedback (RLHF). Zhang et al. [88] formulate the example retrieval problem as a Markov Decision Process (MDP) and propose a reinforcement learning (RL) method to select examples.

2.3   Model/Tool-in-the-Loop

Addressing complex tasks often necessitates the implementation of strategic methodologies that can simplify the process. One such effective strategy is the explicit decomposition of the task into modularized subtasks and then solve these subtasks step by step [89–92]. Alternatively, another strategy involves the implicit decomposition of the task through the division of labor among multiple language model agents. This approach enables a natural and adaptive breakdown of the work, as each agent assumes a specific role in the larger task [93–95]. The procedure of task decomposition not only allows subtask modularization, but also enables subtask composition. Furthermore, by breaking the task into multiple steps, specific steps can be allocated to certain expert models or external tools, such as those specializing in arithmetic computation, web search, counting, and more [60, 61, 96]. Inspired by [61,

Fig. 2.3 Model/Tool-in-the-loop

97], there are primarily three fundamental operations involved in decomposing and solving these subtasks (Fig. 2.3):

1. **Thinking**: The model engages in self-interaction to reason and decompose complex problems into modularized subtasks [61, 91, 97, 98];
2. **Acting**: The model calls tools or models to solve these intermediate subtasks, which may result in effects on the external world [61, 96, 97];
3. **Collaborating**: Multiple models with distinct roles or division of labor communicate and cooperate with each other to achieve a common goal or simulate human social behaviors [94, 99–103].

Thinking. For example, consider the question, *"What is the biggest animal in Africa?"*, which can be decomposed into a chain of three subtasks: *"What animals are in Africa?"* → *"Which of these animals are large?"* → *"Which of these is the largest?"* These three subtasks form a prompt chain (c.f., Sect. 5.3), allowing for the individual solving of each subtask by a single LM, multiple LMs, or even tools. That is, through the process of thinking, the overall task can be decomposed into multiple subtasks that can be efficiently tackled through interactions among language models or tools in a chained manner.

The preliminary instantiation of such a cognitive process is **Chain-of-Thought (CoT)** [89], which seeks to elicit multi-hop complex reasoning capabilities from large language models using a cascading mechanism [91]. Instead of directly producing the answer, multiple thoughts (i.e., reasoning steps) are generated beforehand [89, 90, 104, 105]. Thus, CoT decomposes the task into two sub-tasks: *thought generation* → *answer generation*. However, typical CoT involves solving these subtasks in a single model run [89] without an interaction mechanism.

Derivative works of CoT have shown an increasing tendency to utilize a self-interaction loop that involves iteratively calling the same language model to solve different subtasks [61, 90, 105, 106], also known as **multi-stage CoT** [92, 107]. Furthermore, some other derivative works share similar principles with CoT or multi-stage CoT but employ **different training strategies**, such as bootstrapping [108] (as discussed in Sect. 6.4). Some works go beyond the subtask of *thought generation* and **introduce new subtasks**, including *thought verification* [109], *fact selection and inference* [110], and *self-refinement and self-feedback* [111], among others. Indeed, all of these works can be seen as instantiations of the thinking cognitive process. They employ a self-interaction mechanism, wherein a single language model is utilized iteratively to decompose tasks into subtasks, and effectively solve these subtasks.

Acting. Different from the process of thinking, acting involves the interaction of the LM with external entities, such as other LMs and tools. Since different models or tools can possess specific expertise, the LM can invoke these external entities to perform specific subtasks when the task is decomposed into subtasks. For example, *thought verification* can be accomplished using a discriminative model [112], and *fact selection* may utilize a retriever model [30]. External tools such as calculators [60, 113], simulators [66, 114], search engines [36, 61], code interpreters and executors [115–117], and other APIs [18, 60, 61, 96, 97, 118–121] can also be incorporated into the loop to tackle subtasks that language models typically encounter difficulties with. Generally, tasks emphasizing faithfulness and exactitude (e.g., real facts, complex mathematical operations) and tasks beyond the LM training corpus (e.g., up-to-date information, low-resource languages, awareness of time, image generation) are better solved using external tools than LMs [60, 96, 97, 120–127].

For example, ToolFormer [60] enhances language models with tool-use capabilities by retraining on a tool-use prompted corpus and involving tools such as calculators, calendars, search engines, question-answering systems, and translation systems. ART [128] begins by selecting demonstrations from a task library that involve multi-step reasoning and tool usage. These demonstrations serve as prompts for the frozen LLM to generate intermediate reasoning steps in the form of executable programs. ReAct [61] combines both chain-of-thought reasoning and task-specific tool-use actions to improve the interactive decision-making capabilities of language models. TaskMatrix.AI [120] presents a vision for a new AI ecosystem built on tool-use APIs, proposing an architecture composed of an API platform, API selector, multimodal conversational foundation model, API-based action executor, and integrating RLHF and feedback to API developers to optimize the system. This architecture benefits from its ability to perform digital and physical tasks, its API repository for diverse task experts, its lifelong learning ability, and improved interpretability. HuggingGPT [129] and OpenAGI [130] use ChatGPT as a task controller, planning tasks into multiple subtasks that can be solved by models (tools) selected from the HuggingFace platform.[4]

Moreover, acting can have a tangible impact on the external world through tool-use [97], also referred to as Tool-Oriented Learning [96]. For instance, ChatGPT Plugins[5] empower LLMs to directly utilize tools for tasks such as travel bookings, grocery shopping, and restaurant reservations, among others. LM-Nav [131] leverages a visual navigation model (VNM) to execute the actions planned by the LLM, enabling real-world robotic navigation. In these cases, the overall task is still decomposed into subtasks, but some of which are connected with the external world. By employing specific models or tools to address these subtasks, tangible effects can be realized in the environment. Readers can refer to Sect. 2.4 for additional information related to the interaction between the language model and the environment.

Collaborating. Most of the aforementioned research relies on manual task decomposition. Although some existing works propose automatic task decomposition through distant supervision [132–134] or in-context learning [90, 97, 105, 135, 136], explicit task decomposition is not always straightforward. On the one hand, it requires human expertise or extensive manual effort. On the other hand, in certain cases, different language model agents may share a common goal that is difficult to explicitly decompose [94, 103, 137, 138]. In such scenarios, task decomposition or division of labor may emerge implicitly as different agents with specialized skills assume different roles within the task and interact with one another [94, 95, 99–101, 103, 139–141]. For example, in *MineCraft*, agents with distinct yet complementary recipe skills can communicate and collaborate to synthesize a material, where the specialized agents may automatically discover a potential division of labor [94]. To the best of our knowledge, we can categorize collaboration-based approaches into three clusters:

(1) **Closed-Loop Interaction** refers to a collaborative process where multiple agents interact with each other in a feedback loop [93, 112, 142–145]. In the context of control theory, a closed-loop controller uses feedback to control states or outputs from a dynamical system.[6] Generally, closed-loop controllers are preferred over open-loop controllers as they offer greater adaptability and robustness in changing or uncertain environments. Likewise, closed-loop interaction between language model agents is more effective and robust compared to open-loop interaction [9, 146], making it a primary paradigm for collaboration-based methods. For example, Socratic Models [93] and Inner Monologue [144] enable language models to collaborate with vision-language models, audio-language models, or humans to conduct egocentric perception and robotic manipulation tasks, respectively. The language-based closed-loop feedback is incorporated into LLM planning, significantly improving instruction completion abilities [144]. Planner-Actor-Reporter [145] uses an LLM (Plan-

---

[4] https://huggingface.co/.

[5] https://openai.com/blog/chatgpt-plugins.

[6] https://en.wikipedia.org/wiki/Control_theory.

ner) to generate instructions for a separate RL agent (Actor) to execute in an embodied environment. The state of the environment is reported back to the Planner (via the Reporter) to refine instructions and complete the feedback loop. Note that closed-loop interaction is highly applicable in Environment-in-the-loop scenarios, where closed-loop feedback from the environments can be transferred via a model connected to the environment [93, 144, 147]. Closed-loop interaction is also useful for text generation tasks [111, 148–152]. For example, Self-Refine [111] enhances the output of LLMs through a closed-loop interaction of self-feedback and self-refinement, iteratively improving initial outputs without requiring additional training data, supervised training, or reinforcement learning.

(2) **Theory of Mind** in language models has garnered growing attention in the research community [94, 100, 101, 153–155]. According to [101], *"Theory of Mind (ToM), or the ability to attribute unobservable mental states to others, is central to human social interactions, communication, empathy, self-consciousness, and morality."* Kosinski [101] demonstrates that large language models, like ChatGPT, can successfully tackle 93% of ToM tasks. This finding suggests that ToM-like capabilities may have naturally emerged in large language models. In line with this, MindCraft [94] assigns different material composition tables (sub-skills) to two dialogue agents, enabling them to cooperate and complete the material composition task through mutual communication. Zhu et al. [154] provide a speaker and listener formulation of ToM, where the speaker should model the listener's beliefs (i.e., action possibilities over some instruction candidates). These ToM mechanisms are beneficial for collaborative tasks [155].

(3) **Communicative Agent** perceives language models as agents [156] and delves into the study of multi-agent communication [138]. In addition to Theory of Mind, multi-agent communication also investigates the scenarios of referential game [138], language acquisition [155], language emergence [157], and role playing [103], implying an effort towards LLM society [103]. For example, [157] enable two communicative agents, a speaker and a listener, to learn to play a *Speak, Guess and Draw* game and automatically derive an interaction interface between them, which is so-called machine language. Camel [103] proposes a role-playing framework that involves two cooperative agents, an AI user and an AI assistant. The two language models are prompted with a shared task specifier prompt and different role assignment prompts, which is referred to as *Inception Prompting*. With the condition of Inception Prompting, they communicate with each other without any additional human instruction to solve the specified task. Generative Agents [102] introduces a novel architecture that extends a LLM to enable believable simulations of human behavior in an interactive sandbox environment, demonstrating the agents' ability to autonomously plan and exhibit individual and social behaviors. Yuan and Zhu [158]'s formalism even views existing machine learning paradigms such as passive learning and active learning, as communicative learning, which is in line with [159]'s interactive language modeling. In these paradigms, the language model agents are grouped into teachers and students, where the students learn from the teachers through interaction. They frame learning as a communicative and collaborative process.

Fig. 2.4 Environment-in-the-loop

2.4   Environment-in-the-Loop

A new trend within the NLP community is to harness the power of LMs to address embodied tasks such as robot manipulation, autonomous driving, and egocentric perception, among others [93, 131, 143–145, 160–163]. In these scenarios, the environment is integrated into an interactive loop with language models. The aim of environment-in-the-loop NLP is language grounding, which is to represent language with meaning reference to environments and experiences [164]. It has been argued that only if LMs are put into interaction with real-world or virtual environments can they learn a truly grounded representation of language [164]. During this interaction, the environment assumes the responsibility of furnishing the LM with low-level observations, rewards, and state transitions. Simultaneously, the LM is tasked with generating solutions for environmental tasks, including reasoning, planning, and decision-making [164–166] (Fig. 2.4).

We define two dimensions for language grounding, as shown in Fig. 2.5. The horizontal axis spans from the *concrete* end to the *abstract* end. The term *concrete* refers to models that capture high-dimensional data of the world, such as images, audio, and other similar sensory inputs. On the other hand, the term *abstract* pertains to models that capture low-dimensional data, such as language, code, or other symbolic representations. Compared to a more concrete representation, abstract or bottle-necked representation brings stronger generalization and reasoning ability [167, 168].

The vertical axis ranges from the *low-level* end to the *high-level* end, where *low-level* means a more direct and embodied interaction with the environment, such as perception or manipulation, while *high-level* means a more indirect and conceptual interaction with the

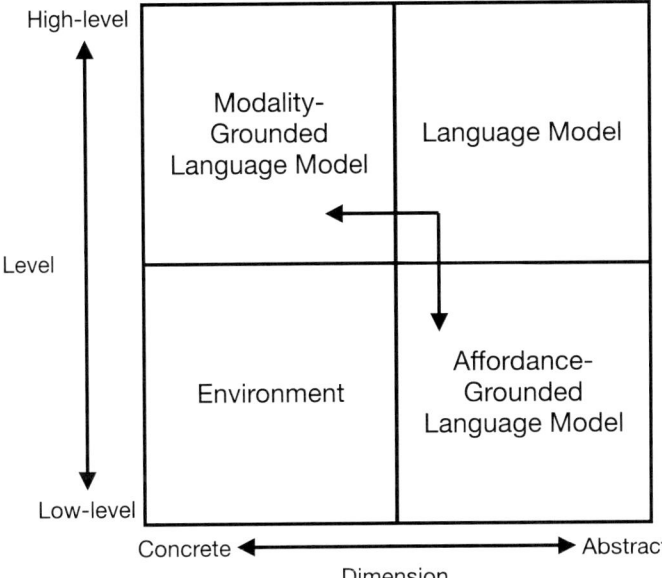

Fig. 2.5 Two directions for language grounding. A third direction for language grounding may be social interaction [138, 155, 164, 169] which is not illustrated in this figure but we have discussed it partly in Sect. 2.3

environment, such as reasoning, planning, and decision-making. This axis can reflect the degree of the model's contextual and situational understanding of the environment.

Generally, the environment can be the real world or virtual world simulated by programs such as MuJoCo [170] and MineCraft.[7] Hence, the environment is in the bottom-left quadrant in Fig. 2.5 with a concrete representation of data and low-level interaction processes. While the language model is in the top-right quadrant in Fig. 2.5 with an abstract representation of data and high-level interaction processes. This discrepancy makes it necessary to ground language models for LM-env interaction. There are mainly two directions: **modality grounding** and **affordance grounding**.

(1) **Modality Grounding** [171] aims to move the language model from the abstract quadrant to the concrete quadrant. It is intuitive to incorporate information in image, audio or other modalities into it. In this way, language models can capture more complete observations from the environment.

(2) **Affordance Grounding** [143] strives to transition language models from the high-level quadrant to the low-level quadrant. The goal is to align the outputs of language models with the contextual scene, ensuring that the generated text corresponds to the surrounding environment rather than being detached from it.

---

[7] https://www.minecraft.net.

It is worth noting that these two goals are not independent processes, and often form a synergy towards the environment. Moreover, other additional requirements such as preference and safety are also possible directions [163], which may further involve human in the loop.

Modality Grounding. Modality-Grounded Language Model (MGLM) is designed to allow language models to process data of more modalities such as vision and audio. In the context of visual grounding (i.e., vision-language pre-trained model), for example, there are three ways: (1) Dual-Tower modeling which trains different encoders for different modalities [172–178]; (2) Single-Tower modeling using the concatenation of multimodal data to train a single model [179–192]; (3) Interaction between pre-trained vision and language models [93, 112, 121, 144, 193–195]. These methods involve the utilization of visual information during both the training and inference stages of a language model. By incorporating visual signals, these approaches enable a visually grounded representation of language. This enhancement in representation facilitates improved interaction efficiency between the language model and the environment, as it allows for increased information throughput.

For example, WebShop [196] and Interactive Language [9] use ResNet [197] and a Transformer model [198] to process visual and linguistic data respectively, and input the fused representations into another Transformer to generate action outputs; VIMA [199] and Gato [184] use one single model to simultaneously process the concatenated multimodal data and predict actions; Socratic Models [93], Inner Monologue [144], and LM-Nav [131] use multimodal language models to convert visual inputs into language captions or phrases and use LLMs for planning, reasoning and question-answering in order to perform embodied tasks. ViperGPT [200] equips the LLM with an API for various perceptual and knowledge modules, along with a Python interpreter, enabling the LLM to generate executable code for visual reasoning tasks.

Another goal of Modality Grounding is to preserve as much high-level knowledge as possible in the language model to ensure that the model is still able to effectively perform tasks such as commonsense reasoning, planning, question answering, code generation, etc. These capabilities become more pronounced and complex as the size of the model increases, known as emergent abilities [201, 202]. These capabilities serve as one of the primary purposes of leveraging language models for embodied tasks. An illustrative example of these capabilities is demonstrated in the context of completing long-horizon navigation tasks. In such tasks, the effective planning of instructions by the LLM is crucial [131].

Affordance Grounding. However, in general, in order to make MGLM knowledge-rich, the model needs to be pre-trained with a large amount of data from open domains, which may result in outputs that are too diverse and therefore do not match the conditions in the real environment [143, 161, 163]. Therefore, some low-level information from the environment is needed to be incorporated into language models, which is referred to as Affordance Grounding [143].

According to [203, 204]: "*Affordances describe the fact that certain states enable an agent to do certain actions, in the context of embodied agents.*". Likewise, according to [143]:

"*The learned affordance functions (Can) provide a world-grounding to determine what is possible to execute upon the plan*". However, [161] argues that [143]'s falls short in providing affordance grounding at the scene-scale, thus limiting the ability to reason about the potential actions a robot can perform within a given environment. Hence, following this thought, there are mainly two requirements for an affordance grounded language model (AGLM): (1) **scene-scale perception**, and (2) **possible action, conditioned on the language-based instructions**. For example, when considering a smart home environment and asking the agent to "*turn off the lights in the living room.*", scene-scale perception aims to make the agent aware of all (or only) the existing and relevant objects, such as "*bedlamps*" and "*droplights*". Secondly, possible action involves the agent to determine the executable actions on the objects that can complete the instructions, such as "***press** the switches.*"

For example, SayCan [143] leverages large language models to generate a list of object-action proposals (i.e., task grounding) which are then scored by a value function connected to the environment (i.e., world-grounding). Similarly, [161] first construct a language queryable scene representation, NLMap, through pre-exploration of a robotic agent and then use the a LLM to generate a list of relevant objects to be filtered and located. The object presence and location are finally used for LLM planning. Abramson et al. [205] train an agent via behavioral cloning on the interactions of paired human players. They then collect human feedback on the learned agent to train a reward model, which is finally used to post-train the agent. That is, they achieve affordance grounding via behavioral cloning and RLHF. Code as Policies [160] enables a language model to generate executable code directly. The generated codes can be executed with a python interpreter for affordance verification [115]. LM-Nav [131] converts the planning results of the language model to image form and then uses a Visual Navigation Model to convert them into executable instructions (i.e., action+distance). Grounded Decoding [163] integrates the high-level semantic understanding of LLMs with the reality-based practicalities of grounded models, enabling the generation of action sequences that are both knowledge-informed and feasible in embodied agent tasks like robotics. Wake et al. [206] provide numerous examples of utilizing ChatGPT for generating executable action sequences to accomplish tasks assigned by users.

Note that KB-in-the-loop, Model/Tool-in-the-loop, or Human-in-the-loop approaches can also be employed for modality grounding or affordance grounding [9, 61, 93, 144, 146, 205]. In these approaches, external objects or entities undertake these functions, such as utilizing humans to describe the visual scene for modality grounding [144].

References

1. Wang, Z.J., Choi, D., Xu, S., Yang, D.: Putting humans in the natural language processing loop: A survey. arXiv preprint arXiv:2103.04044 (2021)
2. Bai, Y., Jones, A., Ndousse, K., Askell, A., Chen, A., DasSarma, N., Drain, D., Fort, S., Ganguli, D., Henighan, T., Joseph, N., Kadavath, S., Kernion, J., Conerly, T., El-Showk, S., Elhage, N., Hatfield-Dodds, Z., Hernandez, D., Hume, T., Johnston, S., Kravec, S., Lovitt, L., Nanda, N.,

Olsson, C., Amodei, D., Brown, T., Clark, J., McCandlish, S., Olah, C., Mann, B., Kaplan, J.: Training a helpful and harmless assistant with reinforcement learning from human feedback. arXiv preprint arXiv:2204.05862 (2022)

3. Kenton, Z., Everitt, T., Weidinger, L., Gabriel, I., Mikulik, V., Irving, G.: Alignment of language agents. arXiv preprint arXiv:2103.14659 (2021)

4. Ouyang, L., Wu, J., Jiang, X., Almeida, D., Wainwright, C.L., Mishkin, P., Zhang, C., Agarwal, S., Slama, K., Ray, A., et al.: Training language models to follow instructions with human feedback. arXiv preprint arXiv:2203.02155 (2022)

5. Leike, J., Krueger, D., Everitt, T., Martic, M., Maini, V., Legg, S.: Scalable agent alignment via reward modeling: a research direction. arXiv preprint arXiv:1811.07871 (2018)

6. Lee, M., Srivastava, M., Hardy, A., Thickstun, J., Durmus, E., Paranjape, A., Gerard-Ursin, I., Li, X.L., Ladhak, F., Rong, F., Wang, R.E., Kwon, M., Park, J.S., Cao, H., Lee, T., Bommasani, R., Bernstein, M.S., Liang, P.: Evaluating human-language model interaction. CoRR **abs/2212.09746** (2022). https://doi.org/10.48550/arXiv.2212.09746

7. Faltings, F., Galley, M., Peng, B., Brantley, K., Cai, W., Zhang, Y., Gao, J., Dolan, B.: Interactive text generation. arXiv preprint arXiv:2303.00908 (2023)

8. Wu, T.S., Terry, M., Cai, C.J.: Ai chains: Transparent and controllable human-ai interaction by chaining large language model prompts. International Conference On Human Factors In Computing Systems (2021). https://doi.org/10.1145/3491102.3517582

9. Lynch, C., Wahid, A., Tompson, J., Ding, T., Betker, J., Baruch, R., Armstrong, T., Florence, P.: Interactive language: Talking to robots in real time (2022)

10. Mehta, N., Teruel, M., Sanz, P.F., Deng, X., Awadallah, A.H., Kiseleva, J.: Improving grounded language understanding in a collaborative environment by interacting with agents through help feedback. arXiv preprint arXiv:2304.10750 (2023)

11. Malmi, E., Dong, Y., Mallinson, J., Chuklin, A., Adamek, J., Mirylenka, D., Stahlberg, F., Krause, S., Kumar, S., Severyn, A.: Text generation with text-editing models. In: Proceedings of the 2022 Conference of the North American Chapter of the Association for Computational Linguistics: Human Language Technologies: Tutorial Abstracts, pp. 1–7. Association for Computational Linguistics, Seattle, United States (2022). https://doi.org/10.18653/v1/2022.naacl-tutorials.1. https://aclanthology.org/2022.naacl-tutorials.1

12. Schick, T., Dwivedi-Yu, J., Jiang, Z., Petroni, F., Lewis, P., Izacard, G., You, Q., Nalmpantis, C., Grave, E., Riedel, S.: Peer: A collaborative language model. arXiv preprint arXiv:2208.11663 (2022)

13. Shi, N., Tang, B., Yuan, B., Huang, L., Pu, Y., Fu, J., Lin, Z.: Text editing as imitation game. In: Findings of the Association for Computational Linguistics: EMNLP 2022, pp. 1583–1594. Association for Computational Linguistics, Abu Dhabi, United Arab Emirates (2022). https://aclanthology.org/2022.findings-emnlp.114

14. Du, W., Kim, Z.M., Raheja, V., Kumar, D., Kang, D.: Read, revise, repeat: A system demonstration for human-in-the-loop iterative text revision. In: Proceedings of the First Workshop on Intelligent and Interactive Writing Assistants (In2Writing 2022), pp. 96–108. Association for Computational Linguistics, Dublin, Ireland (2022). https://doi.org/10.18653/v1/2022.in2writing-1.14. https://aclanthology.org/2022.in2writing-1.14

15. Brown, T., Mann, B., Ryder, N., Subbiah, M., Kaplan, J.D., Dhariwal, P., Neelakantan, A., Shyam, P., Sastry, G., Askell, A., et al.: Language models are few-shot learners. Advances in neural information processing systems **33**, 1877–1901 (2020)

16. Godbole, S., Harpale, A., Sarawagi, S., Chakrabarti, S.: Document classification through interactive supervision of document and term labels. In: Proceedings of the 8th European Conference on Principles and Practice of Knowledge Discovery in Databases, PKDD '04, pp. 185–196. Springer-Verlag, Berlin, Heidelberg (2004)

17. Settles, B.: Closing the loop: Fast, interactive semi-supervised annotation with queries on features and instances. In: Proceedings of the 2011 Conference on Empirical Methods in Natural Language Processing, pp. 1467–1478. Association for Computational Linguistics, Edinburgh, Scotland, UK. (2011). https://aclanthology.org/D11-1136

18. Shuster, K., Xu, J., Komeili, M., Ju, D., Smith, E.M., Roller, S., Ung, M., Chen, M., Arora, K., Lane, J., Behrooz, M., Ngan, W., Poff, S., Goyal, N., Szlam, A., Boureau, Y.L., Kambadur, M., Weston, J.: Blenderbot 3: a deployed conversational agent that continually learns to responsibly engage. arXiv preprint arXiv:2208.03188 (2022)

19. Ramamurthy, R., Ammanabrolu, P., Brantley, K., Hessel, J., Sifa, R., Bauckhage, C., Hajishirzi, H., Choi, Y.: Is reinforcement learning (not) for natural language processing: Benchmarks, baselines, and building blocks for natural language policy optimization. In: The Eleventh International Conference on Learning Representations (2023). https://openreview.net/forum?id=8aHzds2uUyB

20. Fernandes, P., Madaan, A., Liu, E., Farinhas, A., Martins, P.H., Bertsch, A., de Souza, J.G.C., Zhou, S., Wu, T.S., Neubig, G., Martins, A.F.T.: Bridging the gap: A survey on integrating (human) feedback for natural language generation. ARXIV.ORG (2023). https://doi.org/10.48550/arXiv.2305.00955

21. Ji, J., Qiu, T., Chen, B., Zhang, B., Lou, H., Wang, K., Duan, Y., He, Z., Zhou, J., Zhang, Z., Zeng, F., Ng, K.Y., Dai, J., Pan, X., O'Gara, A., Lei, Y., Xu, H., Tse, B., Fu, J., McAleer, S., Yang, Y., Wang, Y., Zhu, S.C., Guo, Y., Gao, W.: Ai alignment: A comprehensive survey (2023)

22. Shen, T., Jin, R., Huang, Y., Liu, C., Dong, W., Guo, Z., Wu, X., Liu, Y., Xiong, D.: Large language model alignment: A survey. arXiv preprint arXiv: 2309.15025 (2023). https://arxiv.org/abs/2309.15025v1

23. Wang, Y., Zhong, W., Li, L., Mi, F., Zeng, X., Huang, W., Shang, L., Jiang, X., Liu, Q.: Aligning large language models with human: A survey. arXiv preprint arXiv:2307.12966 (2023)

24. Wu, T., Jiang, E., Donsbach, A., Gray, J., Molina, A., Terry, M., Cai, C.J.: Promptchainer: Chaining large language model prompts through visual programming. arXiv preprint arXiv:2203.06566 (2022)

25. Holtzman, A., Buys, J., Forbes, M., Choi, Y.: The curious case of neural text degeneration. International Conference On Learning Representations (2019)

26. Vemprala, S., Bonatti, R., Bucker, A., Kapoor, A.: ChatGPT for robotics: Design principles and model abilities. Tech. Rep. MSR-TR-2023-8, Microsoft (2023). https://www.microsoft.com/en-us/research/publication/chatgpt-for-robotics-design-principles-and-model-abilities/

27. Kim, C., Park, J., Shin, J., Lee, H., Abbeel, P., Lee, K.: Preference transformer: Modeling human preferences using transformers for RL. In: The Eleventh International Conference on Learning Representations (2023). https://openreview.net/forum?id=Peot1SFDX0

28. Zhang, T., Kishore, V., Wu, F., Weinberger, K.Q., Artzi, Y.: Bertscore: Evaluating text generation with BERT. In: 8th International Conference on Learning Representations, ICLR 2020, Addis Ababa, Ethiopia, April 26–30, 2020. OpenReview.net (2020). https://openreview.net/forum?id=SkeHuCVFDr

29. Khandelwal, U., Levy, O., Jurafsky, D., Zettlemoyer, L., Lewis, M.: Generalization through memorization: Nearest neighbor language models. In: International Conference on Learning Representations (2020). https://openreview.net/forum?id=HklBjCEKvH

30. Guu, K., Lee, K., Tung, Z., Pasupat, P., Chang, M.W.: Realm: Retrieval-augmented language model pre-training. arXiv preprint arXiv:2002.08909 (2020)

31. Lewis, P., Perez, E., Piktus, A., Petroni, F., Karpukhin, V., Goyal, N., Küttler, H., Lewis, M., tau Yih, W., Rocktäschel, T., Riedel, S., Kiela, D.: Retrieval-augmented generation for knowledge-intensive nlp tasks. arXiv preprint arXiv:2005.11401 (2020)

32. Cheng, Y., Li, S., Liu, B., Zhao, R., Li, S., Lin, C., Zheng, Y.: Guiding the growth: Difficulty-controllable question generation through step-by-step rewriting. In: Proceedings of the 59th Annual Meeting of the Association for Computational Linguistics and the 11th International Joint Conference on Natural Language Processing (Volume 1: Long Papers), pp. 5968–5978. Association for Computational Linguistics, Online (2021). https://doi.org/10.18653/v1/2021.acl-long.465. https://aclanthology.org/2021.acl-long.465

33. Izacard, G., Lewis, P., Lomeli, M., Hosseini, L., Petroni, F., Schick, T., Dwivedi-Yu, J., Joulin, A., Riedel, S., Grave, E.: Atlas: Few-shot learning with retrieval augmented language models. arXiv preprint arXiv:2208.03299 (2022)

34. Menick, J., Trebacz, M., Mikulik, V., Aslanides, J., Song, F., Chadwick, M., Glaese, M., Young, S., Campbell-Gillingham, L., Irving, G., et al.: Teaching language models to support answers with verified quotes. arXiv preprint arXiv:2203.11147 (2022)

35. Borgeaud, S., Mensch, A., Hoffmann, J., Cai, T., Rutherford, E., Millican, K., van den Driessche, G., Lespiau, J., Damoc, B., Clark, A., de Las Casas, D., Guy, A., Menick, J., Ring, R., Hennigan, T., Huang, S., Maggiore, L., Jones, C., Cassirer, A., Brock, A., Paganini, M., Irving, G., Vinyals, O., Osindero, S., Simonyan, K., Rae, J.W., Elsen, E., Sifre, L.: Improving language models by retrieving from trillions of tokens. International Conference On Machine Learning (2021)

36. Nakano, R., Hilton, J., Balaji, S., Wu, J., Ouyang, L., Kim, C., Hesse, C., Jain, S., Kosaraju, V., Saunders, W., et al.: Webgpt: Browser-assisted question-answering with human feedback. arXiv preprint arXiv:2112.09332 (2021)

37. Shuster, K., Poff, S., Chen, M., Kiela, D., Weston, J.: Retrieval augmentation reduces hallucination in conversation. In: M. Moens, X. Huang, L. Specia, S.W. Yih (eds.) Findings of the Association for Computational Linguistics: EMNLP 2021, Virtual Event / Punta Cana, Dominican Republic, 16–20 November, 2021, pp. 3784–3803. Association for Computational Linguistics (2021). https://doi.org/10.18653/v1/2021.findings-emnlp.320

38. Wang, B., Ping, W., Xu, P., McAfee, L., Liu, Z., Shoeybi, M., Dong, Y., Kuchaiev, O., Li, B., Xiao, C., et al.: Shall we pretrain autoregressive language models with retrieval? a comprehensive study. arXiv preprint arXiv:2304.06762 (2023)

39. Lewis, P., Wu, Y., Liu, L., Minervini, P., Küttler, H., Piktus, A., Stenetorp, P., Riedel, S.: Paq: 65 million probably-asked questions and what you can do with them. Transactions of the Association for Computational Linguistics 9, 1098–1115 (2021)

40. Chen, W., Verga, P., de Jong, M., Wieting, J., Cohen, W.: Augmenting pre-trained language models with qa-memory for open-domain question answering. EACL (2023)

41. Lu, Y., Lu, H., Fu, G., Liu, Q.: Kelm: Knowledge enhanced pre-trained language representations with message passing on hierarchical relational graphs. arXiv preprint arXiv:2109.04223 (2021)

42. Liu, W., Zhou, P., Zhao, Z., Wang, Z., Ju, Q., Deng, H., Wang, P.: K-bert: Enabling language representation with knowledge graph. AAAI Conference On Artificial Intelligence (2019). https://doi.org/10.1609/AAAI.V34I03.5681

43. Zhang, Z., Han, X., Liu, Z., Jiang, X., Sun, M., Liu, Q.: ERNIE: Enhanced language representation with informative entities. In: Proceedings of the 57th Annual Meeting of the Association for Computational Linguistics, pp. 1441–1451. Association for Computational Linguistics, Florence, Italy (2019). https://doi.org/10.18653/v1/P19-1139. https://aclanthology.org/P19-1139

44. Sun, Y., Wang, S., Li, Y., Feng, S., Chen, X., Zhang, H., Tian, X., Zhu, D., Tian, H., Wu, H.: Ernie: Enhanced representation through knowledge integration. arXiv preprint arXiv:1904.09223 (2019)

45. Févry, T., Soares, L.B., Fitzgerald, N., Choi, E., Kwiatkowski, T.: Entities as experts: Sparse memory access with entity supervision. In: Proceedings of the 2020 Conference on Empirical Methods in Natural Language Processing (EMNLP), pp. 4937–4951 (2020)

46. Sun, Y., Wang, S., Feng, S., Ding, S., Pang, C., Shang, J., Liu, J., Chen, X., Zhao, Y., Lu, Y., et al.: Ernie 3.0: Large-scale knowledge enhanced pre-training for language understanding and generation. arXiv preprint arXiv:2107.02137 (2021)

47. Xiong, W., Du, J., Wang, W.Y., Stoyanov, V.: Pretrained encyclopedia: Weakly supervised knowledge-pretrained language model. In: International Conference on Learning Representations (2020)

48. Liu, Q., Yogatama, D., Blunsom, P.: Relational memory augmented language models. arXiv preprint arXiv:2201.09680 (2022)

49. Hu, L., Liu, Z., Zhao, Z., Hou, L., Nie, L., Li, J.: A survey of knowledge-enhanced pre-trained language models. arXiv preprint arXiv:2211.05994 (2022)

50. Wikimedia Foundation: Wikimedia downloads. https://dumps.wikimedia.org

51. Vrandečić, D., Krötzsch, M.: Wikidata: A free collaborative knowledgebase. Communications of the ACM **57**, 78–85 (2014). https://doi.org/10.1145/2629489

52. Bollacker, K., Evans, C., Paritosh, P., Sturge, T., Taylor, J.: Freebase: A collaboratively created graph database for structuring human knowledge. In: Proceedings of the 2008 ACM SIGMOD International Conference on Management of Data, SIGMOD '08, pp. 1247–1250. Association for Computing Machinery, New York, NY, USA (2008). https://doi.org/10.1145/1376616.1376746

53. Li, H., Su, Y., Cai, D., Wang, Y., Liu, L.: A survey on retrieval-augmented text generation. arXiv preprint arXiv:2202.01110 (2022)

54. Wu, Y., Zhao, Y., Hu, B., Minervini, P., Stenetorp, P., Riedel, S.: An efficient memory-augmented transformer for knowledge-intensive nlp tasks. EMNLP (2022)

55. Sun, Y., Wang, S., Li, Y., Feng, S., Tian, H., Wu, H., Wang, H.: Ernie 2.0: A continual pre-training framework for language understanding. In: Proceedings of the AAAI conference on artificial intelligence, vol. 34, pp. 8968–8975 (2020)

56. Lazaridou, A., Gribovskaya, E., Stokowiec, W., Grigorev, N.: Internet-augmented language models through few-shot prompting for open-domain question answering. arXiv preprint arXiv:2203.05115 (2022)

57. Kasai, J., Sakaguchi, K., Takahashi, Y., Bras, R.L., Asai, A., Yu, X., Radev, D., Smith, N.A., Choi, Y., Inui, K.: Realtime qa: What's the answer right now? arXiv preprint arXiv:2207.13332 (2022)

58. Gao, L., Dai, Z., Pasupat, P., Chen, A., Chaganty, A.T., Fan, Y., Zhao, V.Y., Lao, N., Lee, H., Juan, D.C., et al.: Attributed text generation via post-hoc research and revision. arXiv preprint arXiv:2210.08726 (2022)

59. Fan, L., Wang, G., Jiang, Y., Mandlekar, A., Yang, Y., Zhu, H., Tang, A., Huang, D.A., Zhu, Y., Anandkumar, A.: Minedojo: Building open-ended embodied agents with internet-scale knowledge. arXiv preprint arXiv:2206.08853 (2022)

60. Schick, T., Dwivedi-Yu, J., Dessì, R., Raileanu, R., Lomeli, M., Zettlemoyer, L., Cancedda, N., Scialom, T.: Toolformer: Language models can teach themselves to use tools. CoRR **abs/2302.04761** (2023). https://doi.org/10.48550/arXiv.2302.04761

61. Yao, S., Zhao, J., Yu, D., Du, N., Shafran, I., Narasimhan, K., Cao, Y.: React: Synergizing reasoning and acting in language models. arXiv preprint arXiv:2210.03629 (2022)

62. Wang, W., Dong, L., Cheng, H., Song, H., Liu, X., Yan, X., Gao, J., Wei, F.: Visually-augmented language modeling. arXiv preprint arXiv:2205.10178 (2022)

63. Saeed, M., Ahmadi, N., Nakov, P., Papotti, P.: Rulebert: Teaching soft rules to pre-trained language models. In: M. Moens, X. Huang, L. Specia, S.W. Yih (eds.) Proceedings of the 2021 Conference on Empirical Methods in Natural Language Processing, EMNLP 2021, Virtual Event/Punta Cana, Dominican Republic, 7–11 November, 2021, pp. 1460–1476. Association for Computational Linguistics (2021). https://doi.org/10.18653/v1/2021.emnlp-main.110

64. Han, X., Zhao, W., Ding, N., Liu, Z., Sun, M.: PTR: prompt tuning with rules for text classification. AI Open **3**, 182–192 (2022). https://doi.org/10.1016/j.aiopen.2022.11.003

65. Wang, R., Tang, D., Duan, N., Wei, Z., Huang, X., Ji, J., Cao, G., Jiang, D., Zhou, M.: K-adapter: Infusing knowledge into pre-trained models with adapters. In: C. Zong, F. Xia, W. Li, R. Navigli (eds.) Findings of the Association for Computational Linguistics: ACL/IJCNLP 2021, Online Event, August 1–6, 2021, *Findings of ACL*, vol. ACL/IJCNLP 2021, pp. 1405–1418. Association for Computational Linguistics (2021). https://doi.org/10.18653/v1/2021.findings-acl.121

66. Liu, R., Wei, J., Gu, S.S., Wu, T.Y., Vosoughi, S., Cui, C., Zhou, D., Dai, A.M.: Mind's eye: Grounded language model reasoning through simulation. arXiv preprint arXiv:2210.05359 (2022)

67. Petroni, F., Rocktäschel, T., Riedel, S., Lewis, P.S.H., Bakhtin, A., Wu, Y., Miller, A.H.: Language models as knowledge bases? In: K. Inui, J. Jiang, V. Ng, X. Wan (eds.) Proceedings of the 2019 Conference on Empirical Methods in Natural Language Processing and the 9th International Joint Conference on Natural Language Processing, EMNLP-IJCNLP 2019, Hong Kong, China, November 3–7, 2019, pp. 2463–2473. Association for Computational Linguistics (2019). https://doi.org/10.18653/v1/D19-1250

68. Li, J., Hui, B., Qu, G., Li, B., Yang, J., Li, B., Wang, B., Qin, B., Cao, R., Geng, R., Huo, N., Ma, C., Chang, K., Huang, F., Cheng, R., Li, Y.: Can LLM already serve as a database interface? a big bench for large-scale database grounded text-to-sqls. ARXIV.ORG (2023). https://doi.org/10.48550/arXiv.2305.03111

69. Zhou, S., Alon, U., Xu, F.F., Wang, Z., Jiang, Z., Neubig, G.: Docprompting: Generating code by retrieving the docs. arXiv preprint arXiv:2207.05987 (2022)

70. Ye, H., Zhang, N., Deng, S., Chen, X., Chen, H., Xiong, F., Chen, X., Chen, H.: Ontology-enhanced prompt-tuning for few-shot learning. arXiv preprint arXiv:2201.11332 (2022)

71. Robertson, S., Zaragoza, H.: The probabilistic relevance framework: Bm25 and beyond. Found. Trends Inf. Retr. **3**(4), 333–389 (2009). https://doi.org/10.1561/1500000019

72. Chen, D., Fisch, A., Weston, J., Bordes, A.: Reading wikipedia to answer open-domain questions. arXiv preprint arXiv:1704.00051 (2017)

73. Zhang, F., Chen, B., Zhang, Y., Liu, J., Zan, D., Mao, Y., Lou, J.G., Chen, W.: Repocoder: Repository-level code completion through iterative retrieval and generation. arXiv preprint arXiv:2303.12570 (2023)

74. Jaccard, P.: The distribution of the flora in the alpine zone. The New Phytologist **11**(2), 37–50 (1912). http://www.jstor.org/stable/2427226

75. Dai, Z., Callan, J.: Context-aware term weighting for first stage passage retrieval. In: Proceedings of the 43rd International ACM SIGIR conference on research and development in Information Retrieval, pp. 1533–1536 (2020)

76. Zhao, T., Lu, X., Lee, K.: Sparta: Efficient open-domain question answering via sparse transformer matching retrieval. arXiv preprint arXiv:2009.13013 (2020)

77. Formal, T., Piwowarski, B., Clinchant, S.: SPLADE: Sparse Lexical and Expansion Model for First Stage Ranking, pp. 2288–2292. Association for Computing Machinery, New York, NY, USA (2021). https://doi.org/10.1145/3404835.3463098

78. Cai, D., Wang, Y., Li, H., Lam, W., Liu, L.: Neural machine translation with monolingual translation memory. In: Proceedings of the 59th Annual Meeting of the Association for Computational Linguistics and the 11th International Joint Conference on Natural Language Processing (Volume 1: Long Papers), pp. 7307–7318. Association for Computational Linguistics, Online (2021). https://doi.org/10.18653/v1/2021.acl-long.567. https://aclanthology.org/2021.acl-long.567

79. Izacard, G., Caron, M., Hosseini, L., Riedel, S., Bojanowski, P., Joulin, A., Grave, E.: Unsupervised dense information retrieval with contrastive learning. arXiv preprint arXiv:2112.09118 (2021)

80. Izacard, G., Grave, E.: Distilling knowledge from reader to retriever for question answering. In: International Conference on Learning Representations (2021)

81. Shi, W., Min, S., Yasunaga, M., Seo, M., James, R., Lewis, M., Zettlemoyer, L., Yih, W.t.: Replug: Retrieval-augmented black-box language models. arXiv preprint arXiv:2301.12652 (2023)

82. Jiang, Z., Xu, F.F., Araki, J., Neubig, G.: How Can We Know What Language Models Know? Transactions of the Association for Computational Linguistics **8**, 423–438 (2020). https://doi.org/10.1162/tacl_a_00324

83. Liu, J., Liu, A., Lu, X., Welleck, S., West, P., Le Bras, R., Choi, Y., Hajishirzi, H.: Generated knowledge prompting for commonsense reasoning. In: Proceedings of the 60th Annual Meeting of the Association for Computational Linguistics (Volume 1: Long Papers), pp. 3154–3169. Association for Computational Linguistics, Dublin, Ireland (2022).https://doi.org/10.18653/v1/2022.acl-long.225. https://aclanthology.org/2022.acl-long.225

84. Tay, Y., Tran, V.Q., Dehghani, M., Ni, J., Bahri, D., Mehta, H., Qin, Z., Hui, K., Zhao, Z., Gupta, J., Schuster, T., Cohen, W.W., Metzler, D.: Transformer memory as a differentiable search index. arXiv preprint arXiv:2202.06991 (2022)

85. Sun, Z., Wang, X., Tay, Y., Yang, Y., Zhou, D.: Recitation-augmented language models. arXiv preprint arXiv:2210.01296 (2022)

86. Yu, W., Iter, D., Wang, S., Xu, Y., Ju, M., Sanyal, S., Zhu, C., Zeng, M., Jiang, M.: Generate rather than retrieve: Large language models are strong context generators. arXiv preprint arXiv:2209.10063 (2022)

87. Ho, N., Schmid, L., Yun, S.Y.: Large language models are reasoning teachers. arXiv preprint arXiv:2212.10071 (2022)

88. Zhang, Y., Feng, S., Tan, C.: Active example selection for in-context learning. arXiv preprint arXiv:2211.04486 (2022)

89. Wei, J., Wang, X., Schuurmans, D., Bosma, M., Chi, E.H., Le, Q., Zhou, D.: Chain of thought prompting elicits reasoning in large language models. CoRR **abs/2201.11903** (2022). https://arxiv.org/abs/2201.11903

90. Zhou, D., Schärli, N., Hou, L., Wei, J., Scales, N., Wang, X., Schuurmans, D., Bousquet, O., Le, Q., Chi, E.: Least-to-most prompting enables complex reasoning in large language models. arXiv preprint arXiv:2205.10625 (2022)

91. Dohan, D., Xu, W., Lewkowycz, A., Austin, J., Bieber, D., Lopes, R.G., Wu, Y., Michalewski, H., Saurous, R.A., Sohl-dickstein, J., Murphy, K., Sutton, C.: Language model cascades. arXiv preprint arXiv:2207.10342 (2022)

92. Qiao, S., Ou, Y., Zhang, N., Chen, X., Yao, Y., Deng, S., Tan, C., Huang, F., Chen, H.: Reasoning with language model prompting: A survey. arXiv preprint arXiv:2212.09597 (2022)

93. Zeng, A., Attarian, M., Ichter, B., Choromanski, K., Wong, A., Welker, S., Tombari, F., Purohit, A., Ryoo, M., Sindhwani, V., Lee, J., Vanhoucke, V., Florence, P.: Socratic models: Composing zero-shot multimodal reasoning with language. arXiv preprint arXiv:2204.00598 (2022)

94. Bara, C.P., CH-Wang, S., Chai, J.: MindCraft: Theory of mind modeling for situated dialogue in collaborative tasks. In: Proceedings of the 2021 Conference on Empirical Methods in Natural Language Processing, pp. 1112–1125. Association for Computational Linguistics, Online and Punta Cana, Dominican Republic (2021). https://doi.org/10.18653/v1/2021.emnlp-main.85. https://aclanthology.org/2021.emnlp-main.85

95. Goyal, A., Didolkar, A.R., Lamb, A., Badola, K., Ke, N.R., Rahaman, N., Binas, J., Blundell, C., Mozer, M.C., Bengio, Y.: Coordination among neural modules through a shared global workspace. In: The Tenth International Conference on Learning Representations, ICLR 2022, Virtual Event, April 25–29, 2022. OpenReview.net (2022). https://openreview.net/forum?id=XzTtHjgPDsT

96. Qin, Y., Hu, S., Lin, Y., Chen, W., Ding, N., Cui, G., Zeng, Z., Huang, Y., Xiao, C., Han, C., Fung, Y., Su, Y., Wang, H., Qian, C., Tian, R., Zhu, K., Liang, S., Shen, X., Xu, B., Zhang, Z., Ye, Y., Li, B., Tang, Z., Yi, J., Zhu, Y., Dai, Z., Yan, L., Cong, X., Lu, Y.T., Zhao, W., Huang, Y., Yan, J.H., Han, X., Sun, X., Li, D., Phang, J., Yang, C., Wu, T., Ji, H., Liu, Z., Sun, M.: Tool learning with foundation models. ARXIV.ORG (2023). https://doi.org/10.48550/arXiv.2304.08354

97. Mialon, G., Dessì, R., Lomeli, M., Nalmpantis, C., Pasunuru, R., Raileanu, R., Roziére, B., Schick, T., Dwivedi-Yu, J., Celikyilmaz, A., Grave, E., LeCun, Y., Scialom, T.: Augmented language models: a survey. arXiv preprint arXiv:2302.07842 (2023)

98. Bubeck, S., Chandrasekaran, V., Eldan, R., Gehrke, J., Horvitz, E., Kamar, E., Lee, P., Lee, Y.T., Li, Y., Lundberg, S., Nori, H., Palangi, H., Ribeiro, M.T., Zhang, Y.: Sparks of artificial general intelligence: Early experiments with GPT-4 (2023). https://www.microsoft.com/en-us/research/publication/sparks-of-artificial-general-intelligence-early-experiments-with-gpt-4/

99. Clark, H.H.: Using Language. 'Using' Linguistic Books. Cambridge University Press (1996). https://doi.org/10.1017/CBO9780511620539

100. Premack, D., Woodruff, G.: Does the chimpanzee have a theory of mind? Behavioral and Brain Sciences **1**(4), 515–526 (1978). https://doi.org/10.1017/S0140525X00076512

101. Kosinski, M.: Theory of mind may have spontaneously emerged in large language models. arXiv preprint arXiv:2302.02083 (2023)

102. Park, J.S., O'Brien, J.C., Cai, C.J., Morris, M.R., Liang, P., Bernstein, M.S.: Generative agents: Interactive simulacra of human behavior. arXiv preprint arXiv:2304.03442 (2023)

103. Li, G., Hammoud, H.A.A.K., Itani, H., Khizbullin, D., Ghanem, B.: Camel: Communicative agents for "mind" exploration of large scale language model society. arXiv preprint arXiv:2303.17760 (2023)

104. Wang, X., Wei, J., Schuurmans, D., Le, Q., Chi, E., Narang, S., Chowdhery, A., Zhou, D.: Self-consistency improves chain of thought reasoning in language models. arXiv preprint arXiv:2203.11171 (2022)

105. Press, O., Zhang, M., Min, S., Schmidt, L., Smith, N.A., Lewis, M.: Measuring and narrowing the compositionality gap in language models. arXiv preprint arXiv:2210.03350 (2022)

106. Wang, B., Deng, X., Sun, H.: Iteratively prompt pre-trained language models for chain of thought. In: Y. Goldberg, Z. Kozareva, Y. Zhang (eds.) Proceedings of the 2022 Conference on Empirical Methods in Natural Language Processing, EMNLP 2022, Abu Dhabi, United Arab Emirates, December 7–11, 2022, pp. 2714–2730. Association for Computational Linguistics (2022). https://aclanthology.org/2022.emnlp-main.174

107. Dong, Q., Li, L., Dai, D., Zheng, C., Wu, Z., Chang, B., Sun, X., Xu, J., Li, L., Sui, Z.: A survey for in-context learning. CoRR **abs/2301.00234** (2023). https://doi.org/10.48550/arXiv.2301.00234

108. Zelikman, E., Mu, J., Goodman, N.D., Wu, Y.T.: Star: Self-taught reasoner bootstrapping reasoning with reasoning (2022)

109. Weng, Y., Zhu, M., He, S., Liu, K., Zhao, J.: Large language models are reasoners with self-verification. arXiv preprint arXiv:2212.09561 (2022)

110. Creswell, A., Shanahan, M., Higgins, I.: Selection-inference: Exploiting large language models for interpretable logical reasoning. arXiv preprint arXiv:2205.09712 (2022)

111. Madaan, A., Tandon, N., Gupta, P., Hallinan, S., Gao, L., Wiegreffe, S., Alon, U., Dziri, N., Prabhumoye, S., Yang, Y., Welleck, S., Majumder, B.P., Gupta, S., Yazdanbakhsh, A., Clark, P.: Self-refine: Iterative refinement with self-feedback. arXiv preprint arXiv:2303.17651 (2023)

112. Chen, Z., Zhou, Q., Shen, Y., Hong, Y., Zhang, H., Gan, C.: See, think, confirm: Interactive prompting between vision and language models for knowledge-based visual reasoning. arXiv preprint arXiv:2301.05226 (2023)

113. Cobbe, K., Kosaraju, V., Bavarian, M., Chen, M., Jun, H., Kaiser, L., Plappert, M., Tworek, J., Hilton, J., Nakano, R., Hesse, C., Schulman, J.: Training verifiers to solve math word problems. arXiv preprint arXiv:2110.14168 (2021)

114. Cranmer, K., Brehmer, J., Louppe, G.: The frontier of simulation-based inference. Proceedings of the National Academy of Sciences **117**(48), 30055–30062 (2020). https://doi.org/10.1073/pnas.1912789117. https://www.pnas.org/doi/abs/10.1073/pnas.1912789117

115. Ni, A., Iyer, S., Radev, D., Stoyanov, V., tau Yih, W., Wang, S.I., Lin, X.V.: Lever: Learning to verify language-to-code generation with execution. arXiv preprint arXiv:2302.08468 (2023)

116. Gao, L., Madaan, A., Zhou, S., Alon, U., Liu, P., Yang, Y., Callan, J., Neubig, G.: Pal: Program-aided language models. arXiv preprint arXiv:2211.10435 (2022)

117. Chen, W., Ma, X., Wang, X., Cohen, W.W.: Program of thoughts prompting: Disentangling computation from reasoning for numerical reasoning tasks. arXiv preprint arXiv:2211.12588 (2022)

118. Parisi, A., Zhao, Y., Fiedel, N.: Talm: Tool augmented language models. arXiv preprint arXiv:2205.12255 (2022)

119. Thoppilan, R., Freitas, D.D., Hall, J., Shazeer, N., Kulshreshtha, A., Cheng, H.T., Jin, A., Bos, T., Baker, L., Du, Y., Li, Y., Lee, H., Zheng, H.S., Ghafouri, A., Menegali, M., Huang, Y., Krikun, M., Lepikhin, D., Qin, J., Chen, D., Xu, Y., Chen, Z., Roberts, A., Bosma, M., Zhao, V., Zhou, Y., Chang, C.C., Krivokon, I., Rusch, W., Pickett, M., Srinivasan, P., Man, L., Meier-Hellstern, K., Morris, M.R., Doshi, T., Santos, R.D., Duke, T., Soraker, J., Zevenbergen, B., Prabhakaran, V., Diaz, M., Hutchinson, B., Olson, K., Molina, A., Hoffman-John, E., Lee, J., Aroyo, L., Rajakumar, R., Butryna, A., Lamm, M., Kuzmina, V., Fenton, J., Cohen, A., Bernstein, R., Kurzweil, R., Aguera-Arcas, B., Cui, C., Croak, M., Chi, E., Le, Q.: Lamda: Language models for dialog applications. arXiv preprint arXiv:2201.08239 (2022)

120. Liang, Y., Wu, C., Song, T., Wu, W., Xia, Y., Liu, Y., Ou, Y., Lu, S., Ji, L., Mao, S., Wang, Y., Shou, L., Gong, M., Duan, N.: Taskmatrix.ai: Completing tasks by connecting foundation models with millions of apis. arXiv preprint arXiv:2303.16434 (2023)

121. Wu, C., Yin, S., Qi, W., Wang, X., Tang, Z., Duan, N.: Visual ChatGPT: Talking, drawing and editing with visual foundation models. arXiv preprint arXiv:2303.04671 (2023)

122. Welleck, S., Kulikov, I., Roller, S., Dinan, E., Cho, K., Weston, J.: Neural text generation with unlikelihood training (2019)

123. Maynez, J., Narayan, S., Bohnet, B., McDonald, R.: On faithfulness and factuality in abstractive summarization (2020)

124. Patel, A., Bhattamishra, S., Goyal, N.: Are NLP models really able to solve simple math word problems? In: Proceedings of the 2021 Conference of the North American Chapter of the Association for Computational Linguistics: Human Language Technologies, pp. 2080–2094. Association for Computational Linguistics, Online (2021). https://doi.org/10.18653/v1/2021.naacl-main.168. https://aclanthology.org/2021.naacl-main.168

125. Komeili, M., Shuster, K., Weston, J.: Internet-augmented dialogue generation. In: Proceedings of the 60th Annual Meeting of the Association for Computational Linguistics (Volume 1: Long Papers), pp. 8460–8478. Association for Computational Linguistics, Dublin, Ireland (2022). https://doi.org/10.18653/v1/2022.acl-long.579. https://aclanthology.org/2022.acl-long.579

126. Lin, X.V., Mihaylov, T., Artetxe, M., Wang, T., Chen, S., Simig, D., Ott, M., Goyal, N., Bhosale, S., Du, J., Pasunuru, R., Shleifer, S., Koura, P.S., Chaudhary, V., O'Horo, B., Wang, J., Zettlemoyer, L., Kozareva, Z., Diab, M., Stoyanov, V., Li, X.: Few-shot learning with multilingual language models (2022)

127. Dhingra, B., Cole, J.R., Eisenschlos, J.M., Gillick, D., Eisenstein, J., Cohen, W.W.: Time-Aware Language Models as Temporal Knowledge Bases. Transactions of the Association for Computational Linguistics **10**, 257–273 (2022). https://doi.org/10.1162/tacl_a_00459

128. Paranjape, B., Lundberg, S., Singh, S., Hajishirzi, H., Zettlemoyer, L., Ribeiro, M.T.: Art: Automatic multi-step reasoning and tool-use for large language models. arXiv preprint arXiv:2303.09014 (2023)

129. Shen, Y., Song, K., Tan, X., Li, D., Lu, W., Zhuang, Y.: Hugginggpt: Solving ai tasks with ChatGPT and its friends in huggingface. arXiv preprint arXiv:2303.17580 (2023)

130. Ge, Y., Hua, W., Ji, J., Tan, J., Xu, S., Zhang, Y.: Openagi: When LLM meets domain experts. arXiv preprint arXiv:2304.04370 (2023)

131. Shah, D., Osinski, B., Ichter, B., Levine, S.: Lm-nav: Robotic navigation with large pre-trained models of language, vision, and action. arXiv preprint arXiv:2207.04429 (2022)

132. Min, S., Zhong, V., Zettlemoyer, L., Hajishirzi, H.: Multi-hop reading comprehension through question decomposition and rescoring. Annual Meeting Of The Association For Computational Linguistics (2019). https://doi.org/10.18653/v1/P19-1613

133. Talmor, A., Berant, J.: The web as a knowledge-base for answering complex questions. In: M.A. Walker, H. Ji, A. Stent (eds.) Proceedings of the 2018 Conference of the North American Chapter of the Association for Computational Linguistics: Human Language Technologies, NAACL-HLT 2018, New Orleans, Louisiana, USA, June 1–6, 2018, Volume 1 (Long Papers), pp. 641–651. Association for Computational Linguistics (2018). https://doi.org/10.18653/v1/n18-1059

134. Perez, E., Lewis, P., Yih, W.t., Cho, K., Kiela, D.: Unsupervised question decomposition for question answering. In: Proceedings of the 2020 Conference on Empirical Methods in Natural Language Processing (EMNLP), pp. 8864–8880. Association for Computational Linguistics, Online (2020). https://doi.org/10.18653/v1/2020.emnlp-main.713. https://aclanthology.org/2020.emnlp-main.713

135. Khot, T., Trivedi, H., Finlayson, M., Fu, Y., Richardson, K., Clark, P., Sabharwal, A.: Decomposed prompting: A modular approach for solving complex tasks. CoRR **abs/2210.02406** (2022). https://doi.org/10.48550/arXiv.2210.02406

136. Dua, D., Gupta, S., Singh, S., Gardner, M.: Successive prompting for decomposing complex questions. In: Y. Goldberg, Z. Kozareva, Y. Zhang (eds.) Proceedings of the 2022 Conference on Empirical Methods in Natural Language Processing, EMNLP 2022, Abu Dhabi, United Arab Emirates, December 7–11, 2022, pp. 1251–1265. Association for Computational Linguistics (2022). https://aclanthology.org/2022.emnlp-main.81

137. Claus, C., Boutilier, C.: The dynamics of reinforcement learning in cooperative multiagent systems. In: Proceedings of the Fifteenth National/Tenth Conference on Artificial Intelligence/Innovative Applications of Artificial Intelligence, AAAI '98/IAAI '98, p. 746–752. American Association for Artificial Intelligence, USA (1998)

138. Lazaridou, A., Peysakhovich, A., Baroni, M.: Multi-agent cooperation and the emergence of (natural) language. In: 5th International Conference on Learning Representations, ICLR 2017, Toulon, France, April 24–26, 2017, Conference Track Proceedings. OpenReview.net (2017). https://openreview.net/forum?id=Hk8N3Sclg

139. Li, Q., Zhou, W.: Connecting the dots between fact verification and fake news detection. In: Proceedings of the 28th International Conference on Computational Linguistics, pp. 1820–1825. International Committee on Computational Linguistics, Barcelona, Spain (Online) (2020). https://doi.org/10.18653/v1/2020.coling-main.165. https://aclanthology.org/2020.coling-main.165

140. Liu, D., Shah, V., Boussif, O., Meo, C., Goyal, A., Shu, T., Mozer, M., Heess, N., Bengio, Y.: Stateful active facilitator: Coordination and environmental heterogeneity in cooperative multi-agent reinforcement learning. arXiv preprint arXiv:2210.03022 (2022)

141. Liu, D., Shah, V., Boussif, O., Meo, C., Goyal, A., Shu, T., Mozer, M., Heess, N., Bengio, Y.: Coordinating policies among multiple agents via an intelligent communication channel. arXiv preprint arXiv:2205.10607 (2022)

142. Freedman, R.G., Fung, Y.R., Ganchin, R., Zilberstein, S.: Responsive planning and recognition for closed-loop interaction. arXiv preprint arXiv:1909.06427 (2019)

143. Ahn, M., Brohan, A., Brown, N., Chebotar, Y., Cortes, O., David, B., Finn, C., Gopalakrishnan, K., Hausman, K., Herzog, A., et al.: Do as i can, not as i say: Grounding language in robotic affordances. arXiv preprint arXiv:2204.01691 (2022)

144. Huang, W., Xia, F., Xiao, T., Chan, H., Liang, J., Florence, P., Zeng, A., Tompson, J., Mordatch, I., Chebotar, Y., et al.: Inner monologue: Embodied reasoning through planning with language models. arXiv preprint arXiv:2207.05608 (2022)

145. Dasgupta, I., Kaeser-Chen, C., Marino, K., Ahuja, A., Babayan, S., Hill, F., Fergus, R.: Collaborating with language models for embodied reasoning. arXiv preprint arXiv:2302.00763 (2023)

146. Huang, W., Abbeel, P., Pathak, D., Mordatch, I.: Language models as zero-shot planners: Extracting actionable knowledge for embodied agents. arXiv preprint arXiv:2201.07207 (2022)

147. Wang, Z., Cai, S., Liu, A., Ma, X., Liang, Y.: Describe, explain, plan and select: Interactive planning with large language models enables open-world multi-task agents. arXiv preprint arXiv:2302.01560 (2023)

148. Shinn, N., Cassano, F., Labash, B., Gopinath, A., Narasimhan, K., Yao, S.: Reflexion: Language agents with verbal reinforcement learning. arXiv preprint arXiv: 2303.11366 (2023)

149. Welleck, S., Lu, X., West, P., Brahman, F., Shen, T., Khashabi, D., Choi, Y.: Generating sequences by learning to self-correct. International Conference on Learning Representations (2022). https://doi.org/10.48550/arXiv.2211.00053

150. Saunders, W., Yeh, C., Wu, J., Bills, S., Ouyang, L., Ward, J., Leike, J.: Self-critiquing models for assisting human evaluators. arXiv preprint arXiv: 2206.05802 (2022)

151. Ye, S., Jo, Y., Kim, D., Kim, S., Hwang, H., Seo, M.: Selfee: Iterative self-revising LLM empowered by self-feedback generation. Blog post (2023). https://kaistai.github.io/SelFee/

152. Li, Y., Wei, F., Zhao, J., Zhang, C., Zhang, H.: Rain: Your language models can align themselves without finetuning. arXiv preprint arXiv: 2309.07124 (2023). https://arxiv.org/abs/2309.07124v2

153. Rabinowitz, N.C., Perbet, F., Song, H.F., Zhang, C., Eslami, S., Botvinick, M.: Machine theory of mind. International Conference On Machine Learning (2018)

154. Zhu, H., Neubig, G., Bisk, Y.: Few-shot language coordination by modeling theory of mind. arXiv preprint arXiv:2107.05697 (2021)

155. Liu, A., Zhu, H., Liu, E., Bisk, Y., Neubig, G.: Computational language acquisition with theory of mind. arXiv preprint arXiv:2303.01502 (2023)

156. Andreas, J.: Language models as agent models. In: Y. Goldberg, Z. Kozareva, Y. Zhang (eds.) Findings of the Association for Computational Linguistics: EMNLP 2022, Abu Dhabi, United Arab Emirates, December 7–11, 2022, pp. 5769–5779. Association for Computational Linguistics (2022). https://aclanthology.org/2022.findings-emnlp.423

157. Wang, Y., Zhang, X.Y., Liu, C.L., Zhang, Z.: Emergence of machine language: Towards symbolic intelligence with neural networks. arXiv preprint arXiv:2201.05489 (2022)

158. Yuan, L., Zhu, S.C.: Communicative learning: A unified learning formalism. Engineering (2023). https://doi.org/10.1016/j.eng.2022.10.017. https://www.sciencedirect.com/science/article/pii/S2095809923001339

159. ter Hoeve, M., Kharitonov, E., Hupkes, D., Dupoux, E.: Towards interactive language modeling. arXiv preprint arXiv:2112.11911 (2021)

160. Liang, J., Huang, W., Xia, F., Xu, P., Hausman, K., Ichter, B., Florence, P., Zeng, A.: Code as policies: Language model programs for embodied control. arXiv preprint arXiv:2209.07753 (2022)

161. Chen, B., Xia, F., Ichter, B., Rao, K., Gopalakrishnan, K., Ryoo, M.S., Stone, A., Kappler, D.: Open-vocabulary queryable scene representations for real world planning. arXiv preprint arXiv:2209.09874 (2022)

162. Carta, T., Romac, C., Wolf, T., Lamprier, S., Sigaud, O., Oudeyer, P.Y.: Grounding large language models in interactive environments with online reinforcement learning. arXiv preprint arXiv:2302.02662 (2023)

163. Huang, W., Xia, F., Shah, D., Driess, D., Zeng, A., Lu, Y., Florence, P., Mordatch, I., Levine, S., Hausman, K., et al.: Grounded decoding: Guiding text generation with grounded models for robot control. arXiv preprint arXiv:2303.00855 (2023)

164. Bisk, Y., Holtzman, A., Thomason, J., Andreas, J., Bengio, Y., Chai, J., Lapata, M., Lazaridou, A., May, J., Nisnevich, A., Pinto, N., Turian, J.: Experience grounds language. arXiv preprint arXiv:2004.10151 (2020)

165. Li, S., Puig, X., Paxton, C., Du, Y., Wang, C., Fan, L., Chen, T., Huang, D.A., Akyürek, E., Anandkumar, A., et al.: Pre-trained language models for interactive decision-making. Advances in Neural Information Processing Systems **35**, 31199–31212 (2022)

166. Yang, S., Nachum, O., Du, Y., Wei, J., Abbeel, P., Schuurmans, D.: Foundation models for decision making: Problems, methods, and opportunities. arXiv preprint arXiv:2303.04129 (2023)

167. Kawaguchi, K., Kaelbling, L.P., Bengio, Y.: Generalization in deep learning. arXiv preprint arXiv:1710.05468 (2017)

168. Liu, D., Lamb, A., Kawaguchi, K., Goyal, A., Sun, C., Mozer, M.C., Bengio, Y.: Discrete-valued neural communication. arXiv preprint arXiv:2107.02367 (2021)

169. Bolotta, S., Dumas, G.: Social neuro ai: Social interaction as the "dark matter" of ai. Frontiers in Computer Science **4** (2022). https://doi.org/10.3389/fcomp.2022.846440. https://www.frontiersin.org/articles/10.3389/fcomp.2022.846440

170. Todorov, E., Erez, T., Tassa, Y.: Mujoco: A physics engine for model-based control. In: 2012 IEEE/RSJ International Conference on Intelligent Robots and Systems, pp. 5026–5033. IEEE (2012). https://doi.org/10.1109/IROS.2012.6386109

171. Beinborn, L., Botschen, T., Gurevych, I.: Multimodal grounding for language processing. In: Proceedings of the 27th International Conference on Computational Linguistics, pp. 2325–2339. Association for Computational Linguistics, Santa Fe, New Mexico, USA (2018). https://aclanthology.org/C18-1197

172. Tan, H., Bansal, M.: Lxmert: Learning cross-modality encoder representations from transformers. IJCNLP (2019)

173. Lu, J., Batra, D., Parikh, D., Lee, S.: Vilbert: Pretraining task-agnostic visiolinguistic representations for vision-and-language tasks. In: H.M. Wallach, H. Larochelle, A. Beygelzimer, F. d'Alché-Buc, E.B. Fox, R. Garnett (eds.) Advances in Neural Information Processing Systems 32: Annual Conference on Neural Information Processing Systems 2019, NeurIPS 2019, December 8-14, 2019, Vancouver, BC, Canada, pp. 13–23 (2019). https://proceedings.neurips.cc/paper/2019/hash/c74d97b01eae257e44aa9d5bade97baf-Abstract.html

174. Radford, A., Kim, J.W., Hallacy, C., Ramesh, A., Goh, G., Agarwal, S., Sastry, G., Askell, A., Mishkin, P., Clark, J., Krueger, G., Sutskever, I.: Learning transferable visual models from natural language supervision. International Conference On Machine Learning (2021)

175. Xu, X., Wu, C., Rosenman, S., Lal, V., Che, W., Duan, N.: Bridgetower: Building bridges between encoders in vision-language representation learning. arXiv preprint arXiv:2206.08657 (2022)

176. Li, J., Selvaraju, R.R., Gotmare, A.D., Joty, S.R., Xiong, C., Hoi, S.: Align before fuse: Vision and language representation learning with momentum distillation. Neural Information Processing Systems (2021)

177. Yu, J., Wang, Z., Vasudevan, V., Yeung, L., Seyedhosseini, M., Wu, Y.: Coca: Contrastive captioners are image-text foundation models. arXiv preprint arXiv:2205.01917 (2022)

178. Zeng, Y., Zhang, X., Li, H., Wang, J., Zhang, J., Zhou, W.: X^2-vlm: All-in-one pre-trained model for vision-language tasks. arXiv preprint arXiv:2211.11402 (2022)

179. Su, W., Zhu, X., Cao, Y., Li, B., Lu, L., Wei, F., Dai, J.: Vl-bert: Pre-training of generic visual-linguistic representations. arXiv preprint arXiv:1908.08530 (2019)

180. Li, L.H., Yatskar, M., Yin, D., Hsieh, C.J., Chang, K.W.: Visualbert: A simple and performant baseline for vision and language. arXiv preprint arXiv:1908.03557 (2019)

181. Chen, Y., Li, L., Yu, L., Kholy, A.E., Ahmed, F., Gan, Z., Cheng, Y., Liu, J.: UNITER: universal image-text representation learning. In: A. Vedaldi, H. Bischof, T. Brox, J. Frahm (eds.) Computer Vision—ECCV 2020—16th European Conference, Glasgow, UK, August 23–28, 2020, Proceedings, Part XXX, *Lecture Notes in Computer Science*, vol. 12375, pp. 104–120. Springer (2020). https://doi.org/10.1007/978-3-030-58577-8_7

182. Li, X., Yin, X., Li, C., Zhang, P., Hu, X., Zhang, L., Wang, L., Hu, H., Dong, L., Wei, F., Choi, Y., Gao, J.: Oscar: Object-semantics aligned pre-training for vision-language tasks. In: A. Vedaldi, H. Bischof, T. Brox, J. Frahm (eds.) Computer Vision—ECCV 2020—16th European Conference, Glasgow, UK, August 23–28, 2020, Proceedings, Part XXX, *Lecture Notes in Computer Science*, vol. 12375, pp. 121–137. Springer (2020). https://doi.org/10.1007/978-3-030-58577-8_8

183. Wang, P., Yang, A., Men, R., Lin, J., Bai, S., Li, Z., Ma, J., Zhou, C., Zhou, J., Yang, H.: OFA: unifying architectures, tasks, and modalities through a simple sequence-to-sequence learning framework. In: K. Chaudhuri, S. Jegelka, L. Song, C. Szepesvári, G. Niu, S. Sabato (eds.) International Conference on Machine Learning, ICML 2022, 17–23 July 2022, Baltimore, Maryland, USA, *Proceedings of Machine Learning Research*, vol. 162, pp. 23318–23340. PMLR (2022). https://proceedings.mlr.press/v162/wang22al.html

184. Reed, S., Zolna, K., Parisotto, E., Colmenarejo, S.G., Novikov, A., Barth-maron, G., Giménez, M., Sulsky, Y., Kay, J., Springenberg, J.T., Eccles, T., Bruce, J., Razavi, A., Edwards, A., Heess, N., Chen, Y., Hadsell, R., Vinyals, O., Bordbar, M., de Freitas, N.: A generalist agent. Transactions on Machine Learning Research (2022). https://openreview.net/forum?id=1ikK0kHjvj. Featured Certification

185. Brohan, A., Brown, N., Carbajal, J., Chebotar, Y., Dabis, J., Finn, C., Gopalakrishnan, K., Hausman, K., Herzog, A., Hsu, J., Ibarz, J., Ichter, B., Irpan, A., Jackson, T., Jesmonth, S., Joshi, N., Julian, R., Kalashnikov, D., Kuang, Y., Leal, I., Lee, K.H., Levine, S., Lu, Y., Malla, U., Manjunath, D., Mordatch, I., Nachum, O., Parada, C., Peralta, J., Perez, E., Pertsch, K., Quiambao, J., Rao, K., Ryoo, M., Salazar, G., Sanketi, P., Sayed, K., Singh, J., Sontakke, S., Stone, A., Tan, C., Tran, H., Vanhoucke, V., Vega, S., Vuong, Q., Xia, F., Xiao, T., Xu, P., Xu, S., Yu, T., Zitkovich, B.: Rt-1: Robotics transformer for real-world control at scale. In: arXiv preprint arXiv:2212.06817 (2022)

186. Koh, J.Y., Salakhutdinov, R., Fried, D.: Grounding language models to images for multimodal generation. arXiv preprint arXiv:2301.13823 (2023)

187. Driess, D., Xia, F., Sajjadi, M.S.M., Lynch, C., Chowdhery, A., Ichter, B., Wahid, A., Tompson, J., Vuong, Q., Yu, T., Huang, W., Chebotar, Y., Sermanet, P., Duckworth, D., Levine, S., Vanhoucke, V., Hausman, K., Toussaint, M., Greff, K., Zeng, A., Mordatch, I., Florence, P.: Palm-e: An embodied multimodal language model. In: arXiv preprint arXiv:2303.03378 (2023)

188. Chen, X., Wang, X., Changpinyo, S., Piergiovanni, A., Padlewski, P., Salz, D., Goodman, S., Grycner, A., Mustafa, B., Beyer, L., Kolesnikov, A., Puigcerver, J., Ding, N., Rong, K., Akbari, H., Mishra, G., Xue, L., Thapliyal, A., Bradbury, J., Kuo, W., Seyedhosseini, M., Jia, C., Ayan, B.K., Riquelme, C., Steiner, A., Angelova, A., Zhai, X., Houlsby, N., Soricut, R.: Pali: A jointly-scaled multilingual language-image model. arXiv preprint arXiv:2209.06794 (2022)

189. Wang, W., Bao, H., Dong, L., Bjorck, J., Peng, Z., Liu, Q., Aggarwal, K., Mohammed, O.K., Singhal, S., Som, S., Wei, F.: Image as a foreign language: Beit pretraining for all vision and vision-language tasks. arXiv preprint arXiv:2208.10442 (2022)

190. Wang, H., Ma, S., Huang, S., Dong, L., Wang, W., Peng, Z., Wu, Y., Bajaj, P., Singhal, S., Benhaim, A., Patra, B., Liu, Z., Chaudhary, V., Song, X., Wei, F.: Foundation transformers. arXiv preprint arXiv:2210.06423 (2022)
191. Diao, S., Zhou, W., Zhang, X., Wang, J.: Write and paint: Generative vision-language models are unified modal learners. In: The Eleventh International Conference on Learning Representations, ICLR 2023, Kigali, Rwanda, May 1–5, 2023. OpenReview.net (2023). https://openreview.net/pdf?id=HgQR0mXQ1_a
192. Huang, S., Dong, L., Wang, W., Hao, Y., Singhal, S., Ma, S., Lv, T., Cui, L., Mohammed, O.K., Patra, B., Liu, Q., Aggarwal, K., Chi, Z., Bjorck, J., Chaudhary, V., Som, S., Song, X., Wei, F.: Language is not all you need: Aligning perception with language models. arXiv preprint arXiv:2302.14045 (2023)
193. Alayrac, J.B., Donahue, J., Luc, P., Miech, A., Barr, I., Hasson, Y., Lenc, K., Mensch, A., Millican, K., Reynolds, M., Ring, R., Rutherford, E., Cabi, S., Han, T., Gong, Z., Samangooei, S., Monteiro, M., Menick, J., Borgeaud, S., Brock, A., Nematzadeh, A., Sharifzadeh, S., Binkowski, M., Barreira, R., Vinyals, O., Zisserman, A., Simonyan, K.: Flamingo: a visual language model for few-shot learning. DEEPMIND (2022)
194. Li, J., Li, D., Savarese, S., Hoi, S.: Blip-2: Bootstrapping language-image pre-training with frozen image encoders and large language models. arXiv preprint arXiv:2301.12597 (2023)
195. Zhu, D., Chen, J., Haydarov, K., Shen, X., Zhang, W., Elhoseiny, M.: ChatGPT asks, blip-2 answers: Automatic questioning towards enriched visual descriptions. arXiv preprint arXiv:2303.06594 (2023)
196. Yao, S., Chen, H., Yang, J., Narasimhan, K.: Webshop: Towards scalable real-world web interaction with grounded language agents. arXiv preprint arXiv:2207.01206 (2022)
197. He, K., Zhang, X., Ren, S., Sun, J.: Deep residual learning for image recognition. Computer Vision And Pattern Recognition (2015). https://doi.org/10.1109/cvpr.2016.90
198. Vaswani, A., Shazeer, N.M., Parmar, N., Uszkoreit, J., Jones, L., Gomez, A.N., Kaiser, L., Polosukhin, I.: Attention is all you need. NIPS (2017)
199. Jiang, Y., Gupta, A., Zhang, Z., Wang, G., Dou, Y., Chen, Y., Fei-Fei, L., Anandkumar, A., Zhu, Y., Fan, L.: Vima: General robot manipulation with multimodal prompts. arXiv preprint arXiv:2210.03094 (2022)
200. Suris, D., Menon, S., Vondrick, C.: Vipergpt: Visual inference via python execution for reasoning. arXiv preprint arXiv:2303.08128 (2023)
201. Kaplan, J., McCandlish, S., Henighan, T., Brown, T.B., Chess, B., Child, R., Gray, S., Radford, A., Wu, J., Amodei, D.: Scaling laws for neural language models. arXiv preprint arXiv:2001.08361 (2020)
202. Wei, J., Tay, Y., Bommasani, R., Raffel, C., Zoph, B., Borgeaud, S., Yogatama, D., Bosma, M., Zhou, D., Metzler, D., Chi, E.H., Hashimoto, T., Vinyals, O., Liang, P., Dean, J., Fedus, W.: Emergent abilities of large language models. arXiv preprint arXiv:2206.07682 (2022)
203. Gibson, J.J.: The ecological approach to visual perception: classic edition. Psychology press (2014)
204. Khetarpal, K., Ahmed, Z., Comanici, G., Abel, D., Precup, D.: What can i do here? a theory of affordances in reinforcement learning. ICML (2020)
205. Abramson, J., Ahuja, A., Carnevale, F., Georgiev, P., Goldin, A., Hung, A., Landon, J., Lhotka, J., Lillicrap, T., Muldal, A., Powell, G., Santoro, A., Scully, G., Srivastava, S., von Glehn, T., Wayne, G., Wong, N., Yan, C., Zhu, R.: Improving multimodal interactive agents with reinforcement learning from human feedback. arXiv preprint arXiv:2211.11602 (2022)
206. Wake, N., Kanehira, A., Sasabuchi, K., Takamatsu, J., Ikeuchi, K.: ChatGPT empowered long-step robot control in various environments: A case application. arXiv preprint arXiv:2304.03893 (2023)

Interaction Interface

3

Zekun Wang, Ge Zhang, Ning Shi and Chenghua Lin

3.1 Natural Language

Natural Language is the most common interaction interface. Communicating via this interface requires that the interactive objects can effectively understand and produce natural language. This interface is therefore commonly used in Model-in-the-loop [1, 2] and Human-in-the-loop [3, 4]. Natural language interaction empowers users to express their needs with inherent expressiveness, enabling effective communication of their requirements without the need for specialized training. Additionally, this interaction interface facilitates a better understanding of the intermediate interaction process, leading to improved debuggability and interpretability of the interaction chain [1, 4, 5]. Crucially, since LMs are primarily pre-trained on natural language, interacting with them through natural language instead of other language is the most effective way to activate and utilize the knowledge encoded in the LMs. This alignment between the LM training data and the interaction interface allows for optimal utilization of the knowledge contained within the LMs.

Z. Wang
Beihang University, Beijing, China
e-mail: zenmoore@buaa.edu.cn

G. Zhang
University of Michigan, Beijing, China
e-mail: gezhang@umich.edu

N. Shi
University of Alberta, Edmonton, Canada
e-mail: ning.shi@ualberta.ca

C. Lin (✉)
Department of Computer Science, University of Manchester, Manchester, UK
e-mail: chenghua.lin@manchester.ac.uk

© The Author(s), under exclusive license to Springer Nature Switzerland AG 2026
Z. Wang et al. (eds.), *Interactive Natural Language Processing*, Synthesis Lectures
on Human Language Technologies, https://doi.org/10.1007/978-3-032-06264-2_3

Fig. 3.1 Interacting via natural language

However, interacting with a language model through natural language heavily relies on the organization and the utterance of the language, often necessitating intricate prompt engineering [6–11]. Organization of the language refers to the structure of a model's prompt, and can be categorized into **unstructural natural language** and **structural natural language**. Utterance, on the other hand, refers to the specific wording or language used to express a given prompt or query. Utterance is more flexible by nature and therefore difficult to determine an optimal one. Different utterances may produce different results as they differ from the activated pattern in the model parameters. Practically, suitable prompts can be discovered through manual or automatic search [6, 10, 12–17]. We refer the readers to Sect. 4.3.3 for more information (Fig. 3.1).

Unstructural Natural Language. Unstructural natural language is a free-form text. When it serves as an output from the language model, it does not have specific categorization, and the content can be free-form responses such as answers to questions and textual feedback. When it serves as an input to the language model, in addition to the main input content, such as interaction messages and queries, it primarily takes three forms of auxiliary context: (1) few-shot examples, (2) task description, and (3) role assignment [18–20]. Thereof,

- Input format example of few-shot prompting: "*[Example-1]; [Example-2]; [Example-3]; [input]*" e.g. "*sea otter → loutre de mer; plush girafe → girafe peluche; cheese →*" for translation task.
- Input format example of task description: "*[task description]: [input]*" e.g. "*translate English to French: cheese →*".

- Input format example of role assignment: *"[role assignment]. [input]"* e.g. *"Act as a python programmer: write codes to detect objects."*.[1]

For example, some recent work, including Natural Instructions [18] and Super Natural Instructions [19], have built comprehensive collections of tasks and their corresponding instructions in natural language. Interactive Language [21] enables humans to provide real-time instructions for the multimodal language model based on the current state of a given environment for robotic manipulation. Camel [20] defines an inception prompt that comprises a task specifier prompt and two role assignment prompts, namely the assistant system prompt and the user system prompt, which are utilized for role-playing tasks.

Structural Natural Language. Structural natural language usually imposes explicit constraints on the text in terms of content or formatting. Such constraints can be imposed on either the input [22] or output [23] of language models. For example, [24–26] define the overall structure of the generated article via an Outline or Plan (e.g., *"(1) Introduction, (2) Related Work, (3) Method, (4) Experimental Results, ..."* or a storyline). References [23, 27] unify the format of a generated text via a Template (e.g., *"pick up [object]"*) to facilitate parsing of the action and the object to be acted upon. ProQA [22] employs a prompt-based input schema that is designed in a structured manner, e.g., *"[Format]: <Extractive QA>; [Task]: <SQuAD>; [Domain]: <Wikipedia>; [Question]: In what Country is Normandy located? [Passage]: ..."*. This schema allows for efficient modeling of knowledge generalization across all QA tasks, while also preserving task-specific knowledge tailored to each individual QA task. Note that although ProQA incorporates certain soft prompts in its input schema (c.f., Sect. 3.4), the main body of its instance still consists of natural language.

While unstructured natural language is a widely used interface for interaction due to its flexibility, simplicity, and readability, it suffers from certain drawbacks, including ambiguity, lack of coherence and parsability. Although these challenges can be partially addressed by employing structural natural language, all forms of natural language are inherently limited by their subjectivity and variability.

3.2 Formal Language

To further unlock the benefits of structural language, such as unambiguity, coherence, and parsability, and to mitigate the inherent limitations of natural language mentioned above, formal language emerges as another important interaction interface. According to Wikipedia,[2] *"a formal language consists of words whose letters are taken from an alphabet and are well-formed according to a specific set of rules."* Formal language is utilized in various domains such as mathematics, logic, linguistics, computer science, as well as other fields

---

[1] Role assignment can be considered a special type of task descriptions.

[2] https://en.wikipedia.org/wiki/Formal_language.

Fig. 3.2 Interacting via formal language

where precise and unambiguous communication is essential. Here are some examples of formal languages (Fig. 3.2):

1. Programming Languages: examples include C, Java, Python, and many others. These programming languages are used to write scripts or commands that computers can execute [28–32].
2. Query Languages: examples include SQL and XQuery, which are used to retrieve and manipulate data stored in databases [28, 33].
3. Mathematical Expressions: examples include Boolean algebra, first-order logic, and equations. They are used to describe mathematical concepts and relationships [34–36].
4. Formal Grammars: examples include context-free grammars, regular grammars, recursive grammars, etc.[3] They are used to describe the syntactic structure of natural language [37–39].
5. Others: for example, knowledge triples [40, 41], and regular expressions (regex, [42]).

The interactive objects that use formal language as an interaction interface usually include knowledge bases [28, 33, 40], environments [29], and models/tools [34, 35, 43–45]. For example, Mind's-Eye [43] uses a text-to-code language model to generate rendering codes for the physical simulation engine. Jiang et al. [44] involve a three-step approach to creating mathematical proofs. This approach includes formulating an initial informal proof, converting it into a formal sketch, and then employing a standard prover to prove the con-

---

[3] https://en.wikipedia.org/wiki/Formal_grammar.

jectures. This allows for the automated transformation of informal mathematical issues into fully formalized proofs using natural and mathematical languages. Binder [28] first parses its input into programs (Python, SQL, etc.) given the questions and knowledge bases, and then executes them to get the results. K-Adapter [39] incorporates linguistic knowledge into PLMs through the use of adapters [46], exemplifying the application of formal grammars as an interaction interface. In some specific cases, other interactive objects may also use formal language. For example, human developers can interact with a code-based language model [47] using formal language. Lahiri et al. [48] create an interactive framework to refine user intents through test case generations and user feedback.

Compared to natural language, formal language offers distinctive advantages as an interaction interface, including: (1) It brings about precision and clarity, eradicating the ambiguity often associated with natural language. (2) Its structured syntax and rules make it directly parsable and easily interpretable by programs, enabling more efficient and accurate interaction with tools, for example. (3) It facilitates complex reasoning and logic-based operations more effectively, as codes or mathematical proofs are data formats that encompass a series of logical reasoning steps, which may provide opportunities to enhance models' reasoning abilities [49, 50]. However, the use of formal language may have certain limitations, including: (1) Limited accessibility: It often requires specialized knowledge or training for proper understanding and usage. And it relies on LMs specifically trained with formal language. (2) High sensitivity: E.g., even small errors in codes can render them non-executable. (3) Lack of expressiveness: It is unable to convey ideas in a nuanced and flexible manner.

3.3 Edits

Text editing aims to reconstruct the textual source input to the target one by applying a set of edits, such as deletion, insertion, and substitution [51]. The motivation behind text editing is the recognition that source and target texts often share significant similarities in various monolingual tasks. Instead of reproducing the source words [52–55], text editing models reduce such copying to predicting a single keep operation. Also, edits are often facilitated with rich metadata about language editing, including the inserted or deleted spans and the word order (Fig. 3.3).

Learning from the editing of textual data is gaining increasing attention, given its success in code pre-training [56], image editing [57], drug design [58], and other areas. Previous cognition-related research has proven that mechanical editing operations require less cognitive effort compared to correcting transfer errors, which have no references to the source text version [59], and that iterative editing procedure plays an important role in improving students' writing abilities [60, 61]. Moreover, these editing-related cognitive phenomena have shined upon various NLP topics. By addressing some limitations of the dominant sequence-to-sequence approaches [62], such as a relatively high computational requirement [63], text editing has found its wide array of applications [63–65] such as automatic post-editing [66,

Fig. 3.3 Interacting via edits

67], data-to-text generation [68], grammatical error correction [69–72], punctuation restoration [73–76], sentence simplification [77, 78], human value alignment [79, 80], style transfer [81], and sequence-to-sequence pre-training [82].

Similar to the pattern of repeated revisions made by humans to a manuscript until it is finalized, a complete process of text editing can be decomposed to multiple iterative rounds of editing, rather than one-pass edit [83–89]. On account of this, edits can be treated as one kind of interaction interface. Typically, text editing can be conducted through interaction between the editing model and itself, with the outputs of the previous iterations as the input of the current one until the text is fully edited to be returned [68, 90–92]. Meanwhile, text editing can also be conducted through interaction among multiple different models or modules [63, 93–95]. For example, an edit can be split into tasks, such as sequence tagging and masked language modeling, for models to cooperate. Specifically, a tagger first attaches an edit operation to each token. Afterwards, a masked language model fills in the placeholders for insertion and substitution operations to complete the edit [63, 95]. Moreover, the participation of a code interpreter [77, 87], environment [88], and user simulator [89] can control the editing better and provide additional supervision signals.

Recent research proves that editing-based models can be expanded to various NLP downstream tasks by retrieving or generating prototypes, i.e., original text to be edited [51, 96]. Additionally, text editing models have shown impressive performance in low-resource settings and can get rid of the typical autoregressive mechanism, thus improving inference speed [63, 69]. However, it is still under-explored how to automatically generate prototypes for general NLG tasks so as to expand the text editing paradigm to them [97], which hinders the broad use of edits as an interaction interface.

3.4　Representation Language

In some cases, the communication language between the language model and interactive objects is not human-readable. This communication interface is referred to as Representation Language, as it employs a human-unreadable language that can be interpreted and processed by computers (e.g., models or tools) to represent signals such as images, models' states, or rewards. We can break down this type of interaction interface into two categories: **discrete representation language** and **continuous representation language** (Fig. 3.4).

Discrete Representation Language. It refers to an interaction interface that is not readable by humans and quantized. For example, OFA [98] and BEiT-3 [99] treat images as a form of "*foreign language*". That is, the sequence of image patches is obtained through image quantization and discretization techniques [100–103]. This process allows the generation or understanding of an image token sequence that cannot be directly readable by humans but can be processed by models such as VQ-VAE [100] or VQGAN [101, 102]. Similarly, the hidden states inside the language model can also be discretized into discrete representation language. For example, [104, 105] have demonstrated that discretized, human-unreadable hidden states can lead to better generalization and robustness.

Continuous Representation Language. It refers to an interaction interface through which the language models communicate with interactive objects using continuous scalars or vectors in a dense space. For example, Flamingo [106] encodes and re-samples images into dense and continuous vectors and then passes them to a language model via cross

Fig. 3.4 Interacting via representation language

attention, BLIP-2 [107] and LLaVA [108] encodes and maps images into numerous soft tokens[4] and passes them to a language model as prefixes of text inputs.

Note that metric signals, such as scalar rewards and ranking scores [3, 109, 110], can also be regarded as a form of representation language employed by language models. In particular, if these signals belong to a discrete set of numbers (e.g., $\in \mathbb{Z}$), they can be classified as a form of discrete representation language. On the other hand, if they are represented as continuous values (e.g., $\in \mathbb{R}$), they can be classified as a form of continuous representation language.

3.5   Shared Memory

The interaction interfaces discussed earlier focus on direct communication between language models and interactive objects. However, there is also a form of indirect communication facilitated through shared information units, commonly referred to as shared memory [2, 111–116]. That is, the message receiver does not directly receive the message from the sender, but instead retrieves it from a memory pool where the message has been pre-written by the sender. Depending on the form in which the message is stored and utilized, this type of interaction interface can be classified into two categories: **hard memory** and **soft memory** (Fig. 3.5).

Hard Memory. Hard memory often utilizes a human-readable history log to store shared information. For example, Socratic Models [2] employs a communication mechanism where Vision-Language Models (VLM), Audio-Language Models (ALM), and Language Models (LM) interact through a history log written in natural language. This log records the complete history of states perceived by each model [2]. MemPrompt [112] edits human prompts to GPT-3 with user feedback memory for better Human-LM interaction. Dalvi et al. [113] augment a question-answering model with a dynamic memory of user feedback for continual learning.

Soft Memory. Soft memory typically employs queryable and human-unreadable memory slots to store shared information. These memory slots utilize a continuous representation language for efficient storage and retrieval of information. For example, Shared Global Workspace [111] stores the information from multiple modules in a shared sequence of working memory slots to facilitate inter-module communication and coordination. Token Turing Machines [114] use an additional memory unit to store historical state information to aid long-horizon robotic manipulation.

Utilizing memory as an indirect interaction interface provides several advantages over direct communication. It allows interactive objects to retrieve messages from earlier moments, enabling them to access past information. Memory enables the storage of a large volume of information, facilitating high-throughput communication. However, memory can become noised or outdated, leading to potential confusion or errors. Retrieval from memory

---

[4] Soft tokens, also known as soft prompts, refer to learnable parameters that are concatenated to the input prompt of a language model. Please refer to [12].

Fig. 3.5 Interacting via shared memory

can be time-consuming, impacting the efficiency of the interaction. It can also introduce unpredictability and uncertainty into the interaction. Therefore, careful design is crucial to ensure the effective and efficient utilization of memory.

References

1. Wu, T.S., Terry, M., Cai, C.J.: Ai chains: Transparent and controllable human-ai interaction by chaining large language model prompts. International Conference On Human Factors In Computing Systems (2021). https://doi.org/10.1145/3491102.3517582
2. Zeng, A., Attarian, M., Ichter, B., Choromanski, K., Wong, A., Welker, S., Tombari, F., Purohit, A., Ryoo, M., Sindhwani, V., Lee, J., Vanhoucke, V., Florence, P.: Socratic models: Composing zero-shot multimodal reasoning with language. arXiv preprint arXiv: Arxiv-2204.00598 (2022)
3. Ouyang, L., Wu, J., Jiang, X., Almeida, D., Wainwright, C.L., Mishkin, P., Zhang, C., Agarwal, S., Slama, K., Ray, A., et al.: Training language models to follow instructions with human feedback. arXiv preprint arXiv:2203.02155 (2022)
4. Lee, M., Srivastava, M., Hardy, A., Thickstun, J., Durmus, E., Paranjape, A., Gerard-Ursin, I., Li, X.L., Ladhak, F., Rong, F., Wang, R.E., Kwon, M., Park, J.S., Cao, H., Lee, T., Bommasani, R., Bernstein, M.S., Liang, P.: Evaluating human-language model interaction. CoRR **abs/2212.09746** (2022). https://doi.org/10.48550/arXiv.2212.09746

5. Wei, J., Wang, X., Schuurmans, D., Bosma, M., Chi, E.H., Le, Q., Zhou, D.: Chain of thought prompting elicits reasoning in large language models. CoRR **abs/2201.11903** (2022). https://arxiv.org/abs/2201.11903

6. Liu, J., Shen, D., Zhang, Y., Dolan, B., Carin, L., Chen, W.: What makes good in-context examples for GPT-3? In: Proceedings of Deep Learning Inside Out (DeeLIO 2022): The 3rd Workshop on Knowledge Extraction and Integration for Deep Learning Architectures, pp. 100–114. Association for Computational Linguistics, Dublin, Ireland and Online (2022). https://doi.org/10.18653/v1/2022.deelio-1.10. https://aclanthology.org/2022.deelio-1.10

7. Gu, J., Xu, H., Nie, L., Yin, W.: Robustness of learning from task instructions. arXiv preprint arXiv: Arxiv-2212.03813 (2022)

8. Zhao, Z., Wallace, E., Feng, S., Klein, D., Singh, S.: Calibrate before use: Improving few-shot performance of language models. In: M. Meila, T. Zhang (eds.) Proceedings of the 38th International Conference on Machine Learning, *Proceedings of Machine Learning Research*, vol. 139, pp. 12697–12706. PMLR (2021). https://proceedings.mlr.press/v139/zhao21c.html

9. Lu, Y., Bartolo, M., Moore, A., Riedel, S., Stenetorp, P.: Fantastically ordered prompts and where to find them: Overcoming few-shot prompt order sensitivity. In: Proceedings of the 60th Annual Meeting of the Association for Computational Linguistics (Volume 1: Long Papers), pp. 8086–8098. Association for Computational Linguistics, Dublin, Ireland (2022). https://doi.org/10.18653/v1/2022.acl-long.556. https://aclanthology.org/2022.acl-long.556

10. Dong, Q., Li, L., Dai, D., Zheng, C., Wu, Z., Chang, B., Sun, X., Xu, J., Li, L., Sui, Z.: A survey for in-context learning. CoRR **abs/2301.00234** (2023). https://doi.org/10.48550/arXiv.2301.00234

11. Chen, Y., Zhao, C., Yu, Z., McKeown, K., He, H.: On the relation between sensitivity and accuracy in in-context learning. arXiv preprint arXiv: Arxiv-2209.07661 (2022)

12. Liu, P., Yuan, W., Fu, J., Jiang, Z., Hayashi, H., Neubig, G.: Pre-train, prompt, and predict: A systematic survey of prompting methods in natural language processing. arXiv preprint arXiv: Arxiv-2107.13586 (2021)

13. Wallace, E., Feng, S., Kandpal, N., Gardner, M., Singh, S.: Universal adversarial triggers for attacking and analyzing NLP. In: Proceedings of the 2019 Conference on Empirical Methods in Natural Language Processing and the 9th International Joint Conference on Natural Language Processing (EMNLP-IJCNLP), pp. 2153–2162. Association for Computational Linguistics, Hong Kong, China (2019). https://doi.org/10.18653/v1/D19-1221. https://aclanthology.org/D19-1221

14. Jiang, Z., Xu, F.F., Araki, J., Neubig, G.: How Can We Know What Language Models Know? Transactions of the Association for Computational Linguistics **8**, 423–438 (2020). https://doi.org/10.1162/tacl_a_00324

15. Li, X.L., Liang, P.: Prefix-tuning: Optimizing continuous prompts for generation. Annual Meeting Of The Association For Computational Linguistics (2021). https://doi.org/10.18653/v1/2021.acl-long.353

16. Zhang, Y., Feng, S., Tan, C.: Active example selection for in-context learning. arXiv preprint arXiv:2211.04486 (2022)

17. Zhou, Y., Muresanu, A.I., Han, Z., Paster, K., Pitis, S., Chan, H., Ba, J.: Large language models are human-level prompt engineers. arXiv preprint arXiv: Arxiv-2211.01910 (2022)

18. Mishra, S., Khashabi, D., Baral, C., Hajishirzi, H.: Cross-task generalization via natural language crowdsourcing instructions. In: ACL (2022)

19. Wang, Y., Mishra, S., Alipoormolabashi, P., Kordi, Y., Mirzaei, A., Arunkumar, A., Ashok, A., Dhanasekaran, A.S., Naik, A., Stap, D., et al.: Super-naturalinstructions:generalization via declarative instructions on 1600+ tasks. In: EMNLP (2022)

20. Li, G., Hammoud, H.A.A.K., Itani, H., Khizbullin, D., Ghanem, B.: Camel: Communicative agents for "mind" exploration of large scale language model society. arXiv preprint arXiv:2303.17760 (2023)
21. Lynch, C., Wahid, A., Tompson, J., Ding, T., Betker, J., Baruch, R., Armstrong, T., Florence, P.: Interactive language: Talking to robots in real time (2022)
22. Zhong, W., Gao, Y., Ding, N., Qin, Y., Liu, Z., Zhou, M., Wang, J., Yin, J., Duan, N.: ProQA: Structural prompt-based pre-training for unified question answering. In: Proceedings of the 2022 Conference of the North American Chapter of the Association for Computational Linguistics: Human Language Technologies, pp. 4230–4243. Association for Computational Linguistics, Seattle, United States (2022). https://doi.org/10.18653/v1/2022.naacl-main.313. https://aclanthology.org/2022.naacl-main.313
23. Ahn, M., Brohan, A., Brown, N., Chebotar, Y., Cortes, O., David, B., Finn, C., Gopalakrishnan, K., Hausman, K., Herzog, A., et al.: Do as i can, not as i say: Grounding language in robotic affordances. arXiv preprint arXiv:2204.01691 (2022)
24. Drissi, M., Watkins, O., Kalita, J.: Hierarchical text generation using an outline. arXiv preprint arXiv: Arxiv-1810.08802 (2018)
25. Sun, X., Fan, C., Sun, Z., Meng, Y., Wu, F., Li, J.: Summarize, outline, and elaborate: Long-text generation via hierarchical supervision from extractive summaries. International Conference On Computational Linguistics (2020)
26. Yang, K., Peng, N., Tian, Y., Klein, D.: Re3: Generating longer stories with recursive reprompting and revision. arXiv preprint arXiv:2210.06774 (2022)
27. Chen, B., Xia, F., Ichter, B., Rao, K., Gopalakrishnan, K., Ryoo, M.S., Stone, A., Kappler, D.: Open-vocabulary queryable scene representations for real world planning. arXiv preprint arXiv: Arxiv-2209.09874 (2022)
28. Cheng, Z., Xie, T., Shi, P., Li, C., Nadkarni, R., Hu, Y., Xiong, C., Radev, D., Ostendorf, M., Zettlemoyer, L., Smith, N.A., Yu, T.: Binding language models in symbolic languages. arXiv preprint arXiv: Arxiv-2210.02875 (2022)
29. Liang, J., Huang, W., Xia, F., Xu, P., Hausman, K., Ichter, B., Florence, P., Zeng, A.: Code as policies: Language model programs for embodied control. arXiv preprint arXiv: Arxiv-2209.07753 (2022)
30. Chen, W., Ma, X., Wang, X., Cohen, W.W.: Program of thoughts prompting: Disentangling computation from reasoning for numerical reasoning tasks. arXiv preprint arXiv:2211.12588 (2022)
31. Schick, T., Dwivedi-Yu, J., Dessì, R., Raileanu, R., Lomeli, M., Zettlemoyer, L., Cancedda, N., Scialom, T.: Toolformer: Language models can teach themselves to use tools. CoRR **abs/2302.04761** (2023). https://doi.org/10.48550/arXiv.2302.04761
32. Paranjape, B., Lundberg, S., Singh, S., Hajishirzi, H., Zettlemoyer, L., Ribeiro, M.T.: Art: Automatic multi-step reasoning and tool-use for large language models. arXiv preprint arXiv: Arxiv-2303.09014 (2023)
33. Li, J., Hui, B., Qu, G., Li, B., Yang, J., Li, B., Wang, B., Qin, B., Cao, R., Geng, R., Huo, N., Ma, C., Chang, K., Huang, F., Cheng, R., Li, Y.: Can LLM already serve as a database interface? a big bench for large-scale database grounded text-to-sqls. ARXIV.ORG (2023). https://doi.org/10.48550/arXiv.2305.03111
34. Wu, Y., Jiang, A.Q., Li, W., Rabe, M.N., Staats, C., Jamnik, M., Szegedy, C.: Autoformalization with large language models. arXiv preprint arXiv: Arxiv-2205.12615 (2022)
35. Lu, P., Gong, R., Jiang, S., Qiu, L., Huang, S., Liang, X., Zhu, S.C.: Inter-gps: Interpretable geometry problem solving with formal language and symbolic reasoning. Annual Meeting Of The Association For Computational Linguistics (2021). https://doi.org/10.18653/v1/2021.acl-long.528

36. Han, S., Schoelkopf, H., Zhao, Y., Qi, Z., Riddell, M., Benson, L., Sun, L., Zubova, E., Qiao, Y., Burtell, M., Peng, D., Fan, J., Liu, Y., Wong, B., Sailor, M., Ni, A., Nan, L., Kasai, J., Yu, T., Zhang, R., Joty, S., Fabbri, A.R., Kryscinski, W., Lin, X.V., Xiong, C., Radev, D.: Folio: Natural language reasoning with first-order logic. arXiv preprint arXiv: Arxiv-2209.00840 (2022)

37. Bai, J., Wang, Y., Chen, Y., Yang, Y., Bai, J., Yu, J., Tong, Y.: Syntax-bert: Improving pre-trained transformers with syntax trees. arXiv preprint arXiv: Arxiv-2103.04350 (2021)

38. Sachan, D.S., Zhang, Y., Qi, P., Hamilton, W.: Do syntax trees help pre-trained transformers extract information? Conference Of The European Chapter Of The Association For Computational Linguistics (2020). https://doi.org/10.18653/v1/2021.eacl-main.228

39. Wang, R., Tang, D., Duan, N., Wei, Z., Huang, X., Ji, J., Cao, G., Jiang, D., Zhou, M.: K-adapter: Infusing knowledge into pre-trained models with adapters. In: C. Zong, F. Xia, W. Li, R. Navigli (eds.) Findings of the Association for Computational Linguistics: ACL/IJCNLP 2021, Online Event, August 1-6, 2021, *Findings of ACL*, vol. ACL/IJCNLP 2021, pp. 1405–1418. Association for Computational Linguistics (2021). https://doi.org/10.18653/v1/2021.findings-acl.121

40. Liu, Q., Yogatama, D., Blunsom, P.: Relational memory augmented language models. arXiv preprint arXiv: Arxiv-2201.09680 (2022)

41. Sun, Y., Wang, S., Feng, S., Ding, S., Pang, C., Shang, J., Liu, J., Chen, X., Zhao, Y., Lu, Y., et al.: Ernie 3.0: Large-scale knowledge enhanced pre-training for language understanding and generation. arXiv preprint arXiv:2107.02137 (2021)

42. Locascio, N., Narasimhan, K., DeLeon, E., Kushman, N., Barzilay, R.: Neural generation of regular expressions from natural language with minimal domain knowledge. Conference On Empirical Methods In Natural Language Processing (2016). https://doi.org/10.18653/v1/D16-1197

43. Liu, R., Wei, J., Gu, S.S., Wu, T.Y., Vosoughi, S., Cui, C., Zhou, D., Dai, A.M.: Mind's eye: Grounded language model reasoning through simulation. arXiv preprint arXiv:2210.05359 (2022)

44. Jiang, A.Q., Welleck, S., Zhou, J.P., Li, W., Liu, J., Jamnik, M., Lacroix, T., Wu, Y., Lample, G.: Draft, sketch, and prove: Guiding formal theorem provers with informal proofs. arXiv preprint arXiv: Arxiv-2210.12283 (2022)

45. Liang, Y., Wu, C., Song, T., Wu, W., Xia, Y., Liu, Y., Ou, Y., Lu, S., Ji, L., Mao, S., Wang, Y., Shou, L., Gong, M., Duan, N.: Taskmatrix.ai: Completing tasks by connecting foundation models with millions of apis. arXiv preprint arXiv: Arxiv-2303.16434 (2023)

46. Houlsby, N., Giurgiu, A., Jastrzebski, S., Morrone, B., De Laroussilhe, Q., Gesmundo, A., Attariyan, M., Gelly, S.: Parameter-efficient transfer learning for NLP. In: K. Chaudhuri, R. Salakhutdinov (eds.) Proceedings of the 36th International Conference on Machine Learning, *Proceedings of Machine Learning Research*, vol. 97, pp. 2790–2799. PMLR (2019). https://proceedings.mlr.press/v97/houlsby19a.html

47. Chen, M., Tworek, J., Jun, H., Yuan, Q., de Oliveira Pinto, H.P., et al.: Evaluating large language models trained on code. arXiv preprint arXiv: Arxiv-2107.03374 (2021)

48. Lahiri, S.K., Naik, A., Sakkas, G., Choudhury, P., von Veh, C., Musuvathi, M., Inala, J.P., Wang, C., Gao, J.: Interactive code generation via test-driven user-intent formalization. arXiv preprint arXiv: Arxiv-2208.05950 (2022)

49. Fu Yao; Peng, H., Khot, T.: How does GPT obtain its ability? tracing emergent abilities of language models to their sources. Yao Fu's Notion (2022). https://yaofu.notion.site/How-does-GPT-Obtain-its-Ability-Tracing-Emergent-Abilities-of-Language-Models-to-their-Sources-b9a57ac0fcf74f30a1ab9e3e36fa1dc1

50. Suris, D., Menon, S., Vondrick, C.: Vipergpt: Visual inference via python execution for reasoning. arXiv preprint arXiv: Arxiv-2303.08128 (2023)

51. Malmi, E., Dong, Y., Mallinson, J., Chuklin, A., Adamek, J., Mirylenka, D., Stahlberg, F., Krause, S., Kumar, S., Severyn, A.: Text generation with text-editing models. In: Proceedings of the 2022 Conference of the North American Chapter of the Association for Computational Linguistics: Human Language Technologies: Tutorial Abstracts, pp. 1–7. Association for Computational Linguistics, Seattle, United States (2022). https://doi.org/10.18653/v1/2022.naacl-tutorials.1. https://aclanthology.org/2022.naacl-tutorials.1

52. Gu, J., Lu, Z., Li, H., Li, V.O.: Incorporating copying mechanism in sequence-to-sequence learning. In: Proceedings of the 54th Annual Meeting of the Association for Computational Linguistics (Volume 1: Long Papers), pp. 1631–1640. Association for Computational Linguistics, Berlin, Germany (2016). https://doi.org/10.18653/v1/P16-1154. https://aclanthology.org/P16-1154

53. See, A., Liu, P.J., Manning, C.D.: Get to the point: Summarization with pointer-generator networks. In: Proceedings of the 55th Annual Meeting of the Association for Computational Linguistics (Volume 1: Long Papers), pp. 1073–1083. Association for Computational Linguistics, Vancouver, Canada (2017). https://doi.org/10.18653/v1/P17-1099. https://aclanthology.org/P17-1099

54. Zhao, W., Wang, L., Shen, K., Jia, R., Liu, J.: Improving grammatical error correction via pre-training a copy-augmented architecture with unlabeled data. In: Proceedings of the 2019 Conference of the North American Chapter of the Association for Computational Linguistics: Human Language Technologies, Volume 1 (Long and Short Papers), pp. 156–165. Association for Computational Linguistics, Minneapolis, Minnesota (2019). https://doi.org/10.18653/v1/N19-1014. https://aclanthology.org/N19-1014

55. Panthaplackel, S., Allamanis, M., Brockschmidt, M.: Copy that! editing sequences by copying spans. Proceedings of the AAAI Conference on Artificial Intelligence **35**(15), 13622–13630 (2021). https://ojs.aaai.org/index.php/AAAI/article/view/17606

56. Zhang, J., Panthaplackel, S., Nie, P., Li, J.J., Gligoric, M.: Coditt5: Pretraining for source code and natural language editing. In: 37th IEEE/ACM International Conference on Automated Software Engineering, pp. 1–12 (2022)

57. Ravi, H., Kelkar, S., Harikumar, M., Kale, A.: Preditor: Text guided image editing with diffusion prior. arXiv preprint arXiv:2302.07979 (2023)

58. Corso, G., Stärk, H., Jing, B., Barzilay, R., Jaakkola, T.: Diffdock: Diffusion steps, twists, and turns for molecular docking. arXiv preprint arXiv:2210.01776 (2022)

59. Lacruz, I., Denkowski, M., Lavie, A.: Cognitive demand and cognitive effort in post-editing. In: Proceedings of the 11th Conference of the Association for Machine Translation in the Americas, pp. 73–84 (2014)

60. Vardi, I.: The impact of iterative writing and feedback on the characteristics of tertiary students' written texts. Teaching in higher education **17**(2), 167–179 (2012)

61. Gollins, A., Gentner, D.: A framework for a cognitive theory of writing. In: Cognitive processes in writing, pp. 51–72. Routledge (2016)

62. Sutskever, I., Vinyals, O., Le, Q.V.: Sequence to sequence learning with neural networks. In: Z. Ghahramani, M. Welling, C. Cortes, N.D. Lawrence, K.Q. Weinberger (eds.) Advances in Neural Information Processing Systems 27, pp. 3104–3112. Curran Associates, Inc. (2014). http://papers.nips.cc/paper/5346-sequence-to-sequence-learning-with-neural-networks.pdf

63. Mallinson, J., Severyn, A., Malmi, E., Garrido, G.: FELIX: Flexible text editing through tagging and insertion. In: Findings of the Association for Computational Linguistics: EMNLP 2020, pp. 1244–1255. Association for Computational Linguistics, Online (2020). https://doi.org/10.18653/v1/2020.findings-emnlp.111. https://aclanthology.org/2020.findings-emnlp.111

64. Malmi, E., Krause, S., Rothe, S., Mirylenka, D., Severyn, A.: Encode, tag, realize: High-precision text editing. In: Proceedings of the 2019 Conference on Empirical Methods in Natural Language Processing and the 9th International Joint Conference on Natural Language

Processing (EMNLP-IJCNLP), pp. 5054–5065. Association for Computational Linguistics, Hong Kong, China (2019)

65. Stahlberg, F., Kumar, S.: Seq2Edits: Sequence transduction using span-level edit operations. In: Proceedings of the 2020 Conference on Empirical Methods in Natural Language Processing (EMNLP), pp. 5147–5159. Association for Computational Linguistics, Online (2020). https://doi.org/10.18653/v1/2020.emnlp-main.418. https://aclanthology.org/2020.emnlp-main.418

66. Bérard, A., Besacier, L., Pietquin, O.: LIG-CRIStAL submission for the WMT 2017 automatic post-editing task. In: Proceedings of the Second Conference on Machine Translation, pp. 623–629. Association for Computational Linguistics, Copenhagen, Denmark (2017). https://doi.org/10.18653/v1/W17-4772. https://aclanthology.org/W17-4772

67. Xu, J., Crego, J., Yvon, F.: Bilingual synchronization: Restoring translational relationships with editing operations. In: Proceedings of the 2022 Conference on Empirical Methods in Natural Language Processing, pp. 8016–8030. Association for Computational Linguistics, Abu Dhabi, United Arab Emirates (2022). https://aclanthology.org/2022.emnlp-main.548

68. Kasner, Z., Dušek, O.: Data-to-text generation with iterative text editing. In: Proceedings of the 13th International Conference on Natural Language Generation, pp. 60–67. Association for Computational Linguistics, Dublin, Ireland (2020). https://aclanthology.org/2020.inlg-1.9

69. Awasthi, A., Sarawagi, S., Goyal, R., Ghosh, S., Piratla, V.: Parallel iterative edit models for local sequence transduction. In: Proceedings of the 2019 Conference on Empirical Methods in Natural Language Processing and the 9th International Joint Conference on Natural Language Processing (EMNLP-IJCNLP), pp. 4260–4270. Association for Computational Linguistics, Hong Kong, China (2019). https://doi.org/10.18653/v1/D19-1435. https://aclanthology.org/D19-1435

70. Zhou, W., Ge, T., Mu, C., Xu, K., Wei, F., Zhou, M.: Improving grammatical error correction with machine translation pairs. In: Findings of the Association for Computational Linguistics: EMNLP 2020, pp. 318–328. Association for Computational Linguistics, Online (2020). https://doi.org/10.18653/v1/2020.findings-emnlp.30. https://aclanthology.org/2020.findings-emnlp.30

71. Hinson, C., Huang, H.H., Chen, H.H.: Heterogeneous recycle generation for Chinese grammatical error correction. In: Proceedings of the 28th International Conference on Computational Linguistics, pp. 2191–2201. International Committee on Computational Linguistics, Barcelona, Spain (Online) (2020). https://doi.org/10.18653/v1/2020.coling-main.199. https://aclanthology.org/2020.coling-main.199

72. Omelianchuk, K., Atrasevych, V., Chernodub, A., Skurzhanskyi, O.: GECToR–grammatical error correction: Tag, not rewrite. In: Proceedings of the Fifteenth Workshop on Innovative Use of NLP for Building Educational Applications, pp. 163–170. Association for Computational Linguistics, Seattle, WA, USA → Online (2020). https://doi.org/10.18653/v1/2020.bea-1.16. https://aclanthology.org/2020.bea-1.16

73. Che, X., Wang, C., Yang, H., Meinel, C.: Punctuation prediction for unsegmented transcript based on word vector. In: Proceedings of the Tenth International Conference on Language Resources and Evaluation (LREC'16), pp. 654–658 (2016)

74. Kim, S.: Deep recurrent neural networks with layer-wise multi-head attentions for punctuation restoration. In: ICASSP 2019-2019 IEEE International Conference on Acoustics, Speech and Signal Processing (ICASSP), pp. 7280–7284. IEEE (2019)

75. Alam, T., Khan, A., Alam, F.: Punctuation restoration using transformer models for high- and low-resource languages. In: Proceedings of the Sixth Workshop on Noisy User-generated Text (W-NUT 2020), pp. 132–142. Association for Computational Linguistics, Online (2020). https://doi.org/10.18653/v1/2020.wnut-1.18. https://aclanthology.org/2020.wnut-1.18

76. Shi, N., Wang, W., Wang, B., Li, J., Liu, X., Lin, Z.: Incorporating External POS Tagger for Punctuation Restoration. In: Proc. Interspeech 2021, pp. 1987–1991 (2021). https://doi.org/10.21437/Interspeech.2021-1708

77. Dong, Y., Li, Z., Rezagholizadeh, M., Cheung, J.C.K.: EditNTS: An neural programmer-interpreter model for sentence simplification through explicit editing. In: Proceedings of the 57th Annual Meeting of the Association for Computational Linguistics, pp. 3393–3402. Association for Computational Linguistics, Florence, Italy (2019)

78. Agrawal, S., Xu, W., Carpuat, M.: A non-autoregressive edit-based approach to controllable text simplification. In: Findings of the Association for Computational Linguistics: ACL-IJCNLP 2021, pp. 3757–3769. Association for Computational Linguistics, Online (2021). https://doi.org/10.18653/v1/2021.findings-acl.330. https://aclanthology.org/2021.findings-acl.330

79. Liu, R., Jia, C., Zhang, G., Zhuang, Z., Liu, T., Vosoughi, S.: Second thoughts are best: Learning to re-align with human values from text edits. Advances in Neural Information Processing Systems **35**, 181–196 (2022)

80. Zhang, G., Li, Y., Wu, Y., Zhang, L., Lin, C., Geng, J., Wang, S., Fu, J.: Corgi-pm: A chinese corpus for gender bias probing and mitigation. arXiv preprint arXiv:2301.00395 (2023)

81. Reid, M., Zhong, V.: LEWIS: Levenshtein editing for unsupervised text style transfer. In: Findings of the Association for Computational Linguistics: ACL-IJCNLP 2021, pp. 3932–3944. Association for Computational Linguistics, Online (2021). https://doi.org/10.18653/v1/2021.findings-acl.344. https://aclanthology.org/2021.findings-acl.344

82. Zhou, W., Ge, T., Xu, C., Xu, K., Wei, F.: Improving sequence-to-sequence pre-training via sequence span rewriting. In: Proceedings of the 2021 Conference on Empirical Methods in Natural Language Processing, pp. 571–582. Association for Computational Linguistics, Online and Punta Cana, Dominican Republic (2021). https://doi.org/10.18653/v1/2021.emnlp-main.45. https://aclanthology.org/2021.emnlp-main.45

83. Ge, T., Wei, F., Zhou, M.: Fluency boost learning and inference for neural grammatical error correction. In: Proceedings of the 56th Annual Meeting of the Association for Computational Linguistics (Volume 1: Long Papers), pp. 1055–1065. Association for Computational Linguistics, Melbourne, Australia (2018). https://doi.org/10.18653/v1/P18-1097. https://aclanthology.org/P18-1097

84. Gu, J., Wang, C., Zhao, J.: Levenshtein transformer. In: H. Wallach, H. Larochelle, A. Beygelzimer, F. d' Alché-Buc, E. Fox, R. Garnett (eds.) Advances in Neural Information Processing Systems 32, pp. 11181–11191. Curran Associates, Inc. (2019)

85. Stern, M., Chan, W., Kiros, J., Uszkoreit, J.: Insertion transformer: Flexible sequence generation via insertion operations. In: K. Chaudhuri, R. Salakhutdinov (eds.) Proceedings of the 36th International Conference on Machine Learning, *Proceedings of Machine Learning Research*, vol. 97, pp. 5976–5985. PMLR (2019). https://proceedings.mlr.press/v97/stern19a.html

86. Kumar, D., Mou, L., Golab, L., Vechtomova, O.: Iterative edit-based unsupervised sentence simplification. In: Proceedings of the 58th Annual Meeting of the Association for Computational Linguistics, pp. 7918–7928. Association for Computational Linguistics, Online (2020). https://doi.org/10.18653/v1/2020.acl-main.707. https://aclanthology.org/2020.acl-main.707

87. Shi, N., Zeng, Z., Zhang, H., Gong, Y.: Recurrent inference in text editing. In: Findings of the Association for Computational Linguistics: EMNLP 2020, pp. 1758–1769. Association for Computational Linguistics, Online (2020). https://doi.org/10.18653/v1/2020.findings-emnlp.159. https://aclanthology.org/2020.findings-emnlp.159

88. Shi, N., Tang, B., Yuan, B., Huang, L., Pu, Y., Fu, J., Lin, Z.: Text editing as imitation game. In: Findings of the Association for Computational Linguistics: EMNLP 2022, pp. 1583–1594. Association for Computational Linguistics, Abu Dhabi, United Arab Emirates (2022). https://aclanthology.org/2022.findings-emnlp.114

89. Faltings, F., Galley, M., Peng, B., Brantley, K., Cai, W., Zhang, Y., Gao, J., Dolan, B.: Interactive text generation. arXiv preprint arXiv:2303.00908 (2023)

90. Schick, T., Dwivedi-Yu, J., Jiang, Z., Petroni, F., Lewis, P., Izacard, G., You, Q., Nalmpantis, C., Grave, E., Riedel, S.: Peer: A collaborative language model. arXiv preprint arXiv: Arxiv-2208.11663 (2022)

91. Kim, Z.M., Du, W., Raheja, V., Kumar, D., Kang, D.: Improving iterative text revision by learning where to edit from other revision tasks. In: Proceedings of the 2022 Conference on Empirical Methods in Natural Language Processing, pp. 9986–9999. Association for Computational Linguistics, Abu Dhabi, United Arab Emirates (2022). https://aclanthology.org/2022.emnlp-main.678

92. Madaan, A., Tandon, N., Gupta, P., Hallinan, S., Gao, L., Wiegreffe, S., Alon, U., Dziri, N., Prabhumoye, S., Yang, Y., Welleck, S., Majumder, B.P., Gupta, S., Yazdanbakhsh, A., Clark, P.: Self-refine: Iterative refinement with self-feedback. arXiv preprint arXiv: Arxiv-2303.17651 (2023)

93. Narayan, S., Gardent, C.: Hybrid simplification using deep semantics and machine translation. In: Proceedings of the 52nd Annual Meeting of the Association for Computational Linguistics (Volume 1: Long Papers), pp. 435–445. Association for Computational Linguistics, Baltimore, Maryland (2014). https://doi.org/10.3115/v1/P14-1041. https://aclanthology.org/P14-1041

94. Mallinson, J., Adamek, J., Malmi, E., Severyn, A.: EdiT5: Semi-autoregressive text editing with t5 warm-start. In: Findings of the Association for Computational Linguistics: EMNLP 2022, pp. 2126–2138. Association for Computational Linguistics, Abu Dhabi, United Arab Emirates (2022). https://aclanthology.org/2022.findings-emnlp.156

95. Malmi, E., Severyn, A., Rothe, S.: Unsupervised text style transfer with padded masked language models. In: Proceedings of the 2020 Conference on Empirical Methods in Natural Language Processing (EMNLP), pp. 8671–8680. Association for Computational Linguistics, Online (2020). https://doi.org/10.18653/v1/2020.emnlp-main.699. https://aclanthology.org/2020.emnlp-main.699

96. Kazemnejad, A., Salehi, M., Soleymani Baghshah, M.: Paraphrase generation by learning how to edit from samples. In: Proceedings of the 58th Annual Meeting of the Association for Computational Linguistics, pp. 6010–6021. Association for Computational Linguistics, Online (2020). https://doi.org/10.18653/v1/2020.acl-main.535. https://aclanthology.org/2020.acl-main.535

97. Guu, K., Hashimoto, T.B., Oren, Y., Liang, P.: Generating sentences by editing prototypes. Transactions of the Association for Computational Linguistics **6**, 437–450 (2018). https://doi.org/10.1162/tacl_a_00030. https://aclanthology.org/Q18-1031

98. Wang, P., Yang, A., Men, R., Lin, J., Bai, S., Li, Z., Ma, J., Zhou, C., Zhou, J., Yang, H.: OFA: unifying architectures, tasks, and modalities through a simple sequence-to-sequence learning framework. In: K. Chaudhuri, S. Jegelka, L. Song, C. Szepesvári, G. Niu, S. Sabato (eds.) International Conference on Machine Learning, ICML 2022, 17–23 July 2022, Baltimore, Maryland, USA, *Proceedings of Machine Learning Research*, vol. 162, pp. 23318–23340. PMLR (2022). https://proceedings.mlr.press/v162/wang22al.html

99. Wang, W., Bao, H., Dong, L., Bjorck, J., Peng, Z., Liu, Q., Aggarwal, K., Mohammed, O.K., Singhal, S., Som, S., Wei, F.: Image as a foreign language: Beit pretraining for all vision and vision-language tasks. arXiv preprint arXiv: Arxiv-2208.10442 (2022)

100. van den Oord, A., Vinyals, O., Kavukcuoglu, K.: Neural discrete representation learning. NIPS (2017)

101. Esser, P., Rombach, R., Ommer, B.: Taming transformers for high-resolution image synthesis. Computer Vision and Pattern Recognition (2020). https://doi.org/10.1109/CVPR46437.2021.01268

102. Yu, J., Li, X., Koh, J.Y., Zhang, H., Pang, R., Qin, J., Ku, A., Xu, Y., Baldridge, J., Wu, Y.: Vector-quantized image modeling with improved vqgan. International Conference On Learning Representations (2021)

103. Peng, Z., Dong, L., Bao, H., Ye, Q., Wei, F.: Beit v2: Masked image modeling with vector-quantized visual tokenizers. arXiv preprint arXiv: Arxiv-2208.06366 (2022)

104. Liu, D., Lamb, A., Kawaguchi, K., Goyal, A., Sun, C., Mozer, M.C., Bengio, Y.: Discrete-valued neural communication. arXiv preprint arXiv: Arxiv-2107.02367 (2021)

105. Wang, Y., Zhang, X.Y., Liu, C.L., Zhang, Z.: Emergence of machine language: Towards symbolic intelligence with neural networks. arXiv preprint arXiv: Arxiv-2201.05489 (2022)

106. Alayrac, J.B., Donahue, J., Luc, P., Miech, A., Barr, I., Hasson, Y., Lenc, K., Mensch, A., Millican, K., Reynolds, M., Ring, R., Rutherford, E., Cabi, S., Han, T., Gong, Z., Samangooei, S., Monteiro, M., Menick, J., Borgeaud, S., Brock, A., Nematzadeh, A., Sharifzadeh, S., Binkowski, M., Barreira, R., Vinyals, O., Zisserman, A., Simonyan, K.: Flamingo: a visual language model for few-shot learning. DEEPMIND (2022)

107. Li, J., Li, D., Savarese, S., Hoi, S.: Blip-2: Bootstrapping language-image pre-training with frozen image encoders and large language models. arXiv preprint arXiv: Arxiv-2301.12597 (2023)

108. Liu, H., Li, C., Wu, Q., Lee, Y.J.: Visual instruction tuning. In: NeurIPS (2023)

109. Christiano, P., Leike, J., Brown, T.B., Martic, M., Legg, S., Amodei, D.: Deep reinforcement learning from human preferences. arXiv preprint arXiv: Arxiv-1706.03741 (2017)

110. Ramamurthy, R., Ammanabrolu, P., Brantley, K., Hessel, J., Sifa, R., Bauckhage, C., Hajishirzi, H., Choi, Y.: Is reinforcement learning (not) for natural language processing: Benchmarks, baselines, and building blocks for natural language policy optimization. In: The Eleventh International Conference on Learning Representations (2023). https://openreview.net/forum?id=8aHzds2uUyB

111. Goyal, A., Didolkar, A.R., Lamb, A., Badola, K., Ke, N.R., Rahaman, N., Binas, J., Blundell, C., Mozer, M.C., Bengio, Y.: Coordination among neural modules through a shared global workspace. In: The Tenth International Conference on Learning Representations, ICLR 2022, Virtual Event, April 25–29, 2022. OpenReview.net (2022). https://openreview.net/forum?id=XzTtHjgPDsT

112. Madaan, A., Tandon, N., Clark, P., Yang, Y.: Memory-assisted prompt editing to improve GPT-3 after deployment. arXiv preprint arXiv:2201.06009 (2022)

113. Dalvi, B., Tafjord, O., Clark, P.: Towards teachable reasoning systems: Using a dynamic memory of user feedback for continual system improvement. Conference On Empirical Methods In Natural Language Processing (2022)

114. Ryoo, M.S., Gopalakrishnan, K., Kahatapitiya, K., Xiao, T., Rao, K., Stone, A., Lu, Y., Ibarz, J., Arnab, A.: Token turing machines. arXiv preprint arXiv: Arxiv-2211.09119 (2022)

115. Yang, Q., Wang, Z., Chen, H., Wang, S., Pu, Y., Gao, X., Huang, W., Song, S., Huang, G.: Psychogat: A novel psychological measurement paradigm through interactive fiction games with LLM agents (2024). https://openreview.net/forum?id=efvlIIRKeO&referrer=%5BAuthor%20Console%5D(%2Fgroup%3Fid%3Daclweb.org%2FACL%2F2024%2FARR_Commitment%2FAuthors%23your-submissions)

116. Park, J.S., O'Brien, J.C., Cai, C.J., Morris, M.R., Liang, P., Bernstein, M.S.: Generative agents: Interactive simulacra of human behavior. arXiv preprint arXiv:2304.03442 (2023)

Part II
Methods for Interactive NLP

In this part, we aim to explore the methodologies employed by language models for understanding and processing interaction messages. We begin with a quick tour through the pre-trained language models, large language models, and multimodal foundation models. Next, we divide interaction methods into six categories: (1) prompting without model training (aka., Inference-Time Scaling or Test-Time Scaling), (2) fine-tuning which involves updating models' parameters, (3) reinforcement learning, (4) agents, (5) active learning, (6) imitation learning, as well as (7) model surgery (including knowledge editing and machine unlearning). Finally, we propose to re-frame and formalize these methods in a unified manner, i.e., interaction message fusion. Additionally, we include current methodologies and trends in LLM reasoning, which is an important instantiation of the Interactive NLP paradigm and a crucial milestone toward human-level artificial general intelligence. This includes Inference-Time Scaling (the Prompting Chapter), as well as RL for Reasoning (a section in the Reinforcement Learning Chapter) and Agentic Workflow for Reasoning (a section in the Agents Chapter).

Foundation Models

4

Zekun Wang, Yizhi Li, Jie Fu and Ge Zhang

Foundation models are large-scale, pre-trained models that serve as the basis for a wide range of tasks [1], and iNLP. These models utilize vast amounts of data and parameters to capture diverse linguistic patterns, world knowledge, commonsense, and knowledge across various domains.

Pre-trained language models (PLMs) are the precursors to current foundation models. Before evolving into large language models (LLMs), the era of PLMs involved extensive exploration of numerous technical designs, including unit modules, modeling architectures, and pre-training strategies. Since the advent of GPT-3 [2], the potential of PLMs has been significantly unlocked through scaling up. In this era, the focus shifts towards data collection and curation, modeling techniques specific to LLMs, pre-training, supervised fine-tuning (SFT), and alignment.

Due to the impressive capabilities demonstrated by LLMs, recent Multimodal Foundation Models (MFMs), also known as Multimodal LLMs (MLLMs), leverage LLMs as their core

Z. Wang
Beihang University, Beijing, China
e-mail: zenmoore@buaa.edu.cn

Y. Li
University of Manchester, Manchester, UK

J. Fu
Hong Kong University of Science and Technology, Hong Kong, China
e-mail: jiefu@ust.hk

G. Zhang (✉)
University of Michigan, Beijing, China
e-mail: gezhang@umich.edu

Table 4.1 Overview of PLMs

Model	Architecture	Strategy	#Parameters	Characteristics
BERT [7]	Enc	MLM, SRC	110M/340M	MLM, NSP
RoBERTa [8]	Enc	MLM	123M/354M	Dynamic Mask, No NSP
XLNet [9]	Enc/Dec	CausalLM	110M/340M	Permutation AR LM
SpanBERT [10]	Enc	MLM	110M/340M	Span Mask
ERNIE [11]	Enc	MLM	110M	Entity/Phrase Mask
ERNIE-2.0 [12]	Enc	MLM, SRC	110M/340M	Learning lexical, syntactic, and semantic information across Multi-Tasks Learning
ALBERT [13]	Enc	MLM, SRC	12M/18M/60M/235M	Embedding Decomposing, Parameters Share, SOP (sentence order prediction)
DistilBERT [14]	Enc	MLM	66M	Teacher-Student, Dynamic Mask, No NSP
ELECTRA [15]	Enc	MLM	14M/110M/335M	Token Generator, Discriminator to predict original or replaced
SqueezeBERT [16]	Enc	MLM, SRC	62M	Replace FC layers with Convolutions
GPT [17]	Dec	CausalLM	117M	Decoder-based Model
GPT-2 [18]	Dec	CausalLM	1.5B	More parameters and data than GPT
BART [19]	Enc-Dec	Seq2Seq	140M, 406M	Arbitrary Noise
PEGASUS [20]	Enc-Dec	Seq2Seq	223M, 568M	GSG (gap-sentences generation)
UniLM [21]	Enc/Dec	PrefixLM	340M	Unified for Bidirectional, Unidirectional, and Seq2Seq LM
T5 [22]	Enc-Dec	Seq2Seq	220M, 60M, 770M, 3B, 11B	Unified NLP tasks with the same input-output format

and align other modalities with language representation. This enables LLMs to process non-textual data such as images and audio. These models can be categorized into those designed solely for understanding non-textual modalities and those capable of both understanding and generating non-textual modalities.

In this chapter, we briefly introduce the PLMs in Sect. 4.1, the LLMs in Sect. 4.2, and the MFMs in Sect. 4.3. We list some representative PLMs in Table 4.1, LLMs in Table 4.2 and 4.3, as well as MFMs in Table 4.4. We refer the readers to the surveys [3–6] for more detailed information.

Table 4.2 Overview of LLMs

Model	Architecture	Pre-training	#Parameters	Characteristics
FLAN-T5 [23]	Enc-Dec	Seq2Seq	8B, 62B, 540B	Scaling and instruction fine-tuning T5
ERNIE-3.0 [24]	Enc-Dec	MLM, CausalLM, SRC	10B	Multi-task learning, External knowledge enhanced
ERNIE-3.0 Titan [25]	Enc-Dec	MLM, CausalLM, SRC	260B	Large scale of ernie 3.0
GPT-3 [2]	Dec	CausalLM	175B	100X parameters compared with GPT-2
PANGU-α [26]	Dec	CausalLM	2.6B, 13B, 200B	Query layer to induce expected output
FLAN [27]	Dec	CausalLM	137B	Instruct tuning
Gopher [28]	Dec	CausalLM	44M, 117M, 417M, 1.4B, 7.1B, 280B	RMSNorm, RoPE
PaLM [29]	Dec	CausalLM	8B, 62B, 540B	SwiGLU, Parallel Layer, Multi-query attention, Shared input-output embeddings, No bias
UL2 [30]	Dec, Enc-Dec	CausalLM, Seq2Seq	1B, 20B	Unified denoising objectives for both Enc-Dec and Dec architecture
PaLM-2 [31]	Dec, Enc-Dec	CausalLM, Seq2Seq	1.04B, 3.35B, 10.7B	Multi-lingual and Multi-domain training data, More efficient model architecture
OPT [32]	Dec	CausalLM	125M, 350M, 1.3B, 2.7B, 6.7B, 13B, 30B, 66B, 175B	Open pre-trained transformer
Galactica [33]	Dec	CausalLM	125M, 1.3B, 6.7B, 30B, 120B	High-quality scientific training data, Prompt pre-training
GLM-130B [34]	Enc-Dec	CausalLM, MLM	130B	2D positional encoding, Autoregressive blank infilling, Multi-task instruction pre-training

4.1   Pre-Trained Language Models

Pre-trained language models (PLMs), especially large language models (LLMs), have demonstrated their tremendous potential to serve as the cornerstone of advancing language intelligence. Transformer [55], BERT [7], GPT-3 [2] and ChatGPT are recognized as four major milestones of utilizing pre-trained language models for various NLP tasks, which also frame the roadmap of AI development. PLM is usually based on Transformer and can be categorized along two dimensions: (1) architectures, (2) pre-training strategies [30].

Table 4.3 Overview of LLMs (continued)

Model	Architecture	Pre-training	#Parameters	Characteristics
Bloom [35]	Dec	CausalLM	560M, 1.1B, 1.7B, 3B, 7.1B, 176B	ALiBi positional embedding, Embedding layerNorm
FLAN-PaLM [23]	Dec	CausalLM	250M, 80M, 780M, 3B, 11B	Scaling and instruction fine-tuning PaLM
LLaMA [36]	Dec	CausalLM	6.7B, 13B, 33B, 65B	Pre-normalization, SwiGLU [37], RoPE
Mistral [38]	Dec	CausalLM	7B	Sliding window attention
Qwen [39]	Dec	CausalLM	1.8B, 7B, 14B, 72B	Dynamic NTK-aware interpolation
Falcon [40]	Dec	CausalLM	7.5B, 40B, 180B	The largest open-source LLM, Data processing pipeline of web corpus
Yi [41]	Dec	CausalLM	7B, 34B	Detailed pretrain corpus processing pipeline, pretrain and SFT data mixture cookbook
DeepSeek [42–46]	Dec	CausalLM	7B, 67B	High-quality data recall pipeline
OLMo [47, 48]	Dec	CausalLM	1B, 7B	Transparent large language model
MAP-Neo [49]	Dec	CausalLM	2B, 7B	Bilingual transparent model and quality data
GPT-4/GPT-4o [50, 51]	Unk	Unk	Unk	Closed-source LLM developed by OpenAI
Gemini [52]	Unk	Unk	Unk	Closed-source LLM developed by Google
Claude [53, 54]	Unk	Unk	Unk	Closed-source LLM developed by anthropic

Architectures. There are overall three types of architectures: (1) **encoder-only**, where the model takes input tokens and produces a fixed-dimensional representation of the input text [7, 8, 11], (2) **encoder-decoder**, where the model first generates a fixed-dimensional representation of the input text with an encoder, and then autoregressively generates tokens based on this representation with a decoder [19, 22], and (3) **decoder-only**, where the model directly generates tokens in an autoregressive manner based on the input text as context, utilizing only a decoder [2, 17, 18]. The encoder-only architecture is especially well-suited for discriminative tasks, such as text classification [56]. On the other hand, the encoder-decoder architecture is particularly suitable for sequence-to-sequence tasks, such as machine translation [57]. Lastly, the decoder-only architecture is particularly well-suited for generative tasks, such as story generation [58].

Table 4.4 Overview of MFMs. "VLA" denotes "vision-language-action" (i.e., embodied MFMs). Underlined models are Any-to-Any MFMs. While boxed models are world models

Model	Modality	#Parameters	Characteristics
RT-1 [95]	VLA	35M	End-to-end robotic Transformer
RT-2 [96]	VLA	5B, 55B	Training visual-language models with large-scale data for zero-shot robotic action control
VIMA [97]	VLA	2M, 4M, 9M, 20M, 43M, 92M, 200M	Leverage text-image prompt to produce motor actions
LAVA [98]	VLA, Audio	N/A	Real-time speech and language guidance to the robots
PALM-E [99]	VLA	562B	Add robotic or object states with image and text
VLMo [100]	Text, Image	130M	Unified various modalities by MoE Transformer, trained jointly with ITC (Image-Text Contrastive), ITM (Image-Text Matching) and MLM
Flamingo [101]	Text, Image	3B, 9B, 80B	In-context learning of visual and textual multi-modal tasks
CoCa [102]	Text, Image	383M, 787M, 2.1B	Unified single-encoder, dual-encoder and encoder-decoder and trained with contrastive and captioning loss
PaLI [103]	Text, Image	3B, 15B, 17B	Joint training on scaled multi-modal and multilingual tasks
OFA [104]	Text, Image	33M, 93M, 182M, 472M, 930M	Unified architectures, tasks, and modalities by instruction based pre-training and fine-tuning
BEiT-3 [105]	Text, Image	1.9B	General multimodal foundation model on text, image and text-image pair with MDM (Masked Data Modeling)
BLIP [106]	Text, Image	446M	Use a synthetic caption producer and a noise caption filter boostrappingly train a unified multi-modal model with ITC, ITM and LM loss

Pre-training Strategies. LMs typically employ self-supervised training objectives for pre-training, including: (1) **CausalLM** (causal language modeling), where the model predicts the next token based on the preceding tokens from left to right [2, 17, 18]. (2) **PrefixLM**

(prefix language modeling), where the model predicts the next token using a bi-directionally encoded prefix as well as the previous tokens from left to right [21]. (3) **MLM** (masked language modeling), where the model predicts the masked span of the input [7]. (4) **Seq2Seq** (sequence-to-sequence), where the model decodes the output from left to right based on the encoded input [19, 22]. (5) **SRC** (sentence relationship capturing), which includes tasks such as Next Sentence Prediction [7] and Sentence Order Prediction [13], aimed at capturing relationships between sentences. Other pre-training objectives, such as Right-to-Left Language Modeling [21] and Permutation Language Modeling [9], are less commonly used.

4.2   Large Language Models

Large Language Models (LLMs) differ from the pre-trained language models primarily due to their scale. By increasing the number of parameters and training data, LLMs achieve significant performance improvements across a wide range of natural language processing tasks [59]. This scaling up enhances the ability to understand and generate human-like text, surpassing previous models in accuracy and versatility. However, with these advancements come new challenges in LLM research: data construction and modeling efficiency.

Data Construction. Constructing datasets for large language models (LLMs) is a meticulous, multi-step process ensuring data quality, diversity, and scalability. Constructing datasets for LLMs is a meticulous and multi-step process ensuring data quality, diversity, and scalability. It begins with data collection, often via large-scale web crawls like Common Crawl [60], providing extensive raw text data. This data undergoes rigorous filtering to eliminate low-quality or harmful content [61], followed by deduplication to avoid redundant information, using techniques like exact and fuzzy matching techniques [40, 62, 63]. Ensuring dataset diversity and quality involves curating high-quality sources like books and technical papers, despite the labor-intensive process [40, 49, 64]. Scalability remains a significant challenge, requiring substantial computational resources and innovative processing methods [65]. Additionally, ethical considerations, including privacy and copyright issues, necessitate careful data handling to comply with legal standards [47]. Overall, constructing datasets for LLMs is a multifaceted process that balances the need for large-scale data, quality, and ethical considerations to support the development of advanced language models.

Modeling Efficiency. Increasing the modeling efficiency of LLMs is essential given the escalating costs associated with larger datasets and more parameters. In LLM architecture and optimization, exploring Transformer models with modified attention mechanisms like sparse attention [66, 67] and grouped-query attention [68], as well as alternative architectures such as state space models [69], can significantly improve efficiency. Techniques like post-training quantization [70–72] and quantization-aware training [73] are key in reducing model size and computational costs by converting weight precision. Training optimization, particularly mixed precision training, accelerates computations without substantial accuracy loss [74]. Efficient distributed training strategies, including data, model, and pipeline parallelism, are essential for scaling across multiple hardware platforms [75–77]. Enhancing data efficiency through methods like data synthesis [78, 79] and scaling law research [80–82] improves data utilization, thereby reducing the need for extensive annotated datasets. These approaches collectively aim to enhance the efficiency and scalability of LLM training and deployment, making advanced language models more accessible and cost-effective.

Mixture of Experts (MoE). MoE architectures significantly enhance the modeling efficiency of Large Language Models by introducing conditional computation, where tokens are dynamically routed to specialized sub-networks, or "experts", based on a gating mechanism [83]. These specialized sub-networks are typically FFN modules. Unlike dense models that activate all parameters for every input, sparse MoE selectively activates a small subset of experts (typically top-1 or top-2) per token, drastically reducing computational costs while scaling up the model parameters [84]. The design of a MoE-based large language model primarily consists of two components: the gating function and the auxiliary loss term specifically introduced to facilitate MoE training. **The gating function**, often referred to as a router, employs strategies like token-choice gating, where methods such as expert prototyping [85] organize experts into groups for efficient routing, or advanced techniques like Batch Prioritized Routing (BPR) [86] prioritize tokens based on gating scores to mitigate routing biases and the drop-towards-the-end phenomenon [87]. **Auxiliary losses**, such as load-balancing loss $\mathcal{L}_{load}$ [83], importance loss $\mathcal{L}_{importance}$ [83], and router z-loss $\mathcal{L}_z$ [88], are critical for balancing the token distribution across experts and improving training stability. Furthermore, innovations like Gating Logit Normalization [89] enhances expert diversification, while non-trainable gating methods, such as hash-based routing [90] or random routed experts (RRE) [91], achieve full load balancing without additional parameters.

To further optimize MoE architectures, techniques like shared experts and dense-to-sparse training schemes improve both scalability and efficiency. **Shared experts**, as implemented in models like DeepSpeed-MoE [92], NLLB [93], and DeepSeekMoE [42], introduce static parameters accessible to all tokens, capturing the shared features of all inputs. For example, DeepSeekMoE [42] leverages multiple shared experts to support its fine-grained expert segmentation. **Dense-to-sparse training schemes**, such as Sparse Upcycling [94], initializes MoE layers from pretrained dense models, duplicating feed-forward network parameters to create experts and fine-tuning the MoE model. While Skywork-MoE [89] notes that scratch-trained MoE models often outperform upcycled ones due to better expert diversification [89] (Tables 4.5 and 4.6).

4.3 Multimodal Foundation Models

Large Language Models (LLMs) are incredibly powerful, yet they predominantly focus on high-level textual data and the information and knowledge embedded within. Although many approaches convert non-textual modality data into textual forms using tools like image captioners or ASR tools [139–142], these approaches have two significant drawbacks. Firstly,

Table 4.5 Overview of MFMs (continued)

Model	Modality	#Parameters	Characteristics
BLIP-2 [107]	Text, Image	474M, 1.2B	Bridge the gap between a frozen image encoder and a frozen LM in two stages by a querying transformer
Kosmos-1 [108]	Text, Image	1.6B	Instruct and multi-modal transformer
Kosmos-2 [109]	Text, Image	1.6B	Enhanced grounding abilities
GPT-4V [50]	Text, Image	N/A	Multi-modal supported ChatGPT
LLaVA [110, 111]	Text, Image	7B, 13B	Leverage generated textual instruction following dataset for visual-language understanding
MiniGPT-4 [112]	Text, Image	N/A	Verify the superiority of a well-trained language model
LLaSM [113]	Text, Audio	N/A	Introduce instruction following in both text and speech
Otter [114]	Text, Image	1.3B	Introduce instruction following and in-context learning for visual-language models
Fuyu [115]	Text, Image	8B	A vision-language model on raw pixels
NeXT-GPT [116]	Text, Image, Audio	98M	Leverage LLMs, modality adapters, and diffusion decoders to deliver an any2any system
Qwen-VL [117]	Text, Image	9.6B	Enhance the visual understanding ability with designed training pipeline, visual receptor, and input-output interface
CM3Leon [118]	Text, Image	350M, 760M, 7B	A retrieval-augmented and token-based model for infilling text and image generation
mPLUG-Owl [119]	Text, Image	200M	Introduce visual knowledge and visual abstractor module for visual-language instruction tuning
Unified-IO 2 [120]	Text, Image, Audio, Action, etc.	1.1B, 3.2B, 6.8B	Support understanding and generating images and audios (spectrograms) with ViT encoders, Audio spectrogram transformer, and VQGANs. Also support bounding boxes, 3D cuboids, points, and discrete robotic actions

much of the rich information present in the original modalities can be lost. Secondly, the vast amount of prior knowledge embedded in non-textual modalities cannot be effectively trained into the LLMs' parameters [122]. Therefore, developing a series of models capable

Table 4.6 Overview of MFMs (continued)

Model	Modality	#Parameters	Characteristics
DeepSeek-VL [121]	Text, Image	7B	Higher-resolution, interleaved image-text data, SAM and SigCLIP image encoders
Cambrian-1 [122]	Text, Image	8B, 13B, 34B	Vision-centric optimization for better perception abilities; A novel connector named spatial vision aggregator to ensemble multiple visual encoders
SEED-LLaMA [123]	Text, Image	7B	Support both image understanding and generation
Emu [124, 125]	Text, Image, Video	13B, 34B	A Continuous-In-Continuous-Out Any-to-Any MFM
Gemini [52]	Text, Image, Video, Audio	N/A	Support understanding images, videos, speechs, texts, and generating images and texts
GPT-4o [51]	Text, Image, Video, Audio	N/A	Support understanding images, speechs, texts, and videos. Support generating images, 3D assets, texts, and speechs. With real-time interaction abilities
iVideoGPT [126]	Video, Action	138M, 436M	Pre-trained on millions of human and robotic manipulation trajectories. An autoregressive transformer that separately maps each step into a sequence of tokens
LWM [127]	Text, Image, Video	7B	The context window is gradually expanded in several training stages to support image and text generation, as well as video comprehension and generation. The video is broken down into image frames. Images are processed using VQGAN
OpenVLA [128]	VLA	7B	Open-source VLA model trained on diverse robot demonstrations, achieving strong results for generalist manipulation
LLaVA-Next [129–133]	Text, Image, Video, 3D	7B, 8B, 13B, 34B, 72B, 110B	Cropping-based dynamic image resolution support. Enhanced OCR, world knowledge, reasoning, and multi-image understanding abilities. Support video and 3D modalities. High performance for real-world scenarios
Genie [134]	Text, Image, Video	11B	Generate action-controllable virtual worlds from unlabeled Internet videos, enabling training of generalist agents
Pandora [135]	Text, Image, Video	7B	Hybrid autoregressive-diffusion model that simulates world states by generating videos and allows real-time control with free-text actions
Video-LLaVA [136]	Text, Image, Video	7B	Unify visual representation into the language feature space for enhanced visual-language understanding and performance. Align images and videos before projection
Chameleon [137]	Text, Image	7B, 34B	Continued from CM3Leon [118] by scaling up and some technical optimizations such as QK-Norm, z-loss, etc.
MIO [138]	Text, Image, Speech, Video	7B	Support images, texts, speechs, and videos with an autoregressive LLM based on discrete tokens

of understanding or even generating multimodal data is beneficial and crucial for advancing towards artificial general intelligence. From the perspective of iNLP, multimodal foundational models (MFMs) can not only provide users with richer interaction interfaces but also more naturally connect foundation model agents and environments, as illustrated in Sect. 2.4.

In this section, we roughly categorize multimodal foundational models based on their functionalities into the following three types:

- **Multimodal Understanding Foundation Models**: This type of MFM focuses solely on enabling the LLM core to understand non-textual modal data, without concern for generating non-textual modalities. They typically involve extensions at the input end.

- **Any-to-Any Multimodal Foundation Models**: These MFMs focus on enabling the LLM core to both understand and generate data in various modalities. They involve extensions at both the input and output ends.
- **Embodied MFMs and World Models**: These are specialized MFMs that can be either understanding-only or any-to-any. The key distinction is their support for specific modalities and functionalities. Specifically, Embodied MFMs support action modality [95, 96, 128, 143], while world models support the coherent input and output of video sequences [126, 127]. Given their significant implications for AGI, they are discussed separately.

4.3.1 Multimodal Understanding Foundation Models

Due to the vast market for Image-Text multimodal foundation models and the moderate processing difficulty compared to modalities like video, most MFMs research has concentrated on incorporating the image modality. These image-text multimodal foundation models often serve as general architectures applied to other modalities as well. Therefore, in this subsection, we focus on image-text multimodal foundation models to introduce the general architecture of multimodal understanding foundation models, and we provide some references in Table 4.4 for other modalities.

Generally, a Multimodal Understanding Foundation Model consists of three components: (1) an image encoder, (2) a connector, and (3) an LLM backbone. The image encoder is responsible for encoding the image, the connector aligns the encoded image features with the language embedding space, and the LLM backbone processes these aligned image features and textual signals to generate textual output. This structure is widely adopted by the most advanced multimodal foundation models [101, 110, 111, 115, 117, 122]. It is noteworthy that in Sect. 2.4, we introduced three basic structures of modality grounding. The mainstream structure of these multimodal foundation models can be seen as an interaction between vision models and language models, facilitated by the connector.

For the image encoder, its core task is to process the image data into high-level representations. According to [122], these encoders can be categorized based on the types of supervision signals:

1. **Class Label Supervised**: These encoders are trained using class labels to supervise the learning process. An example is Vision Transformer (ViT) [144], whose training tasks involve classification.
2. **Language Supervised**: These encoders are trained using paired image-text data, where the supervision comes from the textual descriptions of the images. For example, CLIP [145] and EVA-CLIP [146] align image and text representations with contrastive loss.

3. **SSL-Contrastive**: Self-Supervised Learning (SSL) encoders trained using contrastive methods. For example, DINO [147, 148] employs a teacher model and a student model to encode two different crops of the same image, aligning their resulting features through knowledge distillation.

4. **SSL-Masking**: These encoders are trained using masked image modeling, where parts of the image are masked, and the model learns to reconstruct them. An example is Masked Autoencoders (MAE) [149], which learn effective representations by predicting the missing parts of an image.

5. **Diffusion**: These models use diffusion processes to learn image representations. Stable Diffusion [150] is an example where the model gradually denoises a noisy image to reconstruct the image.

6. **Depth Supervised**: These encoders use image depth information to supervise the learning process. MiDaS [151, 152] is an example.

7. **Segmentation Supervised**: These encoders are trained using image segmentation tasks, where the model learns to identify and delineate different parts of an image. An example is the Segment Anything Model (SAM) [153], which uses a promptable segmentation approach to segment any object in an image with high accuracy.

To ensure training efficiency and maintain the integrity of the image signals, these image encoders are usually frozen during the training of MFMs [110, 111]. Consequently, the image representations produced by these encoders need to be well-aligned with the language embedding space. This is why many advanced MFMs use CLIP [145] as their image encoder. However, Cambrian-1 [122] demonstrated through experiments that unfreezing the image encoder allows other image encoders trained solely with visual signals to achieve competitive performance, especially in tasks emphasizing perception with visual prompts. Moreover, integrating diverse image encoders significantly enhances the overall multimodal understanding performance [122].

In addition to the primary goals of processing visual information and aligning it with the textual space, image encoders can be optimized for various other objectives. For high-resolution images, techniques such as sub-image cropping and concatenation can be employed [129, 154, 155], where the image is divided into smaller sub-images for encoding, and the encoded features are concatenated. To address the diversity in image aspect ratios, using patch tokens of arbitrary lengths [156] can be effective. ConvLLaVA [157] utilizes a hierarchical ConvNeXt [158] to iteratively compress the length of image tokens, mitigating the issue of excessive token length occupying the LLM's context window. Additionally, some work focuses on specific types of images, such as text-rich images [159] and vector graphics [160].

However, while these image encoders can effectively represent images, there remains a significant gap between their output space and the input space of the LLM backbone. Although language-supervised image encoders partially align image and text representations, this alignment typically uses an offline text encoder rather than the LLM backbone's embedding space, limiting its effectiveness. A connector is essential to address this gap.

Connectors aim to convert encoded image features into aligned image features that are compatible with the LLMs' textual embedding space. Common connector architectures include: (1) query-based Transformer, (2) MLP, (3) quantization, and (4) others.

1. **Query-based Transformer**: This method uses learnable query tokens processed through a multi-layer Transformer with cross-attention, where the encoded image features serve as keys and values. Examples include Perceiver Resampler [161], used by Flamingo [101], and Q-Former, leveraged by BLIP-2 [107].
2. **MLP**: Encoded features are processed through an MLP to produce aligned features. A single-layer linear model is also effective [110]. Examples include LLaVA-1.5 [111] and Yi-VL [41].
3. **Quantization**: Image features are transformed through vector quantization (VQ) into aligned image tokens. Unlike other methods that represent images as continuous tokens, these methods represent images as discrete tokens. Typical methods include VQ-VAE [162], VQGAN [163], ViT-VQGAN [164], SEED-Tokenizer [165], and LWM [127]. This approach is highly beneficial for modeling Any-to-Any MFMs discussed in Sect. 4.3.2, but often results in a significant loss of image information.
4. **Others**: For example, Emu [124] uses a Causal Transformer, while Emu-2 [125] employs a more lightweight Average Pooling module as the connectors.

Unlike image encoders, which are typically not trained during the multimodal model training, connectors often require online training with the image encoders and LLM backbone (with image encoders frozen, and with or without LLMs frozen). Initially, the connector is trained using image-text paired data, followed by joint training with the LLM using other data types, such as visual instruction tuning data [6, 110, 111].

Thus, through image encoders and connectors, an image can be transformed into aligned image features. These features are concatenated with textual embeddings along the length dimension, such as "{IMAGE} {QUESTION}", and then input into the language model backbone to produce the desired output.

4.3.2   Any-To-Any Multimodal Foundation Models

In Sect. 4.3.1, we introduced a large number of multimodal foundational models. However, these foundational models are merely extensions of pure text language models that enhance their understanding of multimodal data, without focusing on their multimodal generation capabilities.

A simple solution, as introduced in Sect. 2.3, is to have the language model generate API calls for external tools (such as an image generation tool or a TTS tool) and execute them to produce the required images or audios. However, these tool-based approaches have several drawbacks: one is that using external text-to-image models for image generation restricts the generatable image space to the scope describable by text [52]; another is that world knowledge embedded in other modalities, which cannot be described by text, cannot be effectively learned by the language model.

To remedy these issues, Any-to-Any MFMs are developed. These models not only support non-textual modality input but also non-textual modality output. Among these, the most notable models are GPT-4o [51] and Gemini [52], which exhibit strong capabilities in image-text understanding, text reasoning, and support for end-to-end input and output of multiple modalities. For example, GPT-4o [51] supports input from text, speech, video, and images, and directly outputs images, text, speech, and 3D assets. Notably, it also achieves real-time speech effects. Gemini [52], developed by Google DeepMind, is an advanced foundational model that, beyond its pure text capabilities, can understand long videos, images, and speech, and can generate images based on discrete tokens. These Any-to-Any MFMs extend the functionality of language models, endowing them with greater interactivity, both in interactions with the environment and with human users.

However, these advanced models are closed-source, and their technical details have not been disclosed extensively. Therefore, in this section, we will present some open-source Any-to-Any MFMs.

The structure of these Any-to-Any MFMs is generally consistent with Multimodal Understanding Foundation Models, comprising a modality encoder, a connector, and an LLM backbone. The difference lies in adapting the output layer of the LLM backbone to support multimodal generation, possibly adding some modules [116, 123–125, 166, 167]. Thus, for Any-to-Any MFMs, the key lies in two aspects: handling multimodal input and supporting multimodal output.

Generally, according to the representation forms of multimodal input and output, Any-to-Any MFMs can be divided into three types:

1. **Discrete-In-Discrete-Out (DIDO)**: Non-textual modality data is converted into discrete tokens using vector quantization methods [163] before being processed by LLMs [123, 127, 138, 167].
2. **Continuous-In-Discrete-Out (CIDO)**: LLM backbones take in densely encoded features of non-textual modality data and produce their discretized representations [52, 168].
3. **Continuous-In-Continuous-Out (CICO)**: LLMs both understand and generate non-textual modality data in their densely encoded forms [116, 124, 125, 166, 169].

DIDO. Discrete-In-Discrete-Out MFMs process all modality data into discrete token form, maintaining consistency with textual tokens. Thus, these discrete tokens can be added to the language model's vocabulary. After expanding the dimensions of the LLM's embedding matrix and language modeling head, the standard next-token-prediction method can be used to train the LLM, similar to adding a new "foreign language" [105]. These generated non-textual modality discrete tokens, once reconstructed by an external decoder, can produce outputs like images or speech.

Different modalities often require different discretization methods, with Vector Quantization (VQ) [162, 163] being the most typical and widely applied. For example, for images, ViT-VQGAN [164] uses ViT [144] as an image encoder, and then applies VQ to discretize the encoded image features, followed by a VQGAN decoder to reconstruct the original image from discrete tokens. This method is also adopted and optimized by Parti [170]. SEED-LLaMA [123] compresses long image features encoded by ViT into 32 continuous tokens using a Causal Q-Former and then discretizes them through quantization. CM3Leon [118] leverages an image quantizer with enhanced face modeling abilities. StrokeNVWA [160] introduces a VQ-Stroke method to discretize vector graphic images into stroke tokens. For other modalities like speech, there are different VQ methods. For example, SpeechTokenizer [171] uses multi-layer residual vector quantization for speech discretization, where each layer corresponds to a codebook (i.e., vocabulary), with the first layer containing more content information distilled from HuBERT [172], and the subsequent layers trained with reconstruction loss to capture lower-level information like timbre.

CIDO. A key issue with DIDO MFMs is the significant information loss due to the discretization of the inputs. To mitigate this, the CIDO approach is employed. Unlike DIDO, CIDO processes non-textual modality input in a continuous form without quantization, while adopting the same discretization scheme as DIDO for the output. Since discrete tokens are consistent with textual tokens, CIDO methods in modeling and training objectives are highly similar to those in Sect. 4.3.1, merely adding more "multimodal languages". Gemini [52] mentioned earlier is a typical CIDO-based Any-to-Any MFM. Unified-IO 2 [120] is an open-source model of this type, using two-layer features of ViT to construct image representations and Audio Spectrogram Transformer [173] to understand audio spectrogram data at the input end. At the output end, it generates discrete tokens for images or audio spectrograms, which are decoded using VQGAN [163] or ViT-VQGAN [164]. Notably, Unified-IO 2 also supports the processing of multimodal conversation history and discrete forms of points, bounding boxes, 3D cuboids, and robot actions.

CICO. Continuous-In-Continuous-Out Any-to-Any MFMs reduce information loss inherent in DIDO and, compared to CIDO, maintains input-output consistency. The continuous input part is consistent with the aforementioned multimodal understanding foundation models. The key lies in modifying the language model to support continuous output forms. Unlike discrete token generation, continuous outputs require specific designs or techniques in two aspects: (1) positioning: the model must determine when to switch from discrete (texts) to continuous output (images, etc.); (2) output representation: the model must accurately generate the required continuous outputs.

Regarding positioning methods, there are typically two approaches: (1) using placeholders [116, 169, 174]; and (2) using non-textual modality begin-of-modality (BOM) tokens [124, 125, 166]. Firstly, multiple special tokens can be utilized as placeholders for non-textual modality data. For instance, Mini-GPT5 [169] and GILL [174] use a sequence of image placeholder tokens from [IMG1] to [IMGr], interleaved with textual tokens, which are added to the model's vocabulary. Similarly, NExT-GPT [116] employs 5 image placeholder tokens and introduces 9 audio and 25 video placeholder tokens. Secondly, the numerous placeholder tokens can be reduced to a single BOM token (sometimes accompanied by an end-of-modality (EOM) token) to mark the position of non-textual modality data. For example, DreamLLM [166] introduces a special <dream> token to signal the beginning of modality switching and then inputs a sequence of queries to generate representations in one model run. Emu [124] and Emu2 [125] introduce both BOM and EOM tokens for images, placing encoded image features between them.

Regarding output representation, to enable soft token outputs, modifications to the output layers of LLMs are necessary. There are typically three approaches: (1) change the classification-based language modeling head into the regression head [124, 125]; (2) introduce a novel output head for dense features [166]; and (3) leverage the hidden states near the output layers [169, 174]. These output representations serve as the conditioning latents for a Diffusion Model to decode the images after a reverse connector module such as a MLP [123], a Transformer [116], or a Q-Former [165].

4.3.3 Embodied MFMs and World Models

Embodied MFMs and World Models are two special types of MFMs. Embodied MFMs aim to enable MFMs to output actions for tasks such as robotic manipulation. World Models can simulate the physical world by understanding the current states of the environment and predicting the future [175]. They are intrinsically coherent in the context of iNLP, as actions serve as an interaction interface between MFMs and the environment, and the simulation of the physical world significantly extends the interactive content space of MFMs.

In Table 4.4, we highlight some representative embodied MFMs (with "VLA" as their modality attribute) and world models (marked with boxes). For a comprehensive overview, we refer readers to the surveys by [143] for embodied MFMs and [176] for world models. In this section, we will mention a few examples.

LWM (Large World Models, [127]) is a world model built on discrete image tokens. It first utilizes RingAttention [177] to extend the context window of language foundation models and progressively trains the models to support increasing context lengths. Then, it extends the trained long-context language model with visual data. Initially, it undergoes a first-stage training using high-quality image-text data, followed by the gradual addition of increasingly long video data. Benefiting from the initial LLM's superior support for long text, the final LWM also supports the understanding and generation of very long videos. Notably, LWM segments videos into multiple image frames and processes these using VQGAN [163], following a DIDO approach.

Genie [134], developed by Google DeepMind, is an interactive world model capable of transforming various prompts, such as text-to-image and hand-drawn sketches, into interactive, playable environments that can be easily created, entered, and explored. It consists of three main components: (1) a latent action model that infers the implicit actions between each pair of frames, (2) a video tokenizer that quantizes raw video frames into discrete tokens, and (3) a dynamics model that predicts the next video frame based on the given latent action and preceding frame tokens. After training, users can input an image, choose a latent action as an integer to feed into the model, and the model will generate subsequent video clips based on the initial prompt and action.

Pandora [135] further expands Genie [134]'s action space. It replaces the latent actions limited by several integers in Genie with natural language instructions. Specifically, Pandora is an autoregressive CICO MFM. It takes in images and language instructions and generates videos similarly to DreamLLM [166] (i.e., input some learnable query tokens and let the models generate the output representations of these queries in a parallel manner).

iVideoGPT [126] builds interactive world models based on the DIDO MFMs structure. It is also an autoregressive Transformer, supporting the input of video frames and actions, and generating subsequent video frames, actions, and an additional reward output. This reward output is implemented through a simple reward head.

References

1. Bommasani, R., Hudson, D.A., Adeli, E., Altman, R., Arora, S., et al.: On the opportunities and risks of foundation models. arXiv preprint arXiv: 2108.07258 (2021)
2. Brown, T., Mann, B., Ryder, N., Subbiah, M., Kaplan, J.D., Dhariwal, P., Neelakantan, A., Shyam, P., Sastry, G., Askell, A., et al.: Language models are few-shot learners. Advances in neural information processing systems **33**, 1877–1901 (2020)
3. Liu, P., Yuan, W., Fu, J., Jiang, Z., Hayashi, H., Neubig, G.: Pre-train, prompt, and predict: A systematic survey of prompting methods in natural language processing. arXiv preprint arXiv: Arxiv-2107.13586 (2021)
4. Zhou, C., Li, Q., Li, C., Yu, J., Liu, Y., Wang, G., Zhang, K., Ji, C., Yan, Q., He, L., Peng, H., Li, J., Wu, J., Liu, Z., Xie, P., Xiong, C., Pei, J., Yu, P.S., University, L.S.M.S., University, B., University, L., University, M., University, N.T., of California at San Diego, U., University, D., Chicago, U., Research, S.A.: A comprehensive survey on pretrained foundation models: A history from bert to ChatGPT. ARXIV.ORG (2023). https://doi.org/10.48550/arXiv.2302.09419
5. Zhao, W.X., Zhou, K., Li, J., Tang, T., Wang, X., Hou, Y., Min, Y., Zhang, B., Zhang, J., Dong, Z., Du, Y., Yang, C., Chen, Y., Chen, Z., Jiang, J., Ren, R., Li, Y., Tang, X., Liu, Z., Liu, P., Nie, J., rong Wen, J.: A survey of large language models. ARXIV.ORG (2023). https://doi.org/10.48550/arXiv.2303.18223
6. Yin, S., Fu, C., Zhao, S., Li, K., Sun, X., Xu, T., Chen, E.: A survey on multimodal large language models. arXiv preprint arXiv: 2306.13549 (2023)
7. Devlin, J., Chang, M.W., Lee, K., Toutanova, K.: Bert: Pre-training of deep bidirectional transformers for language understanding. arXiv preprint arXiv:1810.04805 (2018)

8. Liu, Y., Ott, M., Goyal, N., Du, J., Joshi, M., Chen, D., Levy, O., Lewis, M., Zettlemoyer, L., Stoyanov, V.: Roberta: A robustly optimized bert pretraining approach. ARXIV.ORG (2019)

9. Yang, Z., Dai, Z., Yang, Y., Carbonell, J., Salakhutdinov, R.R., Le, Q.V.: Xlnet: Generalized autoregressive pretraining for language understanding. Advances in neural information processing systems **32** (2019)

10. Joshi, M., Chen, D., Liu, Y., Weld, D.S., Zettlemoyer, L., Levy, O.: Spanbert: Improving pretraining by representing and predicting spans. Transactions of the Association for Computational Linguistics **8**, 64–77 (2020)

11. Sun, Y., Wang, S., Li, Y., Feng, S., Chen, X., Zhang, H., Tian, X., Zhu, D., Tian, H., Wu, H.: Ernie: Enhanced representation through knowledge integration. arXiv preprint arXiv:1904.09223 (2019)

12. Sun, Y., Wang, S., Li, Y., Feng, S., Tian, H., Wu, H., Wang, H.: Ernie 2.0: A continual pretraining framework for language understanding. In: Proceedings of the AAAI conference on artificial intelligence, vol. 34, pp. 8968–8975 (2020)

13. Lan, Z., Chen, M., Goodman, S., Gimpel, K., Sharma, P., Soricut, R.: Albert: A lite bert for self-supervised learning of language representations. arXiv preprint arXiv:1909.11942 (2019)

14. Sanh, V., Debut, L., Chaumond, J., Wolf, T.: Distilbert, a distilled version of bert: smaller, faster, cheaper and lighter. arXiv preprint arXiv:1910.01108 (2019)

15. Clark, K., Luong, M.T., Le, Q.V., Manning, C.D.: Electra: Pre-training text encoders as discriminators rather than generators. arXiv preprint arXiv:2003.10555 (2020)

16. Iandola, F.N., Shaw, A.E., Krishna, R., Keutzer, K.W.: Squeezebert: What can computer vision teach nlp about efficient neural networks? arXiv preprint arXiv:2006.11316 (2020)

17. Radford, A., Narasimhan, K., Salimans, T., Sutskever, I., et al.: Improving language understanding by generative pre-training (2018)

18. Radford, A., Wu, J., Child, R., Luan, D., Amodei, D., Sutskever, I.: Language models are unsupervised multitask learners (2019)

19. Lewis, M., Liu, Y., Goyal, N., Ghazvininejad, M., Mohamed, A., Levy, O., Stoyanov, V., Zettlemoyer, L.: Bart: Denoising sequence-to-sequence pre-training for natural language generation, translation, and comprehension (2019)

20. Zhang, J., Zhao, Y., Saleh, M., Liu, P.: Pegasus: Pre-training with extracted gap-sentences for abstractive summarization. In: International Conference on Machine Learning, pp. 11328–11339. PMLR (2020)

21. Dong, L., Yang, N., Wang, W., Wei, F., Liu, X., Wang, Y., Gao, J., Zhou, M., Hon, H.W.: Unified language model pre-training for natural language understanding and generation. Advances in neural information processing systems **32** (2019)

22. Raffel, C., Shazeer, N., Roberts, A., Lee, K., Narang, S., Matena, M., Zhou, Y., Li, W., Liu, P.J.: Exploring the limits of transfer learning with a unified text-to-text transformer. Journal of Machine Learning Research **21**(140), 1–67 (2020). http://jmlr.org/papers/v21/20-074.html

23. Chung, H.W., Hou, L., Longpre, S., Zoph, B., Tay, Y., Fedus, W., Li, E., Wang, X., Dehghani, M., Brahma, S., et al.: Scaling instruction-finetuned language models. arXiv preprint arXiv:2210.11416 (2022)

24. Sun, Y., Wang, S., Feng, S., Ding, S., Pang, C., Shang, J., Liu, J., Chen, X., Zhao, Y., Lu, Y., et al.: Ernie 3.0: Large-scale knowledge enhanced pre-training for language understanding and generation. arXiv preprint arXiv:2107.02137 (2021)

25. Wang, S., Sun, Y., Xiang, Y., Wu, Z., Ding, S., Gong, W., Feng, S., Shang, J., Zhao, Y., Pang, C., et al.: Ernie 3.0 titan: Exploring larger-scale knowledge enhanced pre-training for language understanding and generation. arXiv preprint arXiv:2112.12731 (2021)

26. Zeng, W., Ren, X., Su, T., Wang, H., Liao, Y., Wang, Z., Jiang, X., Yang, Z., Wang, K., Zhang, X., et al.: Pangu-α: Large-scale autoregressive pretrained Chinese language models with auto-parallel computation. arXiv preprint arXiv:2104.12369 (2021)

27. Wei, J., Bosma, M., Zhao, V., Guu, K., Yu, A., Lester, B., Du, N., Dai, A.M., Le, Q.V.: Finetuned language models are zero-shot learners. International Conference On Learning Representations (2021)

28. Rae, J.W., Borgeaud, S., Cai, T., Millican, K., Hoffmann, J., Song, F., Aslanides, J., Henderson, S., Ring, R., Young, S., et al.: Scaling language models: Methods, analysis & insights from training gopher. arXiv preprint arXiv:2112.11446 (2021)

29. Chowdhery, A., Narang, S., Devlin, J., Bosma, M., Mishra, G., Roberts, A., Barham, P., Chung, H.W., Sutton, C., Gehrmann, S., et al.: Palm: Scaling language modeling with pathways. arXiv preprint arXiv:2204.02311 (2022)

30. Tay, Y., Dehghani, M., Tran, V.Q., Garcia, X., Bahri, D., Schuster, T., Zheng, H.S., Houlsby, N., Metzler, D.: Unifying language learning paradigms. arXiv preprint arXiv:2205.05131 (2022)

31. Google: Palm 2 (2023). https://ai.google/discover/palm2

32. Zhang, S., Roller, S., Goyal, N., Artetxe, M., Chen, M., Chen, S., Dewan, C., Diab, M., Li, X., Lin, X.V., et al.: Opt: Open pre-trained transformer language models. arXiv preprint arXiv:2205.01068 (2022)

33. Taylor, R., Kardas, M., Cucurull, G., Scialom, T., Hartshorn, A., Saravia, E., Poulton, A., Kerkez, V., Stojnic, R.: Galactica: A large language model for science (2022)

34. Zeng, A., Liu, X., Du, Z., Wang, Z., Lai, H., Ding, M., Yang, Z., Xu, Y., Zheng, W., Xia, X., et al.: Glm-130b: An open bilingual pre-trained model. arXiv preprint arXiv:2210.02414 (2022)

35. Scao, T.L., Fan, A., Akiki, C., Pavlick, E., Ilić, S., Hesslow, D., Castagné, R., Luccioni, A.S., Yvon, F., Gallé, M., et al.: Bloom: A 176b-parameter open-access multilingual language model. arXiv preprint arXiv:2211.05100 (2022)

36. Touvron, H., Lavril, T., Izacard, G., Martinet, X., Lachaux, M.A., Lacroix, T., Rozière, B., Goyal, N., Hambro, E., Azhar, F., et al.: Llama: Open and efficient foundation language models. arXiv preprint arXiv:2302.13971 (2023)

37. Shazeer, N.: Glu variants improve transformer. arXiv preprint arXiv:2002.05202 (2020)

38. Jiang, A.Q., Sablayrolles, A., Mensch, A., Bamford, C., Chaplot, D.S., Casas, D.d.l., Bressand, F., Lengyel, G., Lample, G., Saulnier, L., et al.: Mistral 7b. arXiv preprint arXiv:2310.06825 (2023)

39. Bai, J., Bai, S., Chu, Y., Cui, Z., Dang, K., Deng, X., Fan, Y., Ge, W., Han, Y., Huang, F., et al.: Qwen technical report. arXiv preprint arXiv:2309.16609 (2023)

40. Penedo, G., Malartic, Q., Hesslow, D., Cojocaru, R., Cappelli, A., Alobeidli, H., Pannier, B., Almazrouei, E., Launay, J.: The refinedweb dataset for falcon LLM: outperforming curated corpora with web data, and web data only. arXiv preprint arXiv:2306.01116 (2023)

41. Young, A., Chen, B., Li, C., Huang, C., Zhang, G., Zhang, G., Li, H., Zhu, J., Chen, J., Chang, J., et al.: Yi: Open foundation models by 01. ai. arXiv preprint arXiv:2403.04652 (2024)

42. Dai, D., Deng, C., Zhao, C., Xu, R.X., Gao, H., Chen, D., Li, J., Zeng, W., Yu, X., Wu, Y., Xie, Z., Li, Y.K., Huang, P., Luo, F., Ruan, C., Sui, Z., Liang, W.: Deepseekmoe: Towards ultimate expert specialization in mixture-of-experts language models. arXiv preprint arXiv: 2401.06066 (2024)

43. Guo, D., Zhu, Q., Yang, D., Xie, Z., Dong, K., Zhang, W., Chen, G., Bi, X., Wu, Y., Li, Y., et al.: Deepseek-coder: When the large language model meets programming–the rise of code intelligence. arXiv preprint arXiv:2401.14196 (2024)

44. Bi, X., Chen, D., Chen, G., Chen, S., Dai, D., Deng, C., Ding, H., Dong, K., Du, Q., Fu, Z., et al.: Deepseek LLM: Scaling open-source language models with longtermism. arXiv preprint arXiv:2401.02954 (2024)

45. Zhu, Q., Guo, D., Shao, Z., Yang, D., Wang, P., Xu, R., Wu, Y., Li, Y., Gao, H., Ma, S., et al.: Deepseek-coder-v2: Breaking the barrier of closed-source models in code intelligence. arXiv preprint arXiv:2406.11931 (2024)

46. Xin, H., Guo, D., Shao, Z., Ren, Z., Zhu, Q., Liu, B., Ruan, C., Li, W., Liang, X.: Deepseek-prover: Advancing theorem proving in LLMs through large-scale synthetic data. arXiv preprint arXiv:2405.14333 (2024)
47. Soldaini, L., Kinney, R., Bhagia, A., Schwenk, D., Atkinson, D., Authur, R., Bogin, B., Chandu, K., Dumas, J., Elazar, Y., et al.: Dolma: An open corpus of three trillion tokens for language model pretraining research. arXiv preprint arXiv:2402.00159 (2024)
48. Groeneveld, D., Beltagy, I., Walsh, P., Bhagia, A., Kinney, R., Tafjord, O., Jha, A.H., Ivison, H., Magnusson, I., Wang, Y., et al.: Olmo: Accelerating the science of language models. arXiv preprint arXiv:2402.00838 (2024)
49. Zhang, G., Qu, S., Liu, J., Zhang, C., Lin, C., Yu, C.L., Pan, D., Cheng, E., Liu, J., Lin, Q., et al.: Map-neo: Highly capable and transparent bilingual large language model series. arXiv preprint arXiv:2405.19327 (2024)
50. OpenAI: GPT-4 technical report. PREPRINT (2023)
51. OpenAI: hello-gpt-4o (2024). https://openai.com/index/hello-gpt-4o/
52. Gemini Team Google: Gemini: A family of highly capable multimodal models. arXiv preprint arXiv:2312.11805 (2023)
53. Anthropic: Model card and evaluations for claude models (2023). https://www.anthropic.com/news/claude-2
54. Anthropic: Introducing the next generation of claude (2024). https://www.anthropic.com/news/claude-3-family
55. Vaswani, A., Shazeer, N.M., Parmar, N., Uszkoreit, J., Jones, L., Gomez, A.N., Kaiser, L., Polosukhin, I.: Attention is all you need. NIPS (2017)
56. Adhikari, A., Ram, A., Tang, R., Lin, J.: Docbert: Bert for document classification. arXiv preprint arXiv: Arxiv-1904.08398 (2019)
57. Liu, Y., Gu, J., Goyal, N., Li, X., Edunov, S., Ghazvininejad, M., Lewis, M., Zettlemoyer, L.: Multilingual denoising pre-training for neural machine translation. arXiv preprint arXiv: Arxiv-2001.08210 (2020)
58. Guan, J., Huang, F., Zhao, Z., Zhu, X., Huang, M.: A Knowledge-Enhanced Pretraining Model for Commonsense Story Generation. Transactions of the Association for Computational Linguistics **8**, 93–108 (2020). https://doi.org/10.1162/tacl_a_00302.
59. Wei, J., Tay, Y., Bommasani, R., Raffel, C., Zoph, B., Borgeaud, S., Yogatama, D., Bosma, M., Zhou, D., Metzler, D., Chi, E.H., Hashimoto, T., Vinyals, O., Liang, P., Dean, J., Fedus, W.: Emergent abilities of large language models. arXiv preprint arXiv: Arxiv-2206.07682 (2022)
60. Common Crawl Foundation: Common crawl (2012). https://commoncrawl.org/
61. Longpre, S., Yauney, G., Reif, E., Lee, K., Roberts, A., Zoph, B., Zhou, D., Wei, J., Robinson, K., Mimno, D., et al.: A pretrainer's guide to training data: Measuring the effects of data age, domain coverage, quality, & toxicity. arXiv preprint arXiv:2305.13169 (2023)
62. Zhou, T., Chen, Y., Cao, P., Liu, K., Zhao, J., Liu, S.: Oasis: Data curation and assessment system for pretraining of large language models. arXiv preprint arXiv:2311.12537 (2023)
63. Du, X., Yu, Z., Gao, S., Pan, D., Cheng, Y., Ma, Z., Yuan, R., Qu, X., Liu, J., Zheng, T., et al.: Chinese tiny LLM: Pretraining a Chinese-centric large language model. arXiv preprint arXiv:2404.04167 (2024)
64. Zhang, B., Nagesh, A., Knight, K.: Parallel corpus filtering via pre-trained language models. In: D. Jurafsky, J. Chai, N. Schluter, J. Tetreault (eds.) Proceedings of the 58th Annual Meeting of the Association for Computational Linguistics, pp. 8545–8554. Association for Computational Linguistics, Online (2020). https://doi.org/10.18653/v1/2020.acl-main.756. https://aclanthology.org/2020.acl-main.756
65. Penedo, G., Kydlíček, H., Lozhkov, A., Mitchell, M., Raffel, C., Von Werra, L., Wolf, T., et al.: The fineweb datasets: Decanting the web for the finest text data at scale. arXiv preprint arXiv:2406.17557 (2024)

66. Beltagy, I., Peters, M.E., Cohan, A.: Longformer: The long-document transformer. arXiv preprint arXiv:2004.05150 (2020)
67. Choromanski, K., Likhosherstov, V., Dohan, D., Song, X., Gane, A., Sarlos, T., Hawkins, P., Davis, J., Mohiuddin, A., Kaiser, L., et al.: Rethinking attention with performers. arXiv preprint arXiv:2009.14794 (2020)
68. Ainslie, J., Lee-Thorp, J., de Jong, M., Zemlyanskiy, Y., Lebrón, F., Sanghai, S.: Gqa: Training generalized multi-query transformer models from multi-head checkpoints. arXiv preprint arXiv:2305.13245 (2023)
69. Gu, A., Goel, K., Ré, C.: Efficiently modeling long sequences with structured state spaces. arXiv preprint arXiv:2111.00396 (2021)
70. Xiao, G., Lin, J., Seznec, M., Wu, H., Demouth, J., Han, S.: Smoothquant: Accurate and efficient post-training quantization for large language models. In: International Conference on Machine Learning, pp. 38087–38099. PMLR (2023)
71. Yao, Z., Yazdani Aminabadi, R., Zhang, M., Wu, X., Li, C., He, Y.: Zeroquant: Efficient and affordable post-training quantization for large-scale transformers. Advances in Neural Information Processing Systems **35**, 27168–27183 (2022)
72. Frantar, E., Ashkboos, S., Hoefler, T., Alistarh, D.: Gptq: Accurate post-training quantization for generative pre-trained transformers. arXiv preprint arXiv:2210.17323 (2022)
73. Liu, Z., Oguz, B., Zhao, C., Chang, E., Stock, P., Mehdad, Y., Shi, Y., Krishnamoorthi, R., Chandra, V.: LLM-QAT: Data-free quantization aware training for large language models. arXiv preprint arXiv:2305.17888 (2023)
74. Micikevicius, P., Narang, S., Alben, J., Diamos, G., Elsen, E., Garcia, D., Ginsburg, B., Houston, M., Kuchaiev, O., Venkatesh, G., et al.: Mixed precision training. arXiv preprint arXiv:1710.03740 (2017)
75. Shoeybi, M., Patwary, M., Puri, R., LeGresley, P., Casper, J., Catanzaro, B.: Megatron-lm: Training multi-billion parameter language models using model parallelism. arXiv preprint arXiv:1909.08053 (2019)
76. Rajbhandari, S., Rasley, J., Ruwase, O., He, Y.: Zero: Memory optimizations toward training trillion parameter models. In: SC20: International Conference for High Performance Computing, Networking, Storage and Analysis, pp. 1–16. IEEE (2020)
77. Das, P., Ivkin, N., Bansal, T., Rouesnel, L., Gautier, P., Karnin, Z., Dirac, L., Ramakrishnan, L., Perunicic, A., Shcherbatyi, I., et al.: Amazon sagemaker autopilot: a white box automl solution at scale. In: Proceedings of the fourth international workshop on data management for end-to-end machine learning, pp. 1–7 (2020)
78. Maini, P., Seto, S., Bai, H., Grangier, D., Zhang, Y., Jaitly, N.: Rephrasing the web: A recipe for compute and data-efficient language modeling. arXiv preprint arXiv:2401.16380 (2024)
79. Adler, B., Agarwal, N., Aithal, A., Anh, D.H., Bhattacharya, P., Brundyn, A., Casper, J., Catanzaro, B., Clay, S., Cohen, J., et al.: Nemotron-4 340b technical report. arXiv preprint arXiv:2406.11704 (2024)
80. Hoffmann, J., Borgeaud, S., Mensch, A., Buchatskaya, E., Cai, T., Rutherford, E., Casas, D.d.L., Hendricks, L.A., Welbl, J., Clark, A., et al.: Training compute-optimal large language models. arXiv preprint arXiv:2203.15556 (2022)
81. Muennighoff, N., Rush, A., Barak, B., Le Scao, T., Tazi, N., Piktus, A., Pyysalo, S., Wolf, T., Raffel, C.A.: Scaling data-constrained language models. Advances in Neural Information Processing Systems **36** (2024)
82. Yang, C., Li, J., Niu, X., Du, X., Gao, S., Zhang, H., Chen, Z., Qu, X., Yuan, R., Li, Y., et al.: The fine line: Navigating large language model pretraining with down-streaming capability analysis. arXiv preprint arXiv:2404.01204 (2024)

83. Shazeer, N.M., Mirhoseini, A., Maziarz, K., Davis, A., Le, Q.V., Hinton, G.E., Dean, J.: Outrageously large neural networks: The sparsely-gated mixture-of-experts layer. International Conference on Learning Representations (2017)

84. Fedus, W., Zoph, B., Shazeer, N.: Switch transformers: Scaling to trillion parameter models with simple and efficient sparsity. arXiv preprint arXiv: 2101.03961 (2021)

85. Yang, A., Lin, J., Men, R., Zhou, C., Jiang, L., Jia, X., Wang, A., Zhang, J., Wang, J., Li, Y., Zhang, D., Lin, W., Qu, L., Zhou, J., Yang, H.: M6-t: Exploring sparse expert models and beyond. arXiv preprint arXiv: 2105.15082 (2021)

86. Riquelme, C., Puigcerver, J., Mustafa, B., Neumann, M., Jenatton, R., Pinto, A.S., Keysers, D., Houlsby, N.: Scaling vision with sparse mixture of experts. arXiv preprint arXiv: 2106.05974 (2021)

87. Xue, F., Zheng, Z.A., Fu, Y., Ni, J., Zheng, Z., Zhou, W., You, Y.: Openmoe: An early effort on open mixture-of-experts language models. International Conference on Machine Learning (2024). https://doi.org/10.48550/arXiv.2402.01739

88. Zoph, B., Bello, I., Kumar, S., Du, N., Huang, Y., Dean, J., Shazeer, N., Fedus, W.: St-moe: Designing stable and transferable sparse expert models. arXiv preprint arXiv: 2202.08906 (2022)

89. Wei, T., Zhu, B., Zhao, L., Cheng, C., Li, B., Lü, W., Cheng, P., Zhang, J., Zhang, X., Zeng, L., Wang, X., Ma, Y., Hu, R., Yan, S., Fang, H., Zhou, Y.: Skywork-moe: A deep dive into training techniques for mixture-of-experts language models. arXiv preprint arXiv: 2406.06563 (2024)

90. Roller, S., Sukhbaatar, S., szlam, a., Weston, J.: Hash layers for large sparse models. In: M. Ranzato, A. Beygelzimer, Y. Dauphin, P. Liang, J.W. Vaughan (eds.) Advances in Neural Information Processing Systems, vol. 34, pp. 17555–17566. Curran Associates, Inc. (2021). https://proceedings.neurips.cc/paper_files/paper/2021/file/92bf5e6240737e0326ea59846a83e076-Paper.pdf

91. Ren, X., Zhou, P., Meng, X., Huang, X., Wang, Y., Wang, W., Li, P., Zhang, X., Podolskiy, A., Arshinov, G., Bout, A., Piontkovskaya, I., Wei, J., Jiang, X., Su, T., Liu, Q., Yao, J.: Pangu-: Towards trillion parameter language model with sparse heterogeneous computing. arXiv preprint arXiv: 2303.10845 (2023)

92. Rajbhandari, S., Li, C., Yao, Z., Zhang, M., Aminabadi, R.Y., Awan, A.A., Rasley, J., He, Y.: Deepspeed-moe: Advancing mixture-of-experts inference and training to power next-generation ai scale. International Conference on Machine Learning (2022)

93. Team, N., Costa-jussà, M.R., Cross, J., Çelebi, O., Elbayad, M., Heafield, K., Heffernan, K., Kalbassi, E., Lam, J., Licht, D., Maillard, J., Sun, A., Wang, S., Wenzek, G., Youngblood, A., Akula, B., Barrault, L., Gonzalez, G.M., Hansanti, P., Hoffman, J., Jarrett, S., Sadagopan, K.R., Rowe, D., Spruit, S., Tran, C., Andrews, P., Ayan, N.F., Bhosale, S., Edunov, S., Fan, A., Gao, C., Goswami, V., Guzmán, F., Koehn, P., Mourachko, A., Ropers, C., Saleem, S., Schwenk, H., Wang, J.: No language left behind: Scaling human-centered machine translation. arXiv preprint arXiv: 2207.04672 (2022)

94. Komatsuzaki, A., Puigcerver, J., Lee-Thorp, J., Ruiz, C.R., Mustafa, B., Ainslie, J., Tay, Y., Dehghani, M., Houlsby, N.: Sparse upcycling: Training mixture-of-experts from dense checkpoints. arXiv preprint arXiv: 2212.05055 (2022)

95. Brohan, A., Brown, N., Carbajal, J., Chebotar, Y., Dabis, J., Finn, C., Gopalakrishnan, K., Hausman, K., Herzog, A., Hsu, J., Ibarz, J., Ichter, B., Irpan, A., Jackson, T., Jesmonth, S., Joshi, N., Julian, R., Kalashnikov, D., Kuang, Y., Leal, I., Lee, K.H., Levine, S., Lu, Y., Malla, U., Manjunath, D., Mordatch, I., Nachum, O., Parada, C., Peralta, J., Perez, E., Pertsch, K., Quiambao, J., Rao, K., Ryoo, M., Salazar, G., Sanketi, P., Sayed, K., Singh, J., Sontakke, S., Stone, A., Tan, C., Tran, H., Vanhoucke, V., Vega, S., Vuong, Q., Xia, F., Xiao, T., Xu, P., Xu, S., Yu, T., Zitkovich, B.: Rt-1: Robotics transformer for real-world control at scale. In: arXiv preprint arXiv:2212.06817 (2022)

96. Chebotar, Y., Yu, T., Brohan, A., Brown, N., Carbajal, J., Chen, X., Choromanski, K., Ding, T., Driess, D., Dubey, A., Finn, C., Florence, P., Fu, C., Arenas, M.G., Gopalakrishnan, K., Han, K., Hausman, K., Herzog, A., Hsu, J., Ichter, B., Irpan, A., Joshi, N., Julian, R., Kalashnikov, D., Kuang, Y., Leal, I., Lee, L., Lee, T.W.E., Levine, S., Lu, Y., Michalewski, H., Mordatch, I., Pertsch, K., Rao, K., Reymann, K., Ryoo, M., Salazar, G., Sanketi, P., Sermanet, P., Singh, J., Singh, A., Soricut, R., Tran, H., Vanhoucke, V., Vuong, Q., Wahid, A., Welker, S., Wohlhart, P., Wu, J., Xia, F., Xiao, T., Xu, P., Xu, S., Yu, T., Zitkovich, B.: Rt-2: New model trans-lates vision and language into action (2023). https://deepmind.google/discover/blog/rt-2-new-model-translates-vision-and-language-into-action/

97. Jiang, Y., Gupta, A., Zhang, Z., Wang, G., Dou, Y., Chen, Y., Fei-Fei, L., Anandkumar, A., Zhu, Y., Fan, L.: Vima: General robot manipulation with multimodal prompts. arXiv preprint arXiv: Arxiv-2210.03094 (2022)

98. Lynch, C., Wahid, A., Tompson, J., Ding, T., Betker, J., Baruch, R., Armstrong, T., Florence, P.: Interactive language: Talking to robots in real time (2022)

99. Driess, D., Xia, F., Sajjadi, M.S.M., Lynch, C., Chowdhery, A., Ichter, B., Wahid, A., Tompson, J., Vuong, Q., Yu, T., Huang, W., Chebotar, Y., Sermanet, P., Duckworth, D., Levine, S., Van-houcke, V., Hausman, K., Toussaint, M., Greff, K., Zeng, A., Mordatch, I., Florence, P.: Palm-e: An embodied multimodal language model. In: arXiv preprint arXiv:2303.03378 (2023)

100. Bao, H., Wang, W., Dong, L., Liu, Q., Mohammed, O.K., Aggarwal, K., Som, S., Piao, S., Wei, F.: Vlmo: Unified vision-language pre-training with mixture-of-modality-experts. Advances in Neural Information Processing Systems 35, 32897–32912 (2022)

101. Alayrac, J.B., Donahue, J., Luc, P., Miech, A., Barr, I., Hasson, Y., Lenc, K., Mensch, A., Millican, K., Reynolds, M., Ring, R., Rutherford, E., Cabi, S., Han, T., Gong, Z., Samangooei, S., Monteiro, M., Menick, J., Borgeaud, S., Brock, A., Nematzadeh, A., Sharifzadeh, S., Binkowski, M., Barreira, R., Vinyals, O., Zisserman, A., Simonyan, K.: Flamingo: a visual language model for few-shot learning. DEEPMIND (2022)

102. Yu, J., Wang, Z., Vasudevan, V., Yeung, L., Seyedhosseini, M., Wu, Y.: Coca: Contrastive captioners are image-text foundation models. arXiv preprint arXiv: Arxiv-2205.01917 (2022)

103. Chen, X., Wang, X., Changpinyo, S., Piergiovanni, A., Padlewski, P., Salz, D., Goodman, S., Grycner, A., Mustafa, B., Beyer, L., Kolesnikov, A., Puigcerver, J., Ding, N., Rong, K., Akbari, H., Mishra, G., Xue, L., Thapliyal, A., Bradbury, J., Kuo, W., Seyedhosseini, M., Jia, C., Ayan, B.K., Riquelme, C., Steiner, A., Angelova, A., Zhai, X., Houlsby, N., Soricut, R.: Pali: A jointly-scaled multilingual language-image model. arXiv preprint arXiv: Arxiv-2209.06794 (2022)

104. Wang, P., Yang, A., Men, R., Lin, J., Bai, S., Li, Z., Ma, J., Zhou, C., Zhou, J., Yang, H.: OFA: unifying architectures, tasks, and modalities through a simple sequence-to-sequence learning framework. In: K. Chaudhuri, S. Jegelka, L. Song, C. Szepesvári, G. Niu, S. Sabato (eds.) International Conference on Machine Learning, ICML 2022, 17–23 July 2022, Baltimore, Maryland, USA, *Proceedings of Machine Learning Research*, vol. 162, pp. 23318–23340. PMLR (2022). https://proceedings.mlr.press/v162/wang22al.html

105. Wang, W., Bao, H., Dong, L., Bjorck, J., Peng, Z., Liu, Q., Aggarwal, K., Mohammed, O.K., Singhal, S., Som, S., Wei, F.: Image as a foreign language: Beit pretraining for all vision and vision-language tasks. arXiv preprint arXiv: Arxiv-2208.10442 (2022)

106. Li, J., Li, D., Xiong, C., Hoi, S.: Blip: Bootstrapping language-image pre-training for unified vision-language understanding and generation. In: ICML (2022)

107. Li, J., Li, D., Savarese, S., Hoi, S.: Blip-2: Bootstrapping language-image pre-training with frozen image encoders and large language models. arXiv preprint arXiv: Arxiv-2301.12597 (2023)

108. Huang, S., Dong, L., Wang, W., Hao, Y., Singhal, S., Ma, S., Lv, T., Cui, L., Mohammed, O.K., Patra, B., Liu, Q., Aggarwal, K., Chi, Z., Bjorck, J., Chaudhary, V., Som, S., Song, X., Wei, F.:

Language is not all you need: Aligning perception with language models. arXiv preprint arXiv: Arxiv-2302.14045 (2023)

109. Peng, Z., Wang, W., Dong, L., Hao, Y., Huang, S., Ma, S., Wei, F.: Kosmos-2: Grounding multimodal large language models to the world. arXiv preprint arXiv: 2306.14824 (2023)

110. Liu, H., Li, C., Wu, Q., Lee, Y.J.: Visual instruction tuning. In: NeurIPS (2023)

111. Liu, H., Li, C., Li, Y., Lee, Y.J.: Improved baselines with visual instruction tuning (2023)

112. Zhu, D., Chen, J., Shen, X., Li, X., Elhoseiny, M.: Minigpt-4: Enhancing vision-language understanding with advanced large language models. arXiv preprint arXiv:2304.10592 (2023)

113. Shu, Y., Dong, S., Chen, G., Huang, W., Zhang, R., Shi, D., Xiang, Q., Shi, Y.: Llasm: Large language and speech model. arXiv preprint arXiv: 2308.15930 (2023)

114. Li, B., Zhang, Y., Chen, L., Wang, J., Yang, J., Liu, Z.: Otter: A multi-modal model with in-context instruction tuning. arXiv preprint arXiv: Arxiv-2305.03726 (2023)

115. Bavishi, R., Elsen, E., Hawthorne, C., Nye, M., Odena, A., Somani, A., Taşırlar, S.: Introducing our multimodal models (2023). https://www.adept.ai/blog/fuyu-8b

116. Wu, S., Fei, H., Qu, L., Ji, W., Chua, T.S.: NExT-GPT: Any-to-any multimodal LLM. CoRR **abs/2309.05519** (2023)

117. Bai, J., Bai, S., Yang, S., Wang, S., Tan, S., Wang, P., Lin, J., Zhou, C., Zhou, J.: Qwen-vl: A versatile vision-language model for understanding, localization, text reading, and beyond. arXiv preprint arXiv:2308.12966 (2023)

118. Yu, L., Shi, B., Pasunuru, R., Muller, B., Golovneva, O., Wang, T., Babu, A., Tang, B., Karrer, B., Sheynin, S., Ross, C., Polyak, A., Howes, R., Sharma, V., Xu, P., Tamoyan, H., Ashual, O., Singer, U., Li, S.W., Zhang, S., James, R., Ghosh, G., Taigman, Y., Fazel-Zarandi, M., Celikyilmaz, A., Zettlemoyer, L., Aghajanyan, A.: Scaling autoregressive multi-modal models: Pretraining and instruction tuning. arXiv preprint arXiv: 2309.02591 (2023)

119. Ye, Q., Xu, H., Xu, G., Ye, J., Yan, M., Zhou, Y., Wang, J., Hu, A., Shi, P., Shi, Y., Li, C., Xu, Y., Chen, H., Tian, J., Qi, Q., Zhang, J., Huang, F.: mplug-owl: Modularization empowers large language models with multimodality. arXiv preprint arXiv: 2304.14178 (2023)

120. Lu, J., Clark, C., Lee, S., Zhang, Z., Khosla, S., Marten, R., Hoiem, D., Kembhavi, A.: Unified-io 2: Scaling autoregressive multimodal models with vision, language, audio, and action. arXiv preprint arXiv: 2312.17172 (2023)

121. Lu, H., Liu, W., Zhang, B., Wang, B., Dong, K., Liu, B., Sun, J., Ren, T., Li, Z., Yang, H., Sun, Y., Deng, C., Xu, H., Xie, Z., Ruan, C.: Deepseek-vl: Towards real-world vision-language understanding. arXiv preprint arXiv: 2403.05525 (2024)

122. Tong, S., Brown, E., Wu, P., Woo, S., Middepogu, M., Akula, S.C., Yang, J., Yang, S., Iyer, A., Pan, X., Wang, A., Fergus, R., LeCun, Y., Xie, S.: Cambrian-1: A fully open, vision-centric exploration of multimodal LLMs. arXiv preprint arXiv: 2406.16860 (2024)

123. Ge, Y., Zhao, S., Zeng, Z., Ge, Y., Li, C., Wang, X., Shan, Y.: Making llama see and draw with seed tokenizer. arXiv preprint arXiv:2310.01218 (2023)

124. Sun, Q., Yu, Q., Cui, Y., Zhang, F., Zhang, X., Wang, Y., Gao, H., Liu, J., Huang, T., Wang, X.: Generative pretraining in multimodality (2023)

125. Sun, Q., Cui, Y., Zhang, X., Zhang, F., Yu, Q., Luo, Z., Wang, Y., Rao, Y., Liu, J., Huang, T., Wang, X.: Generative multimodal models are in-context learners. CVPR (2024)

126. Wu, J., Yin, S., Feng, N., He, X., Li, D., Hao, J., Long, M.: ivideogpt: Interactive videogpts are scalable world models. arXiv preprint arXiv: 2405.15223 (2024)

127. Liu, H., Yan, W., Zaharia, M., Abbeel, P.: World model on million-length video and language with blockwise ringattention. arXiv preprint arXiv: 2402.08268 (2024)

128. Kim, M.J., Pertsch, K., Karamcheti, S., Xiao, T., Balakrishna, A., Nair, S., Rafailov, R., Foster, E., Lam, G., Sanketi, P., Vuong, Q., Kollar, T., Burchfiel, B., Tedrake, R., Sadigh, D., Levine, S., Liang, P., Finn, C.: Openvla: An open-source vision-language-action model. arXiv preprint arXiv: 2406.09246 (2024)

129. Liu, H., Li, C., Li, Y., Li, B., Zhang, Y., Shen, S., Lee, Y.J.: Llava-next: Improved reasoning, ocr, and world knowledge (2024). https://llava-vl.github.io/blog/2024-01-30-llava-next/
130. Li, B., Zhang, K., Zhang, H., Guo, D., Zhang, R., Li, F., Zhang, Y., Liu, Z., Li, C.: Llava-next: Stronger LLMs supercharge multimodal capabilities in the wild (2024). https://llava-vl.github.io/blog/2024-05-10-llava-next-stronger-llms/
131. Li, B., Zhang, H., Zhang, K., Guo, D., Zhang, Y., Zhang, R., Li, F., Liu, Z., Li, C.: Llava-next: What else influences visual instruction tuning beyond data? (2024). https://llava-vl.github.io/blog/2024-05-25-llava-next-ablations/
132. Zhang, Y., Li, B., Liu, h., Lee, Y.j., Gui, L., Fu, D., Feng, J., Liu, Z., Li, C.: Llava-next: A strong zero-shot video understanding model (2024). https://llava-vl.github.io/blog/2024-04-30-llava-next-video/
133. Li, F., Zhang, R., Zhang, H., Zhang, Y., Li, B., Li, W., Ma, Z., Li, C.: Llava-next: Tackling multi-image, video, and 3d in large multimodal models (2024). https://llava-vl.github.io/blog/2024-06-16-llava-next-interleave/
134. Bruce, J., Dennis, M., Edwards, A., Parker-Holder, J., Shi, Y., Hughes, E., Lai, M., Mavalankar, A., Steigerwald, R., Apps, C., Aytar, Y., Bechtle, S., Behbahani, F., Chan, S., Heess, N., Gonzalez, L., Osindero, S., Ozair, S., Reed, S., Zhang, J., Zolna, K., Clune, J., de Freitas, N., Singh, S., Rocktäschel, T.: Genie: Generative interactive environments. arXiv preprint arXiv: 2402.15391 (2024)
135. Xiang, J., Liu, G., Gu, Y., Gao, Q., Ning, Y., Zha, Y., Feng, Z., Tao, T., Hao, S., Shi, Y., Liu, Z., Xing, E.P., Hu, Z.: Pandora: Towards general world model with natural language actions and video states. arXiv preprint arXiv: 2406.09455 (2024)
136. Lin, B., Ye, Y., Zhu, B., Cui, J., Ning, M., Jin, P., Yuan, L.: Video-llava: Learning united visual representation by alignment before projection. arXiv preprint arXiv: 2311.10122 (2023)
137. Team, C.: Chameleon: Mixed-modal early-fusion foundation models. arXiv preprint arXiv: 2405.09818 (2024)
138. Wang, Z., Zhu, K., Xu, C., Zhou, W., Liu, J., Zhang, Y., Wang, J., Shi, N., Li, S., Li, Y., Que, H., Zhang, Z., Zhang, Y., Zhang, G., Xu, K., Fu, J., Huang, W.: Mio: A foundation model on multimodal tokens. arXiv preprint arXiv: 2409.17692 (2024)
139. Lu, P., Peng, B., Cheng, H., Galley, M., Chang, K.W., Wu, Y., Zhu, S.C., Gao, J.: Chameleon: Plug-and-play compositional reasoning with large language models. Neural Information Processing Systems (2023). https://doi.org/10.48550/arXiv.2304.09842. https://arxiv.org/abs/2304.09842v3
140. Shen, Y., Song, K., Tan, X., Li, D., Lu, W., Zhuang, Y.: Hugginggpt: Solving ai tasks with ChatGPT and its friends in huggingface. arXiv preprint arXiv: Arxiv-2303.17580 (2023)
141. Yu, L., Cheng, Y., Wang, Z., Kumar, V., Macherey, W., Huang, Y., Ross, D.A., Essa, I., Bisk, Y., Yang, M., Murphy, K.P., Hauptmann, A.G., Jiang, L.: SPAE: semantic pyramid autoencoder for multimodal generation with frozen LLMs. In: A. Oh, T. Naumann, A. Globerson, K. Saenko, M. Hardt, S. Levine (eds.) Advances in Neural Information Processing Systems 36: Annual Conference on Neural Information Processing Systems 2023, NeurIPS 2023, New Orleans, LA, USA, December 10–16, 2023 (2023). http://papers.nips.cc/paper_files/paper/2023/hash/a526cc8f6ffb74bedb6ff313e3fdb450-Abstract-Conference.html
142. Zhu, L., Wei, F., Lu, Y.: Beyond text: Frozen large language models in visual signal comprehension. CVPR (2024)
143. Ma, Y., Song, Z., Zhuang, Y., Hao, J., King, I.: A survey on vision-language-action models for embodied ai. arXiv preprint arXiv: 2405.14093 (2024)
144. Dosovitskiy, A., Beyer, L., Kolesnikov, A., Weissenborn, D., Zhai, X., Unterthiner, T., Dehghani, M., Minderer, M., Heigold, G., Gelly, S., Uszkoreit, J., Houlsby, N.: An image is worth 16x16 words: Transformers for image recognition at scale. International Conference on Learning Representations (2020)

145. Radford, A., Kim, J.W., Hallacy, C., Ramesh, A., Goh, G., Agarwal, S., Sastry, G., Askell, A., Mishkin, P., Clark, J., Krueger, G., Sutskever, I.: Learning transferable visual models from natural language supervision. International Conference On Machine Learning (2021)

146. Sun, Q., Fang, Y., Wu, L., Wang, X., Cao, Y.: Eva-clip: Improved training techniques for clip at scale. arXiv preprint arXiv: 2303.15389 (2023)

147. Zhang, H., Li, F., Liu, S., Zhang, L., Su, H., Zhu, J., Ni, L.M., Shum, H.Y.: Dino: Detr with improved denoising anchor boxes for end-to-end object detection. arXiv preprint arXiv: 2203.03605 (2022)

148. Oquab, M., Darcet, T., Moutakanni, T., Vo, H., Szafraniec, M., Khalidov, V., Fernandez, P., Haziza, D., Massa, F., El-Nouby, A., Assran, M., Ballas, N., Galuba, W., Howes, R., Huang, P.Y., Li, S.W., Misra, I., Rabbat, M., Sharma, V., Synnaeve, G., Xu, H., Jegou, H., Mairal, J., Labatut, P., Joulin, A., Bojanowski, P.: Dinov2: Learning robust visual features without supervision. arXiv preprint arXiv: 2304.07193 (2023)

149. He, K., Chen, X., Xie, S., Li, Y., Dollár, P., Girshick, R.: Masked autoencoders are scalable vision learners. arXiv preprint arXiv: 2111.06377 (2021)

150. Rombach, R., Blattmann, A., Lorenz, D., Esser, P., Ommer, B.: High-resolution image synthesis with latent diffusion models. CVPR (2022)

151. Ranftl, R., Lasinger, K., Hafner, D., Schindler, K., Koltun, V.: Towards robust monocular depth estimation: Mixing datasets for zero-shot cross-dataset transfer. IEEE Transactions on Pattern Analysis and Machine Intelligence (2019). https://doi.org/10.1109/TPAMI.2020.3019967

152. Birkl, R., Wofk, D., Müller, M.: Midas v3.1—a model zoo for robust monocular relative depth estimation. arXiv preprint arXiv: 2307.14460 (2023)

153. Kirillov, A., Mintun, E., Ravi, N., Mao, H., Rolland, C., Gustafson, L., Xiao, T., Whitehead, S., Berg, A.C., Lo, W.Y., Dollár, P., Girshick, R.: Segment anything. ICCV (2023)

154. Chen, Z., Wu, J., Wang, W., Su, W., Chen, G., Xing, S., Zhong, M., Zhang, Q., Zhu, X., Lu, L., Li, B., Luo, P., Lu, T., Qiao, Y., Dai, J.: Internvl: Scaling up vision foundation models and aligning for generic visual-linguistic tasks. arXiv preprint arXiv:2312.14238 (2023)

155. Chen, Z., Wang, W., Tian, H., Ye, S., Gao, Z., Cui, E., Tong, W., Hu, K., Luo, J., Ma, Z., et al.: How far are we to GPT-4v? closing the gap to commercial multimodal models with open-source suites. arXiv preprint arXiv:2404.16821 (2024)

156. Dehghani, M., Mustafa, B., Djolonga, J., Heek, J., Minderer, M., Caron, M., Steiner, A., Puigcerver, J., Geirhos, R., Alabdulmohsin, I.M., Oliver, A., Padlewski, P., Gritsenko, A., Luvci'c, M., Houlsby, N.: Patch n' pack: Navit, a vision transformer for any aspect ratio and resolution. Neural Information Processing Systems (2023). https://doi.org/10.48550/arXiv.2307.06304

157. Ge, C., Cheng, S., Wang, Z., Yuan, J., Gao, Y., Song, J., Song, S., Huang, G., Zheng, B.: Convllava: Hierarchical backbones as visual encoder for large multimodal models. arXiv preprint arXiv: 2405.15738 (2024)

158. Liu, Z., Mao, H., Wu, C., Feichtenhofer, C., Darrell, T., Xie, S.: A convnet for the 2020s. Computer Vision and Pattern Recognition (2022). https://doi.org/10.1109/CVPR52688.2022.01167. https://arxiv.org/abs/2201.03545v2

159. Wang, J., Zhang, Y., Ji, Y., Zhang, Y., Jiang, C., Wang, Y., Zhu, K., Wang, Z., Wang, T., Huang, W., Fu, J., Chen, B., Lin, Q., Liu, M., Zhang, G., Chen, W.: Pin: A knowledge-intensive dataset for paired and interleaved multimodal documents. arXiv preprint arXiv: 2406.13923 (2024)

160. Tang, Z., Wu, C., Zhang, Z., Ni, M., Yin, S., Liu, Y., Yang, Z., Wang, L., Liu, Z., Li, J., Duan, N.: Strokenuwa: Tokenizing strokes for vector graphic synthesis. arXiv preprint arXiv: 2401.17093 (2024)

161. Jaegle, A., Gimeno, F., Brock, A., Zisserman, A., Vinyals, O., Carreira, J.: Perceiver: General perception with iterative attention. International Conference on Machine Learning (2021). https://arxiv.org/abs/2103.03206v2

162. van den Oord, A., Vinyals, O., Kavukcuoglu, K.: Neural discrete representation learning. NIPS (2017)

163. Esser, P., Rombach, R., Ommer, B.: Taming transformers for high-resolution image synthesis. Computer Vision And Pattern Recognition (2020). https://doi.org/10.1109/CVPR46437.2021. 01268

164. Yu, J., Li, X., Koh, J.Y., Zhang, H., Pang, R., Qin, J., Ku, A., Xu, Y., Baldridge, J., Wu, Y.: Vector-quantized image modeling with improved vqgan. International Conference On Learning Representations (2021)

165. Ge, Y., Ge, Y., Zeng, Z., Wang, X., Shan, Y.: Planting a seed of vision in large language model. arXiv preprint arXiv:2307.08041 (2023)

166. Dong, R., Han, C., Peng, Y., Qi, Z., Ge, Z., Yang, J., Zhao, L., Sun, J., Zhou, H., Wei, H., Kong, X., Zhang, X., Ma, K., Yi, L.: Dreamllm: Synergistic multimodal comprehension and creation. arXiv preprint arXiv:2309.11499 (2023)

167. Zhan, J., Dai, J., Ye, J., Zhou, Y., Zhang, D., Liu, Z., Zhang, X., Yuan, R., Zhang, G., Li, L., Yan, H., Fu, J., Gui, T., Sun, T., Jiang, Y., Qiu, X.: Anygpt: Unified multimodal LLM with discrete sequence modeling. arXiv preprint arXiv: 2402.12226 (2024)

168. Diao, S., Zhou, W., Zhang, X., Wang, J.: Write and paint: Generative vision-language models are unified modal learners. In: The Eleventh International Conference on Learning Representations, ICLR 2023, Kigali, Rwanda, May 1–5, 2023. OpenReview.net (2023). https://openreview.net/pdf?id=HgQR0mXQ1_a

169. Zheng, K., He, X., Wang, X.E.: Minigpt-5: Interleaved vision-and-language generation via generative vokens. arXiv preprint arXiv: 2310.02239 (2023)

170. Yu, J., Xu, Y., Koh, J.Y., Luong, T., Baid, G., Wang, Z., Vasudevan, V., Ku, A., Yang, Y., Ayan, B.K., Hutchinson, B., Han, W., Parekh, Z., Li, X., Zhang, H., Baldridge, J., Wu, Y.: Scaling autoregressive models for content-rich text-to-image generation. arXiv preprint arXiv: 2206.10789 (2022)

171. Zhang, X., Zhang, D., Li, S., Zhou, Y., Qiu, X.: Speechtokenizer: Unified speech tokenizer for speech language models (2023)

172. Hsu, W.N., Bolte, B., Tsai, Y.H.H., Lakhotia, K., Salakhutdinov, R., rahman Mohamed, A.: Hubert: Self-supervised speech representation learning by masked prediction of hidden units. IEEE/ACM Transactions on Audio Speech and Language Processing (2021). https://doi.org/10.1109/taslp.2021.3122291

173. Gong, Y., Chung, Y.A., Glass, J.R.: Ast: Audio spectrogram transformer. INTERSPEECH (2021). https://doi.org/10.21437/interspeech.2021-698

174. Koh, J.Y., Fried, D., Salakhutdinov, R.: Generating images with multimodal language models. NeurIPS (2023)

175. Garrido, Q., Assran, M., Ballas, N., Bardes, A., Najman, L., LeCun, Y.: Learning and leveraging world models in visual representation learning. arXiv preprint arXiv: 2403.00504 (2024)

176. Zhu, Z., Wang, X., Zhao, W., Min, C., Deng, N., Dou, M., Wang, Y., Shi, B., Wang, K., Zhang, C., You, Y., Zhang, Z., Zhao, D., Xiao, L., Zhao, J., Lu, J., Huang, G.: Is sora a world simulator? a comprehensive survey on general world models and beyond. arXiv preprint arXiv: 2405.03520 (2024)

177. Liu, H., Zaharia, M., Abbeel, P.: Ring attention with blockwise transformers for near-infinite context. arXiv preprint arXiv: 2310.01889 (2023)

Prompting

Zekun Wang, Ruibo Liu and Jie Fu

According to [1], prompting refers to the interaction methods that focus on calling a model via prompts, without involving any parameter updating.[2] This line of research stems from in-context learning [3, 4], a significant capability of large language models. In-Context Learning (ICL) refers to the approach that allows large language models to learn from examples provided in context [3]. Moreover, the task description can also be incorporated within the context, accompanied with few-shot examples [5–8]. Prompting is one of the simplest ways to incorporate interactive messages. However, making it effective can still be tricky, as we will discuss below.

Note that in this section, the discussion focuses on large-scale generative language models, as prompting is challenging to implement with small language models, which may necessitate fine-tuning with prompts [9].

---

[1] This definition is a bit different from that of [2]. We align this definition as "Tuning-free Prompting" in [2]'s categorization. Additionally, we put "Promptless Fine-tuning", "Fixed-prompt LM Tuning", "Prompt+LM Tuning" in Sect. 5.3 and "Fixed-LM Prompt Tuning" in Sect. 6.3.

Z. Wang (✉)
Beihang University, Beijing, China
e-mail: zenmoore@buaa.edu.cn

R. Liu
Department of Computer Science, Dartmouth College, Hanover, USA
e-mail: ruibo.liu.gr@dartmouth.edu

J. Fu
Hong Kong University of Science and Technology, Hong Kong, China
e-mail: jie.fu@polymtl.ca; jiefu@ust.hk

© The Author(s), under exclusive license to Springer Nature Switzerland AG 2026
Z. Wang et al. (eds.), *Interactive Natural Language Processing*, Synthesis Lectures on Human Language Technologies, https://doi.org/10.1007/978-3-032-06264-2_5

In the following sections, prompting methods are classified into three categories according to their characteristics and objectives: (1) Standard Prompting with straightforward task descriptions and demonstrations (i.e., examples) as context for instruction-following; (2) Elicitive Prompting with the context which can stimulate the language model to generate intermediate steps for reasoning; and (3) Prompt Chaining, which cascades multiple language model runs for complex reasoning and pipelined tasks.

Note that in recent literature, the term Inference-Time Scaling (also known as Test-Time Scaling) has emerged as another name for prompting research. Inference-Time Scaling refers to the family of techniques that improve or extend model capabilities during inference, without any parameter updates–encompassing not only prompt design, but also output planning, branching, task decomposition, ensemble answer aggregation, among other inference strategies. Thus, the prompting techniques discussed in this chapter are essentially inference-time scaling methods. For a comprehensive overview of inference-time scaling in large language models, readers are referred to the survey [10].

5.1   Standard Prompting

Standard prompting represents the most elementary form of In-Context Learning. The prompting context primarily comprises a concise, answer-focused task description, along with few-shot examples, as elucidated in Sect. 3.1. In Natural Instructions [7] and Super-Natural Instructions [8], the fundamental structure of a context, or instruction, is composed of: task definition, several positive examples accompanied by explanations (demonstrations), and numerous negative examples with clarifications. Despite its simplicity, various approaches to standard prompting continue to be proposed, as large language models tend to be context-sensitive, often resulting in a lack of robustness [4, 11–15].

This line of research endeavors to enhance the organization of instructions to improve the performance of ICL [4], which enables a language model to better understand and respond to the interaction messages. In accordance with [4], this primarily entails optimizing the subsequent factors: (1) instance selection; (2) instance processing; and (3) instance combination (Fig. 5.1).

Instance Selection. In order to find useful examples, various unsupervised prompt retrieval methods can be utilized, including distance metrics [11], mutual information [16], and n-gram overlap [17], which have been discussed in [4]. Additionally, [18, 19] utilize learned retrievers to identify the most relevant demonstrations to the input. Zhang et al. [20] select demonstrations using reinforcement learning. Li and Qiu [21] propose *InfoScore*, a metric designed to evaluate the informativeness of examples, which facilitates example selection using feedback from language models. It employs an iterative diversity-guided search algorithm to improve and assess the examples. Most studies along this line build upon the premise that an increased relevance of demonstrations directly correlates with enhanced ICL performance [11]. However, [22] find that using randomly sampled demon-

Fig. 5.1 Standard prompting

strations leads to similar results with GPT-3 [3] compared to in-distribution demonstrations. Li et al. [23] reveal that controllability and robustness in Large Language Models (LLMs) can be improved by incorporating counterfactual and irrelevant contexts during fine-tuning.

Instance Processing. The processing of context involves four main types: expansion, filtering, edit and formatting. For example, SuperICL [24] expands in-context examples by incorporating labels, predicted by a small plug-in model, and their associated confidence scores to augment the context for large language models. Zhou et al. [25] employ LLMs for instruction generation, example generation, and filtering through a scoring model. Honovich et al. [26] generate task descriptions based on examples. GrIPS [27] employs a gradient-free, edit-based approach to conduct instruction search (processing). In particular, it follows an iterative process of modifying the base instruction at the phrase-level and subsequently evaluating the candidate instructions to identify the optimal one. ProQA [28] uses a structured schema to format the context.

Instance Combination. The order and structure of demonstrations in a given context also play a crucial role [4, 11, 14, 29]. For example, [11, 14] sort examples in the context according to their distance and entropy metrics with the input, respectively, as mentioned in [4]. Batch prompting [30] enables LLMs to perform inference on multiple samples in a batch, thus reducing token and time costs while maintaining the overall performance. Structured prompting [31] involves encoding multiple groups of examples into multiple LM replica, which are then merged using rescaled attention. This process allows LMs to incorporate and contextualize 1000+ examples. ICIL [29] puts multiple task instructions composed of task definitions and groups of examples together in the context to improve LLMs' zero-shot task generalization performance.

Note that although the diverse approaches mentioned in this part are mainly designed for general-purpose in-context learning, they can be used as methods for interaction message communication. During the interaction with language models, determining the most appropriate way to organize context for interaction messages via elaborate prompt engineering is crucial for performance gain. For example, in the scope of KB-in-the-loop, [32–34] work on how to feed the retrieved knowledge into language models via ICL; in the scope of env-in-the-loop, [35] demonstrate how to generate task instructions and enable cross-environment transfer to help agents generalize their execution.

5.2   Elicitive Prompting

Extending standard prompting, elicitive prompting improves the abilities of LLMs, such as reasoning and planning, by providing them with extra step-by-step guidance in context.

Few-Shot Demonstrations. Typical chain-of-thought [36] uses few-shot examples with reasoning steps to elicit reasoning as shown below (Fig. 5.2):

Fig. 5.2 Elicitive prompting

> **Question**: If a rectangle has a width of 5 units and a length of 8 units, what is its perimeter?
> **Answer**: The perimeter of a rectangle is the sum of the lengths of all its sides. In this case, the rectangle has two sides with a length of 5 units and two sides with a length of 8 units. Therefore, its perimeter is 2 x (5 + 8) = 26 units.
> **Question**: If I need to be at work by 9:00 am, and it takes me 20 min to drive there, what time should I leave my house?
> **Answer**: [to be generated]

For example, Scratchpads [37] and CoT [36] are two representative techniques for elicitive prompting. They explicitly describe the reasoning steps in the few-shot examples, significantly improving math reasoning abilities compared with standard prompting. Least-to-most prompting [38] aims to tackle complex tasks that CoT struggles with. It achieves this by decomposing a complex problem into smaller and more manageable ones with few-shot demonstrations. Other follow-up works focus on how to improve the robustness of CoT, such as majority voting on results [39], perplexity check [40], or retrieving CoTs from pre-defined clusters [41].

Other Forms of Instructions. According to recent studies, it may not be necessary to rely solely on human-written, step-by-step rationales for eliciting prompts, as other forms of instructions may be useful. For example, zero-shot CoT [42] uses a simple phrase "*Let's think step by step.*" to induce the CoT-style reasoning in zero-shot settings:

> **Question**: If I need to be at work by 9:00 am, and it takes me 20 min to drive there, what time should I leave my house?
> **Answer**: Let's think step by step: [to be generated]

The format of answers [43] and task descriptions [7] have also been explored to serve as elicitive prompts. In addition to text-form CoT, Program-of-Thought (PoT) [44], Program-aided Language Model (PAL) [45], and ViperGPT [46] leverage program-form CoTs to obtain reliable reasoning performance in many tasks that programs can solve. PoT, PAL, and ViperGPT offer advantages over text-based CoT since they deliver verified, stepwise results by executing the programs. Vanilla CoT, on the other hand, cannot verify results. Furthermore, through specially-designed prompts (e.g., "*Search[query]*", "*<API> Calculator(735 / 499) → 1.47 </API>*"), humans can unlock tool-using abilities of language models, such as web-searching [47, 48], calculators [47], physical simulation [49], etc. (c.f., Sect. 2.3).

Note that in the scope of interactive natural language processing, elicitive prompting can be used to enhance reasoning and planning capabilities of language models during interactions with other objects [41, 48, 50–53]. Furthermore, the idea of elicitive prompting is usually instantiated within the scope of model/tool-in-the-loop (Sect. 2.3), which will be discussed in detail in the next part (Sect. 5.3).

5.3 Prompt Chaining

An increasing number of studies are using multi-stage chain-of-thought to improve multi-hop reasoning capabilities. In this approach, LMs are cascaded and can be prompted via different contexts, allowing for more complex reasoning. This is in contrast to typical elicitive prompting, which generally only performs one stage of chain-of-thought via In-Context Learning for reasoning. This approach is intuitive as it can aid in generating precise reasoning steps by conducting multiple model runs with different yet interdependent prompts [53]. In contrast, elicitive prompt relies on a single model run with only one context [53] (Fig. 5.3).

By decomposing the task and cascading language models for different reasoning steps or sub-tasks, prompt chaining can not only perform multi-hop reasoning [53, 54], but also work well in pipelined tasks such as peer review writing [50] and advertisement generation [51]. Prompt chaining is one of the fundamental methods for model/tool-in-the-loop natural language processing (c.f., Sect. 2.3).

LM Cascades [54] has presented some works in this line, including sequential reasoning mechanisms [36, 37, 55], reasoning procedures with verifiers or tools [49, 56, 57], and multi-agent interacting question-answering [58].[3] Qiao et al. [53] investigated the enhancement of reasoning through language model prompting, focusing on strategy enhancement and knowledge enhancement. The study explored various aspects, such as prompt engi-

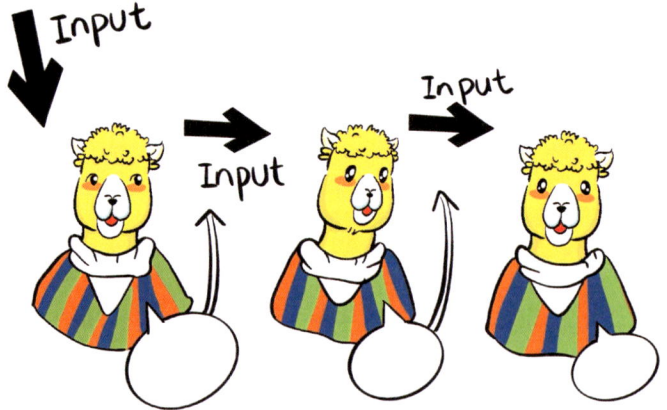

Fig. 5.3 Prompt chaining

---

[3] https://github.com/google/BIG-bench/tree/main/bigbench/benchmark_tasks/twenty_questions.

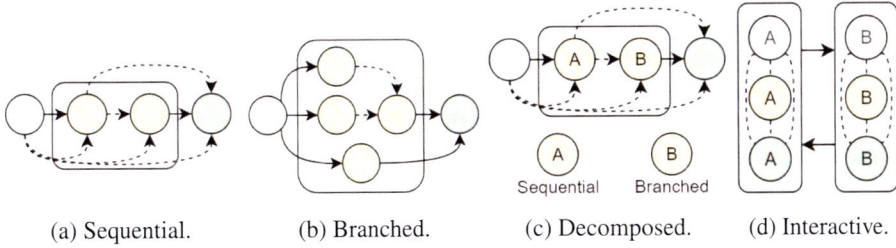

(a) Sequential.	(b) Branched.	(c) Decomposed.	(d) Interactive.

Fig. 5.4 Examples of prompt chaining schemes. Gray, yellow, and green circles refer to input queries, intermediate reasoning steps, and final responses, respectively. Solid and dashed arrows represent indispensable and optional (i.e., none, single, or multiple) conditional probabilities ($P(Y|X)$), respectively. The rounded rectangle refers to a looping block

neering, process optimization, external engines, and both implicit and explicit knowledge. However, to the best of our knowledge, no survey has systematically examined the structure of prompt chaining. Hence, in this section, we categorize the prompt chaining schemes into four distinct types based on their topologies, as illustrated in Fig. 5.4. It is important to note that we do not include graph topology in our categorization. This is because, at the scale of execution nodes, any graph structure must be transformed into one of the configurations defined in Fig. 5.4, or it must be nested or combined with them. For further details, readers are referred to [59]. Additionally, instead of adhering to fixed prompt chaining schemes, users have the flexibility to customize these schemes as discussed in Sect. 5.3. They can also be automatically constructed, as elaborated in Sect. 5.3.

Sequential. The nodes are arranged in a straight line, where each node takes as input the specific outputs of the previous nodes, including the initial input query. For example, Self-Ask [60] and Successive Prompting [61] construct the reasoning chain via sequential question generation (QG) and question answering (QA) nodes. Wang et al. [9] further enable smaller language models to construct the similar QG-QA chain via a learned context-aware prompter. Selection-Inference [55] begins by utilizing the selection module to choose a group of relevant facts based on the given question. Subsequently, the inference module generates new facts by utilizing this subset of facts. Multimodal-CoT [62] addresses the visual reasoning problem through a two-step processing consisting of rationale generation and answer inference conditioned on the image, question, and the generated rationale. Mind's-Eye [49] first generates the rendering codes for an intermediate environment simulator to get grounded rationales, and then generates the final answer based on the simulation results. Note that when the number of looping blocks becomes zero, it is reduced to standard prompting (Sect. 5.1). When the intermediate reasoning steps and the final answer are simultaneously prompted in a single model run, it is reduced to single-stage CoT (Sect. 5.2).

Branched. The nodes are arranged in a tree-like structure, where a node's output may serve as input to multiple other nodes, and a node's input may come from multiple other nodes' outputs. For example, [63] decompose one multi-hop question into many single-hop sub-questions and then aggregates their answers to get the final answer. Wang et al. [39] first generate a range of reasoning paths through sampling from the language model's decoder and then aggregates the most consistent answer in the final answer set by computing the likelihood of the reasoning paths. Ask Me Anything [64] first reformats the question into diverse possible ones with different in-context demonstrations, which are then answered respectively. Finally, the answers are aggregated into the final answer via a learned probabilistic graphical model. Tree of Thoughts [65] constructs the language model's reasoning chain in a tree form, enabling evaluation of states via heuristic approaches, and exploration of potential solutions via Breadth-first search (BFS) or Depth-first search (DFS), significantly improving the model's problem-solving capabilities. Similar approaches involve Graph-of-Thoughts [59], Algorithm-of-Thought [66], Forest-of-Thought [67], and Monte Carlo Tree Search (MCTS) [68]. Liu et al. [2] have investigated multi-prompt learning, including prompt ensembling, prompt composition, and prompt decomposition, which can also be viewed as in this line.

Decomposed. The overall process is linear, but some nodes can be broken down further into nested or hierarchical chains recursively that follow any of the four schemes described. For example, [38] first call an LM to reduce the problem into multiple sub-problems and then iteratively call the language model to solve these sub-problems step-by-step (i.e., the first node serves for problem reduction while the second node is another sequential chain). Decomposed Prompting [1] decomposes the question into a sequence of sub-questions, each being answered immediately after generation by utilizing another prompt chaining block or just standard prompting. This approach facilitates the hierarchical and recursive decomposition of tasks.

Interactive. The nodes are split into multiple groups, each with their own functions. These groups communicate with each other in an alternating fashion to construct a chain of interactions. For example, Socratic Models [69] and Inner Monologue [70] partition the groups of nodes according to the modality. They utilize LLMs for planning and reasoning while making use of VLM for observation. *ChatGPT Asks, BLIP-2 Answers* [71] use Chat-GPT to ask questions about an image, while BLIP-2 [72] is used to answer these questions. This communication through question-answering is performed interactively to generate the image captions. The visual reasoning path in [73] can be divided into two groups, one for answer and explanation generation, and another for explanation verification via a multimodal classifier. The frameworks of [56, 74] can be viewed as interaction between the thought generation group and the thought verification group. MindCraft [75] first assigns two agents with different skills (i.e., recipe knowledge) and then lets them interact through question-answering to accomplish the task in the *MineCraft* environment. Such prompt chaining scheme can also be effectively utilized for alignment. For example, Reflexion [76], Self-Correct [77], RL4F [78], SELF [79], Selfee [80], Self-Refine [81], RAIN [82], and Self-Critique [83], among others, categorize prompt groups into generation, feedback (i.e., evaluation), and refinement (i.e., correction, revision) groups. The interaction among these groups enables the models' responses to iteratively improve, eventually leading to an aligned response.

Customization. The prompt chaining scheme depends on several factors related to the nature of the problem, such as its complexity, structure, and the availability of relevant data. Hence, due to their diversity and variability, a fixed prompt chaining scheme may not satisfy the users' needs. An intuitive solution is to enable users to create or modify the prompt chaining scheme on their own accord, which can further enhance the debuggability and configurability of the system [50, 51]. Generally, this feature is more important for pipelined tasks such as peer review writing, brainstorming, personalized flashcard creation, writing assistant, etc. [50, 51]. For example, AI Chains [50] defines a set of primitive operations such as classification, factual query, information extraction, split points, compose points, etc. These primitive operations are implemented with different instructions fed into the language model. An interactive interface then shows the prompt chaining schemes, which users can customize to construct a pipeline. PromptChainer [51] also introduces an interactive interface that facilitates the visual programming of chains. It divides the nodes into three types: LLM nodes, such as generic LLM and LLM classifier; Helper nodes, such as model output evaluation, data processing, and generic JavaScript; and Communication nodes, such as data inputs, user actions, and API calls. The workflow defined by the graph of nodes is also transparent and configurable. It has been shown that this approach can assist users in building a satisfactory pipeline for applications such as music chatbots, advertisement generators, image query generators, and writing assistants [51].

Automatization. The prompt chaining schemes can also be constructed automatically. For example, ReAct [48] and MM-ReAct [52] determine whether a thought should be generated or a tool should be called automatically based on the context. ToolFormer [47] can determine which tools to utilize or whether to use tools based on the context, as it is trained on tool-use prompted data. AutoAgents [84] automates the generation of different LLM agents' profiles, which are then cascaded in a sequential and interactive manner to accomplish user instructions.

References

1. Khot, T., Trivedi, H., Finlayson, M., Fu, Y., Richardson, K., Clark, P., Sabharwal, A.: Decomposed prompting: A modular approach for solving complex tasks. CoRR **abs/2210.02406** (2022). https://doi.org/10.48550/arXiv.2210.02406
2. Liu, P., Yuan, W., Fu, J., Jiang, Z., Hayashi, H., Neubig, G.: Pre-train, prompt, and predict: A systematic survey of prompting methods in natural language processing. arXiv preprint arXiv: Arxiv-2107.13586 (2021)
3. Brown, T., Mann, B., Ryder, N., Subbiah, M., Kaplan, J.D., Dhariwal, P., Neelakantan, A., Shyam, P., Sastry, G., Askell, A., et al.: Language models are few-shot learners. Advances in neural information processing systems **33**, 1877–1901 (2020)
4. Dong, Q., Li, L., Dai, D., Zheng, C., Wu, Z., Chang, B., Sun, X., Xu, J., Li, L., Sui, Z.: A survey for in-context learning. CoRR **abs/2301.00234** (2023). https://doi.org/10.48550/arXiv.2301.00234

5. Sanh, V., Webson, A., Raffel, C., Bach, S.H., Sutawika, L., Alyafeai, Z., Chaffin, A., Stiegler, A., Scao, T.L., Raja, A., Dey, M., Bari, M.S., Xu, C., Thakker, U., Sharma, S., Szczechla, E., Kim, T., Chhablani, G., Nayak, N.V., Datta, D., Chang, J., Jiang, M.T.J., Wang, H., Manica, M., Shen, S., Yong, Z.X., Pandey, H., Bawden, R., Wang, T., Neeraj, T., Rozen, J., Sharma, A., Santilli, A., Févry, T., Fries, J.A., Teehan, R., Biderman, S.R., Gao, L., Bers, T., Wolf, T., Rush, A.M.: Multitask prompted training enables zero-shot task generalization. International Conference On Learning Representations (2021)

6. Wei, J., Bosma, M., Zhao, V., Guu, K., Yu, A., Lester, B., Du, N., Dai, A.M., Le, Q.V.: Finetuned language models are zero-shot learners. International Conference On Learning Representations (2021)

7. Mishra, S., Khashabi, D., Baral, C., Hajishirzi, H.: Cross-task generalization via natural language crowdsourcing instructions. In: ACL (2022)

8. Wang, Y., Mishra, S., Alipoormolabashi, P., Kordi, Y., Mirzaei, A., Arunkumar, A., Ashok, A., Dhanasekaran, A.S., Naik, A., Stap, D., et al.: Super-naturalinstructions:generalization via declarative instructions on 1600+ tasks. In: EMNLP (2022)

9. Wang, B., Deng, X., Sun, H.: Iteratively prompt pre-trained language models for chain of thought. In: Y. Goldberg, Z. Kozareva, Y. Zhang (eds.) Proceedings of the 2022 Conference on Empirical Methods in Natural Language Processing, EMNLP 2022, Abu Dhabi, United Arab Emirates, December 7-11, 2022, pp. 2714–2730. Association for Computational Linguistics (2022). https://aclanthology.org/2022.emnlp-main.174

10. Zhang, Q., Lyu, F., Sun, Z., Wang, L., Zhang, W., Hua, W., Wu, H., Guo, Z., Wang, Y., Muennighoff, N., King, I., Liu, X., Ma, C.: A survey on test-time scaling in large language models: What, how, where, and how well? arXiv preprint arXiv: 2503.24235 (2025)

11. Liu, J., Shen, D., Zhang, Y., Dolan, B., Carin, L., Chen, W.: What makes good in-context examples for GPT-3? In: Proceedings of Deep Learning Inside Out (DeeLIO 2022): The 3rd Workshop on Knowledge Extraction and Integration for Deep Learning Architectures, pp. 100–114. Association for Computational Linguistics, Dublin, Ireland and Online (2022). https://doi.org/10.18653/v1/2022.deelio-1.10. https://aclanthology.org/2022.deelio-1.10

12. Gu, J., Xu, H., Nie, L., Yin, W.: Robustness of learning from task instructions. arXiv preprint arXiv: Arxiv-2212.03813 (2022)

13. Zhao, Z., Wallace, E., Feng, S., Klein, D., Singh, S.: Calibrate before use: Improving few-shot performance of language models. In: M. Meila, T. Zhang (eds.) Proceedings of the 38th International Conference on Machine Learning, *Proceedings of Machine Learning Research*, vol. 139, pp. 12697–12706. PMLR (2021). https://proceedings.mlr.press/v139/zhao21c.html

14. Lu, Y., Bartolo, M., Moore, A., Riedel, S., Stenetorp, P.: Fantastically ordered prompts and where to find them: Overcoming few-shot prompt order sensitivity. In: Proceedings of the 60th Annual Meeting of the Association for Computational Linguistics (Volume 1: Long Papers), pp. 8086–8098. Association for Computational Linguistics, Dublin, Ireland (2022). https://doi.org/10.18653/v1/2022.acl-long.556. https://aclanthology.org/2022.acl-long.556

15. Chen, Y., Zhao, C., Yu, Z., McKeown, K., He, H.: On the relation between sensitivity and accuracy in in-context learning. arXiv preprint arXiv: Arxiv-2209.07661 (2022)

16. Sorensen, T., Robinson, J., Rytting, C., Shaw, A., Rogers, K., Delorey, A., Khalil, M., Fulda, N., Wingate, D.: An information-theoretic approach to prompt engineering without ground truth labels. In: Proceedings of the 60th Annual Meeting of the Association for Computational Linguistics (Volume 1: Long Papers), pp. 819–862. Association for Computational Linguistics, Dublin, Ireland (2022). https://doi.org/10.18653/v1/2022.acl-long.60. https://aclanthology.org/2022.acl-long.60

17. Agrawal, S., Zhou, C., Lewis, M., Zettlemoyer, L., Ghazvininejad, M.: In-context examples selection for machine translation. arXiv preprint arXiv: Arxiv-2212.02437 (2022)

18. Rubin, O., Herzig, J., Berant, J.: Learning to retrieve prompts for in-context learning. In: M. Carpuat, M. de Marneffe, I.V.M. Ruíz (eds.) Proceedings of the 2022 Conference of the North American Chapter of the Association for Computational Linguistics: Human Language Technologies, NAACL 2022, Seattle, WA, United States, July 10-15, 2022, pp. 2655–2671. Association for Computational Linguistics (2022). https://doi.org/10.18653/v1/2022.naacl-main.191

19. Cheng, D., Huang, S., Bi, J., Zhan, Y., Liu, J., Wang, Y., Sun, H., Wei, F., Deng, D., Zhang, Q.: Uprise: Universal prompt retrieval for improving zero-shot evaluation. arXiv preprint arXiv: Arxiv-2303.08518 (2023)

20. Zhang, Y., Feng, S., Tan, C.: Active example selection for in-context learning. arXiv preprint arXiv:2211.04486 (2022)

21. Li, X., Qiu, X.: Finding supporting examples for in-context learning. arXiv preprint arXiv: Arxiv-2302.13539 (2023)

22. Si, C., Gan, Z., Yang, Z., Wang, S., Wang, J., Boyd-Graber, J., Wang, L.: Prompting GPT-3 to be reliable. arXiv preprint arXiv: Arxiv-2210.09150 (2022)

23. Li, D., Rawat, A.S., Zaheer, M., Wang, X., Lukasik, M., Veit, A., Yu, F., Kumar, S.: Large language models with controllable working memory. arXiv preprint arXiv: Arxiv-2211.05110 (2022)

24. Xu, C., Xu, Y., Wang, S., Liu, Y., Zhu, C., McAuley, J.: Small models are valuable plug-ins for large language models. arXiv preprint arXiv: Arxiv-2305.08848 (2023)

25. Zhou, Y., Muresanu, A.I., Han, Z., Paster, K., Pitis, S., Chan, H., Ba, J.: Large language models are human-level prompt engineers. arXiv preprint arXiv: Arxiv-2211.01910 (2022)

26. Honovich, O., Shaham, U., Bowman, S.R., Levy, O.: Instruction induction: From few examples to natural language task descriptions. arXiv preprint arXiv: Arxiv-2205.10782 (2022)

27. Prasad, A., Hase, P., Zhou, X., Bansal, M.: Grips: Gradient-free, edit-based instruction search for prompting large language models. In: A. Vlachos, I. Augenstein (eds.) Proceedings of the 17th Conference of the European Chapter of the Association for Computational Linguistics, EACL 2023, Dubrovnik, Croatia, May 2-6, 2023, pp. 3827–3846. Association for Computational Linguistics (2023). https://aclanthology.org/2023.eacl-main.277

28. Zhong, W., Gao, Y., Ding, N., Qin, Y., Liu, Z., Zhou, M., Wang, J., Yin, J., Duan, N.: ProQA: Structural prompt-based pre-training for unified question answering. In: Proceedings of the 2022 Conference of the North American Chapter of the Association for Computational Linguistics: Human Language Technologies, pp. 4230–4243. Association for Computational Linguistics, Seattle, United States (2022). https://doi.org/10.18653/v1/2022.naacl-main.313. https://aclanthology. org/2022.naacl-main.313

29. Ye, S., Hwang, H., Yang, S., Yun, H., Kim, Y., Seo, M.: In-context instruction learning. arXiv preprint arXiv:2302.14691 (2023)

30. Cheng, Z., Kasai, J., Yu, T.: Batch prompting: Efficient inference with large language model apis. arXiv preprint arXiv: Arxiv-2301.08721 (2023)

31. Hao, Y., Sun, Y., Dong, L., Han, Z., Gu, Y., Wei, F.: Structured prompting: Scaling in-context learning to 1,000 examples. arXiv preprint arXiv: Arxiv-2212.06713 (2022)

32. Lazaridou, A., Gribovskaya, E., Stokowiec, W., Grigorev, N.: Internet-augmented language models through few-shot prompting for open-domain question answering. arXiv preprint arXiv:2203.05115 (2022)

33. Izacard, G., Lewis, P., Lomeli, M., Hosseini, L., Petroni, F., Schick, T., Dwivedi-Yu, J., Joulin, A., Riedel, S., Grave, E.: Atlas: Few-shot learning with retrieval augmented language models. arXiv preprint arXiv: Arxiv-2208.03299 (2022)

34. Ram, O., Levine, Y., Dalmedigos, I., Muhlgay, D., Shashua, A., Leyton-Brown, K., Shoham, Y.: In-context retrieval-augmented language models. arXiv preprint arXiv: Arxiv-2302.00083 (2023)

35. Weir, N., Yuan, X., Côté, M.A., Hausknecht, M., Laroche, R., Momennejad, I., Seijen, H.V., Durme, B.V.: One-shot learning from a demonstration with hierarchical latent language. arXiv preprint arXiv: Arxiv-2203.04806 (2022)

36. Wei, J., Wang, X., Schuurmans, D., Bosma, M., Chi, E.H., Le, Q., Zhou, D.: Chain of thought prompting elicits reasoning in large language models. CoRR **abs/2201.11903** (2022). https://arxiv.org/abs/2201.11903

37. Nye, M., Andreassen, A.J., Gur-Ari, G., Michalewski, H., Austin, J., Bieber, D., Dohan, D., Lewkowycz, A., Bosma, M., Luan, D., et al.: Show your work: Scratchpads for intermediate computation with language models. arXiv preprint arXiv:2112.00114 (2021)

38. Zhou, D., Schärli, N., Hou, L., Wei, J., Scales, N., Wang, X., Schuurmans, D., Bousquet, O., Le, Q., Chi, E.: Least-to-most prompting enables complex reasoning in large language models. arXiv preprint arXiv:2205.10625 (2022)

39. Wang, X., Wei, J., Schuurmans, D., Le, Q., Chi, E., Narang, S., Chowdhery, A., Zhou, D.: Self-consistency improves chain of thought reasoning in language models. arXiv preprint arXiv: Arxiv-2203.11171 (2022)

40. Fu, Y., Peng, H., Sabharwal, A., Clark, P., Khot, T.: Complexity-based prompting for multi-step reasoning. arXiv preprint arXiv:2210.00720 (2022)

41. Zhang, Z., Zhang, A., Li, M., Smola, A.: Automatic chain of thought prompting in large language models. arXiv preprint arXiv:2210.03493 (2022)

42. Kojima, T., Gu, S.S., Reid, M., Matsuo, Y., Iwasawa, Y.: Large language models are zero-shot reasoners. arXiv preprint arXiv:2205.11916 (2022)

43. Marasović, A., Beltagy, I., Downey, D., Peters, M.E.: Few-shot self-rationalization with natural language prompts. ArXiv preprint **abs/2111.08284** (2021). https://arxiv.org/abs/2111.08284

44. Chen, W., Ma, X., Wang, X., Cohen, W.W.: Program of thoughts prompting: Disentangling computation from reasoning for numerical reasoning tasks. arXiv preprint arXiv:2211.12588 (2022)

45. Gao, L., Madaan, A., Zhou, S., Alon, U., Liu, P., Yang, Y., Callan, J., Neubig, G.: Pal: Program-aided language models. arXiv preprint arXiv:2211.10435 (2022)

46. Suris, D., Menon, S., Vondrick, C.: Vipergpt: Visual inference via python execution for reasoning. arXiv preprint arXiv: Arxiv-2303.08128 (2023)

47. Schick, T., Dwivedi-Yu, J., Dessì, R., Raileanu, R., Lomeli, M., Zettlemoyer, L., Cancedda, N., Scialom, T.: Toolformer: Language models can teach themselves to use tools. CoRR **abs/2302.04761** (2023). https://doi.org/10.48550/arXiv.2302.04761

48. Yao, S., Zhao, J., Yu, D., Du, N., Shafran, I., Narasimhan, K., Cao, Y.: React: Synergizing reasoning and acting in language models. arXiv preprint arXiv: Arxiv-2210.03629 (2022)

49. Liu, R., Wei, J., Gu, S.S., Wu, T.Y., Vosoughi, S., Cui, C., Zhou, D., Dai, A.M.: Mind's eye: Grounded language model reasoning through simulation. arXiv preprint arXiv:2210.05359 (2022)

50. Wu, T.S., Terry, M., Cai, C.J.: Ai chains: Transparent and controllable human-ai interaction by chaining large language model prompts. International Conference On Human Factors In Computing Systems (2021). https://doi.org/10.1145/3491102.3517582

51. Wu, T., Jiang, E., Donsbach, A., Gray, J., Molina, A., Terry, M., Cai, C.J.: Promptchainer: Chaining large language model prompts through visual programming. arXiv preprint arXiv: Arxiv-2203.06566 (2022)

52. Yang, Z., Li, L., Wang, J., Lin, K., Azarnasab, E., Ahmed, F., Liu, Z., Liu, C., Zeng, M., Wang, L.: Mm-react: Prompting ChatGPT for multimodal reasoning and action (2023)

53. Qiao, S., Ou, Y., Zhang, N., Chen, X., Yao, Y., Deng, S., Tan, C., Huang, F., Chen, H.: Reasoning with language model prompting: A survey. arXiv preprint arXiv: Arxiv-2212.09597 (2022)

54. Dohan, D., Xu, W., Lewkowycz, A., Austin, J., Bieber, D., Lopes, R.G., Wu, Y., Michalewski, H., Saurous, R.A., Sohl-dickstein, J., Murphy, K., Sutton, C.: Language model cascades. arXiv preprint arXiv: Arxiv-2207.10342 (2022)

55. Creswell, A., Shanahan, M., Higgins, I.: Selection-inference: Exploiting large language models for interpretable logical reasoning. arXiv preprint arXiv: Arxiv-2205.09712 (2022)

56. Cobbe, K., Kosaraju, V., Bavarian, M., Chen, M., Jun, H., Kaiser, L., Plappert, M., Tworek, J., Hilton, J., Nakano, R., Hesse, C., Schulman, J.: Training verifiers to solve math word problems. arXiv preprint arXiv: Arxiv-2110.14168 (2021)

57. Nakano, R., Hilton, J., Balaji, S., Wu, J., Ouyang, L., Kim, C., Hesse, C., Jain, S., Kosaraju, V., Saunders, W., et al.: Webgpt: Browser-assisted question-answering with human feedback. arXiv preprint arXiv:2112.09332 (2021)

58. Srivastava, A., Rastogi, A., Rao, A., Shoeb, A.A.M., Abid, A., Fisch, A., Brown, A.R., Santoro, A., Gupta, A., Garriga-Alonso, A., Kluska, A., Lewkowycz, A., Agarwal, A., Power, A., Ray, A., Warstadt, A., Kocurek, A.W., Safaya, A., Tazarv, A., Xiang, A., Parrish, A., Nie, A., Hussain, A., Askell, A., Dsouza, A., Slone, A., Rahane, A., Iyer, A.S., Andreassen, A., Madotto, A., Santilli, A., Stuhlmüller, A., Dai, A., La, A., Lampinen, A., Zou, A., Jiang, A., Chen, A., Vuong, A., Gupta, A., Gottardi, A., Norelli, A., Venkatesh, A., Gholamidavoodi, A., Tabassum, A., Menezes, A., Kirubarajan, A., Mullokandov, A., Sabharwal, A., Herrick, A., Efrat, A., Erdem, A., Karakaş, A., Roberts, B.R., Loe, B.S., Zoph, B., Bojanowski, B., Özyurt, B., Hedayatnia, B., Neyshabur, B., Inden, B., Stein, B., Ekmekci, B., Lin, B.Y., Howald, B., Diao, C., Dour, C., Stinson, C., Argueta, C., Ramírez, C.F., Singh, C., Rathkopf, C., Meng, C., Baral, C., Wu, C., Callison-Burch, C., Waites, C., Voigt, C., Manning, C.D., Potts, C., Ramirez, C., Rivera, C.E., Siro, C., Raffel, C., Ashcraft, C., Garbacea, C., Sileo, D., Garrette, D., Hendrycks, D., Kilman, D., Roth, D., Freeman, D., Khashabi, D., Levy, D., Gonzßlez, D.M., Perszyk, D., Hernandez, D., Chen, D., Ippolito, D., Gilboa, D., Dohan, D., Drakard, D., Jurgens, D., Datta, D., Ganguli, D., Emelin, D., Kleyko, D., Yuret, D., Chen, D., Tam, D., Hupkes, D., Misra, D., Buzan, D., Mollo, D.C., Yang, D., Lee, D.H., Shutova, E., Cubuk, E.D., Segal, E., Hagerman, E., Barnes, E., Donoway, E., Pavlick, E., Rodola, E., Lam, E., Chu, E., Tang, E., Erdem, E., Chang, E., Chi, E.A., Dyer, E., Jerzak, E., Kim, E., Manyasi, E.E., Zheltonozhskii, E., Xia, F., Siar, F., Martínez-Plumed, F., Happé, F., Chollet, F., Rong, F., Mishra, G., Winata, G.I., de Melo, G., Kruszewski, G., Parascandolo, G., Mariani, G., Wang, G., Jaimovitch-López, G., Betz, G., Gur-Ari, G., Galijasevic, H., Kim, H., Rashkin, H., Hajishirzi, H., Mehta, H., Bogar, H., Shevlin, H., Schütze, H., Yakura, H., Zhang, H., Wong, H.M., Ng, I., Noble, I., Jumelet, J., Geissinger, J., Kernion, J., Hilton, J., Lee, J., Fisac, J.F., Simon, J.B., Koppel, J., Zheng, J., Zou, J., Kocoń, J., Thompson, J., Kaplan, J., Radom, J., Sohl-Dickstein, J., Phang, J., Wei, J., Yosinski, J., Novikova, J., Bosscher, J., Marsh, J., Kim, J., Taal, J., Engel, J., Alabi, J., Xu, J., Song, J., Tang, J., Waweru, J., Burden, J., Miller, J., Balis, J.U., Berant, J., Frohberg, J., Rozen, J., Hernandez-Orallo, J., Boudeman, J., Jones, J., Tenenbaum, J.B., Rule, J.S., Chua, J., Kanclerz, K., Livescu, K., Krauth, K., Gopalakrishnan, K., Ignatyeva, K., Markert, K., Dhole, K.D., Gimpel, K., Omondi, K., Mathewson, K., Chiafullo, K., Shkaruta, K., Shridhar, K., McDonell, K., Richardson, K., Reynolds, L., Gao, L., Zhang, L., Dugan, L., Qin, L., Contreras-Ochando, L., Morency, L.P., Moschella, L., Lam, L., Noble, L., Schmidt, L., He, L., Colón, L.O., Metz, L., Şenel, L.K., Bosma, M., Sap, M., ter Hoeve, M., Farooqi, M., Faruqui, M., Mazeika, M., Baturan, M., Marelli, M., Maru, M., Quintana, M.J.R., Tolkiehn, M., Giulianelli, M., Lewis, M., Potthast, M., Leavitt, M.L., Hagen, M., Schubert, M., Baitemirova, M.O., Arnaud, M., McElrath, M., Yee, M.A., Cohen, M., Gu, M., Ivanitskiy, M., Starritt, M., Strube, M., Swędrowski, M., Bevilacqua, M., Yasunaga, M., Kale, M., Cain, M., Xu, M., Suzgun, M., Tiwari, M., Bansal, M., Aminnaseri, M., Geva, M., Gheini, M., T, M.V., Peng, N., Chi, N., Lee, N., Krakover, N.G.A., Cameron, N., Roberts, N., Doiron, N., Nangia, N., Deckers, N., Muennighoff, N., Keskar, N.S., Iyer, N.S., Constant, N., Fiedel, N., Wen, N.,

Zhang, O., Agha, O., Elbaghdadi, O., Levy, O., Evans, O., Casares, P.A.M., Doshi, P., Fung, P., Liang, P.P., Vicol, P., Alipoormolabashi, P., Liao, P., Liang, P., Chang, P., Eckersley, P., Htut, P.M., Hwang, P., Miłkowski, P., Patil, P., Pezeshkpour, P., Oli, P., Mei, Q., Lyu, Q., Chen, Q., Banjade, R., Rudolph, R.E., Gabriel, R., Habacker, R., Delgado, R.R., Millière, R., Garg, R., Barnes, R., Saurous, R.A., Arakawa, R., Raymaekers, R., Frank, R., Sikand, R., Novak, R., Sitelew, R., LeBras, R., Liu, R., Jacobs, R., Zhang, R., Salakhutdinov, R., Chi, R., Lee, R., Stovall, R., Teehan, R., Yang, R., Singh, S., Mohammad, S.M., Anand, S., Dillavou, S., Shleifer, S., Wiseman, S., Gruetter, S., Bowman, S.R., Schoenholz, S.S., Han, S., Kwatra, S., Rous, S.A., Ghazarian, S., Ghosh, S., Casey, S., Bischoff, S., Gehrmann, S., Schuster, S., Sadeghi, S., Hamdan, S., Zhou, S., Srivastava, S., Shi, S., Singh, S., Asaadi, S., Gu, S.S., Pachchigar, S., Toshniwal, S., Upadhyay, S., Shyamolima, Debnath, Shakeri, S., Thormeyer, S., Melzi, S., Reddy, S., Makini, S.P., Lee, S.H., Torene, S., Hatwar, S., Dehaene, S., Divic, S., Ermon, S., Biderman, S., Lin, S., Prasad, S., Piantadosi, S.T., Shieber, S.M., Misherghi, S., Kiritchenko, S., Mishra, S., Linzen, T., Schuster, T., Li, T., Yu, T., Ali, T., Hashimoto, T., Wu, T.L., Desbordes, T., Rothschil: Beyond the imitation game: Quantifying and extrapolating the capabilities of language models. arXiv preprint arXiv: Arxiv-2206.04615 (2022)

59. Besta, M., Blach, N., Kubicek, A., Gerstenberger, R., Podstawski, M., Gianinazzi, L., Gajda, J., Lehmann, T., Niewiadomski, H., Nyczyk, P., Hoefler, T.: Graph of thoughts: Solving elaborate problems with large language models. Proceedings of the AAAI Conference on Artificial Intelligence **38**(16), 17682–17690 (2024). https://doi.org/10.1609/aaai.v38i16.29720. https://ojs.aaai.org/index.php/AAAI/article/view/29720

60. Press, O., Zhang, M., Min, S., Schmidt, L., Smith, N.A., Lewis, M.: Measuring and narrowing the compositionality gap in language models. arXiv preprint arXiv: Arxiv-2210.03350 (2022)

61. Dua, D., Gupta, S., Singh, S., Gardner, M.: Successive prompting for decomposing complex questions. In: Y. Goldberg, Z. Kozareva, Y. Zhang (eds.) Proceedings of the 2022 Conference on Empirical Methods in Natural Language Processing, EMNLP 2022, Abu Dhabi, United Arab Emirates, December 7-11, 2022, pp. 1251–1265. Association for Computational Linguistics (2022). https://aclanthology.org/2022.emnlp-main.81

62. Zhang, Z., Zhang, A., Li, M., Zhao, H., Karypis, G., Smola, A.: Multimodal chain-of-thought reasoning in language models. arXiv preprint arXiv: Arxiv-2302.00923 (2023)

63. Perez, E., Lewis, P., Yih, W.t., Cho, K., Kiela, D.: Unsupervised question decomposition for question answering. In: Proceedings of the 2020 Conference on Empirical Methods in Natural Language Processing (EMNLP), pp. 8864–8880. Association for Computational Linguistics, Online (2020). https://doi.org/10.18653/v1/2020.emnlp-main.713. https://aclanthology.org/2020.emnlp-main.713

64. Arora, S., Narayan, A., Chen, M.F., Orr, L., Guha, N., Bhatia, K., Chami, I., Sala, F., Ré, C.: Ask me anything: A simple strategy for prompting language models. arXiv preprint arXiv: Arxiv-2210.02441 (2022)

65. Yao, S., Yu, D., Zhao, J., Shafran, I., Griffiths, T.L., Cao, Y., Narasimhan, K.: Tree of thoughts: Deliberate problem solving with large language models. arXiv preprint arXiv: 2305.10601 (2023)

66. Sel, B., Al-Tawaha, A.S., Khattar, V., Wang, L., Jia, R., Jin, M.: Algorithm of thoughts: Enhancing exploration of ideas in large language models. International Conference on Machine Learning (2023). https://doi.org/10.48550/arXiv.2308.10379

67. Bi, Z., Han, K., Liu, C., Tang, Y., Wang, Y.: Forest-of-thought: Scaling test-time compute for enhancing LLM reasoning. arXiv preprint arXiv: 2412.09078 (2024)

68. Lin, Q., Xu, B., Hu, G., Li, Z., Hao, Z., Zhang, K., Cai, R.: Cmcts: A constrained monte carlo tree search framework for mathematical reasoning in large language model. arXiv preprint arXiv: 2502.11169 (2025)

69. Zeng, A., Attarian, M., Ichter, B., Choromanski, K., Wong, A., Welker, S., Tombari, F., Purohit, A., Ryoo, M., Sindhwani, V., Lee, J., Vanhoucke, V., Florence, P.: Socratic models: Composing zero-shot multimodal reasoning with language. arXiv preprint arXiv: Arxiv-2204.00598 (2022)

70. Huang, W., Xia, F., Xiao, T., Chan, H., Liang, J., Florence, P., Zeng, A., Tompson, J., Mordatch, I., Chebotar, Y., et al.: Inner monologue: Embodied reasoning through planning with language models. arXiv preprint arXiv:2207.05608 (2022)

71. Zhu, D., Chen, J., Haydarov, K., Shen, X., Zhang, W., Elhoseiny, M.: ChatGPT asks, blip-2 answers: Automatic questioning towards enriched visual descriptions. arXiv preprint arXiv: Arxiv-2303.06594 (2023)

72. Li, J., Li, D., Savarese, S., Hoi, S.: Blip-2: Bootstrapping language-image pre-training with frozen image encoders and large language models. arXiv preprint arXiv: Arxiv-2301.12597 (2023)

73. Chen, Z., Zhou, Q., Shen, Y., Hong, Y., Zhang, H., Gan, C.: See, think, confirm: Interactive prompting between vision and language models for knowledge-based visual reasoning. arXiv preprint arXiv: Arxiv-2301.05226 (2023)

74. Weng, Y., Zhu, M., He, S., Liu, K., Zhao, J.: Large language models are reasoners with self-verification. arXiv preprint arXiv: Arxiv-2212.09561 (2022)

75. Bara, C.P., CH-Wang, S., Chai, J.: MindCraft: Theory of mind modeling for situated dialogue in collaborative tasks. In: Proceedings of the 2021 Conference on Empirical Methods in Natural Language Processing, pp. 1112–1125. Association for Computational Linguistics, Online and Punta Cana, Dominican Republic (2021). https://doi.org/10.18653/v1/2021.emnlp-main.85. https://aclanthology.org/2021.emnlp-main.85

76. Shinn, N., Cassano, F., Labash, B., Gopinath, A., Narasimhan, K., Yao, S.: Reflexion: Language agents with verbal reinforcement learning. arXiv preprint arXiv: 2303.11366 (2023)

77. Welleck, S., Lu, X., West, P., Brahman, F., Shen, T., Khashabi, D., Choi, Y.: Generating sequences by learning to self-correct. International Conference on Learning Representations (2022). https://doi.org/10.48550/arXiv.2211.00053

78. Akyurek, A.F., Akyürek, E., Madaan, A., Kalyan, A., Clark, P., Wijaya, D., Tandon, N.: Rl4f: Generating natural language feedback with reinforcement learning for repairing model outputs. Annual Meeting of the Association for Computational Linguistics (2023). https://doi.org/10.48550/arXiv.2305.08844

79. Lu, J., Zhong, W., Huang, W., Wang, Y., Mi, F., Wang, B., Wang, W., Shang, L., Liu, Q.: Self: Language-driven self-evolution for large language model. arXiv preprint arXiv: 2310.00533 (2023)

80. Ye, S., Jo, Y., Kim, D., Kim, S., Hwang, H., Seo, M.: Selfee: Iterative self-revising LLM empowered by self-feedback generation. Blog post (2023). https://kaistai.github.io/SelFee/

81. Madaan, A., Tandon, N., Gupta, P., Hallinan, S., Gao, L., Wiegreffe, S., Alon, U., Dziri, N., Prabhumoye, S., Yang, Y., Welleck, S., Majumder, B.P., Gupta, S., Yazdanbakhsh, A., Clark, P.: Self-refine: Iterative refinement with self-feedback. arXiv preprint arXiv: Arxiv-2303.17651 (2023)

82. Li, Y., Wei, F., Zhao, J., Zhang, C., Zhang, H.: Rain: Your language models can align themselves without finetuning. arXiv preprint arXiv: 2309.07124 (2023)

83. Saunders, W., Yeh, C., Wu, J., Bills, S., Ouyang, L., Ward, J., Leike, J.: Self-critiquing models for assisting human evaluators. arXiv preprint arXiv: 2206.05802 (2022)

84. Chen, G., Dong, S., Shu, Y., Zhang, G., Sesay, J., Karlsson, B.F., Fu, J., Shi, Y.: Autoagents: A framework for automatic agent generation. arXiv preprint arXiv: 2309.17288 (2023)

Fine-Tuning

Zekun Wang, Qingqing Zhu, Xiuying Chen, Mong Yuan Sim,
Wangchunshu Zhou, Shaochun Hao and Jie Fu

Fine-tuning refers to the process of updating the parameters of a model. The ongoing interaction provides an increasing amount of interaction messages that can be used to update
the models' parameters, resulting in better language model adaptation to interactions like
instruction following [1–6] and grounding [7–10]. This line of research also explores how to

Z. Wang
Beihang University, Beijing, China
e-mail: zenmoore@buaa.edu.cn

Q. Zhu
Peking University, Beijing, China
e-mail: zhuqingqing@pku.edu.cn

X. Chen
King Abdullah University Of Science And Technology, Thuwal, Saudi Arabia
e-mail: xiuying.chen@kaust.edu.sa

M. Y. Sim
University of Adelaide, Adelaide, Australia
e-mail: mongyuan.sim@student.adelaide.edu.au

W. Zhou
Hangzhou, China
e-mail: chunshu@aiwaves.cn

S. Hao
Xi'an Jiaotong University, Hangzhou, China
e-mail: haoshaochun@stu.xjtu.edu.cn

J. Fu (✉)
Hong Kong University of Science and Technology, Hong Kong, China
e-mail: jie.fu@polymtl.ca

Z. Wang et al. (eds.), *Interactive Natural Language Processing*, Synthesis Lectures
on Human Language Technologies, https://doi.org/10.1007/978-3-032-06264-2_6

make effective use of this new data without catastrophic forgetting [11–16], how to ensure generalization to new tasks [1–6], and how to adapt the language model more efficiently [17–22].

In this section, we discuss four commonly employed fine-tuning-based methods: (1) Supervised Instruction Tuning, which aims to adapt language models for instruction following and to enhance their task generalization abilities (Sect. 6.1); (2) Continual Learning, which aims to infuse new data into language models without catastrophic forgetting (Sect. 6.2); (3) Parameter-Efficient Fine-Tuning, which focuses on the efficient adaptation of language models (Sect. 6.3); and (4) Semi-Supervised Fine-Tuning, which further tackles the problem of unlabeled data, as in some cases, the interaction message may not provide adequate supervision to train the model [23–27] (Sect. 6.4).

6.1   Supervised Instruction Tuning

Supervised instruction tuning involves fine-tuning a pre-trained language model using data that provides task instruction supervision. Various studies [1–5, 7, 28–35] have been conducted in this area. These methods fine-tune a pre-trained model by using supervised instructions on a multitask mixture, covering various tasks and inducing zero-shot generalization capabilities.

The first line of work, investigated by researchers such as [1–4, 7, 28–30, 36, 37], focuses on providing instructions to language models as part of the input. Typically, these instructions are prepended to the input and contain specific details about the task the model is expected to perform. These models explore different aspects, such as training and evaluation data, model architectures (decoder-only v.s. encoder-decoder), instruction formatting, task mixtures, and other related factors (Fig. 6.1).

The discussed studies offer conclusive evidence that fine-tuning language models on multiple NLP tasks and incorporating instructions allow these models to generalize to unseen tasks and better understand and respond to user queries [6]. As demonstrated in [38, 39], scaling up language models leads to improvements in performance. The researchers also study the impact of different scales of instruction data, aiming to better understand the influence of the amount and diversity of this kind of training data [35].

Subsequently, OpenAI releases the InstructGPT [5] and develops a series of GPT-3.5 variants, all of which are built upon the foundation of GPT-3 [40]. These variants include *code-davinci-002* and *text-davinci-002*, which only involve supervised instruction tuning, as well as *text-davinci-003* and *gpt-3.5-turbo*, which are refined through both supervised fine-tuning and reinforcement learning from human feedback (RLHF). These modifications enhance the models' alignment with human intent, resulting in more truthful and less toxic responses from the language models. Along with this line, DeepMind's Sparrow [34] and Anthropic's Claude[1] also use instruction tuning and RLHF to teach models to produce

---

[1] https://www.anthropic.com/index/introducing-claude.

Fig. 6.1 Supervised instruction tuning

answers that align with human values [41]. Furthermore, apart from scaling up the instructional fine-tuning process by increasing the number of tasks and the model size, [31, 32] improve the process by jointly integrating chain-of-thought data during instruction tuning. They fine-tune the T5 [28] and PaLM [42] into FLAN-T5 and FLAN-PaLM models [32], resulting in robust performance across a diverse range of natural language processing tasks, including translation, reasoning, and question answering.

Apart from supervised instruction fine-tuning using existing instruction datasets or human-annotated instruction datasets, recent studies have also highlighted semi-supervised approaches for creating instruction-following data generated by LLMs [24, 43–47]. This synthetic data can be utilized for fine-tuning PLMs [24, 44]. For example, Baize [44] first generates some dialogue data using GPT, and then uses these data to fine-tune another model. UltraChat [48] generates dialogue data in accordance with the practices established by Camel [49], utilizing these datasets to enhance the open-ended chatting capabilities of models. Such data augmentation and distillation techniques are increasingly employed in the development of domain-specific language models. These models are typically fine-tuned using synthetic data relevant to their specific domain and incorporate domain-specific inductive biases. For instance, MAmmoTH [50] is partly fine-tuned with synthetic mathematical reasoning data generated by LLMs and employs the "Program of Thoughts" (PoT) [51] technique as an inductive bias to enhance math reasoning skills. RoleLLM [52] is fine-tuned using synthetic role-playing data. It utilizes a technique known as "Context-Instruct" as an inductive bias, which helps in infusing role-specific knowledge into the model. TIGER-Score [53] is designed to improve models' evaluation capabilities. Its inductive bias is the "Error Breakdown Analysis", which acts to elicit more accurate evaluation scores by analyzing errors in a structured manner. This line is partly overlapped with knowledge distillation, for which we refer the readers to Sect. 6.4 for more information.

Recently, in the field of alignment, there has been a shift away from RLHF due to its inefficiency in training [54–57]. Instead, more research is focusing on supervised fine-tuning (SFT) using preferred responses [54, 57, 58] or employing contrastive or ranking-based supervised fine-tuning on preference-contrastive response pairs or preference-ranked response lists [55, 56, 59–62]. For example, Dromendary [57] generates preferred responses based on manually designed principle prompts and then trains models on these synthetic responses, known as Self-Alignment. Chain of Hindsight [61] trains models with both better and worse responses but concatenates these contrastive pairs with templates like "[response-1] is better than [response-2]", decoding only the better response during inference. SLiC [62] calibrates sequence likelihood using various ranking functions, including rank loss, margin loss, list rank loss, and expected rank loss, given preference-ranked responses. RRHF [55] simplifies this approach by omitting margin terms and focusing on list rank loss. DPO [56] introduces a novel contrastive-learning-based loss function, $\mathcal{L}_{\text{DPO}}(\pi_\theta; \pi_{\text{ref}}) = -\mathbb{E}_{(x, y_w, y_l) \sim \mathcal{D}}\left[\log \sigma\left(\beta \log \frac{\pi_\theta(y_w|x)}{\pi_{\text{ref}}(y_w|x)} - \beta \log \frac{\pi_\theta(y_l|x)}{\pi_{\text{ref}}(y_l|x)}\right)\right]$, deducted from PPO [63]'s formula, for training with preference-contrastive response pairs (y_w, y_l). PRO [59] utilizes $\mathcal{L}_{\text{PRO}} = -\sum_{k=1}^{n-1} \log \frac{\exp(\pi_\theta(y^k|x))}{\sum_{i=k}^{n} \exp(\pi_\theta(y^i|x))}$ for training with preference-ranked responses. Note that these contrastive or rank-based training objectives are often combined with vanilla supervised instruction tuning losses to prevent training collapse [55, 59].

Supervised instruction tuning can be regarded as one of the crucial steps in interactive natural language processing. By fine-tuning language models with supervised instructions, their ability to comprehend and respond to a diverse range of queries can be enhanced, enabling them to perform tasks such as question-answering and task completion with greater precision and efficiency. Recent developments in contrastive or ranking-based supervised instruction tuning have also proven to be effective methods for alignment (Fig. 6.2).

6.2   Continual Learning

LMs that have been pre-trained on static data may become outdated and no longer aligned with new domains or tasks [16, 64]. Therefore, it is beneficial to utilize interaction messages accumulated over time to fine-tune LMs. This guarantees that the LMs are up-to-date with the newest information and perform optimally in novel scenarios [65]. Although typical fine-tuning is an effective approach to updating an LM, it can suffer from catastrophic forgetting [66]. As data size increases, attempting to incorporate new knowledge into a fixed-sized LM may result in losing previous knowledge. Continual Learning (CL) is a promising solution to this problem. It seeks to continuously integrate knowledge from novel sources without expunging prior learning [67].

Fig. 6.2 Continual learning

To begin, we present a mathematical formalization of continual learning. Consider a language model that is presented with a sequence of n tasks $(T_1, ..., T_n)$. For each task T_k, the model is provided with a set of N i.i.d. training examples $(x_i, y_i)_{i=1}^{N}$. Assuming that the language model is parameterized by θ and is aware of the task identity during both training and inference, the overall learning objective across all tasks is:

$$\max_{\theta} \sum_{k=1}^{n} \sum_{(x,y) \in T_k} \log p(y \mid x; \theta) \tag{6.1}$$

In the continual learning setting, it involves sequentially optimizing the loss for each task T_k via fine-tuning:

$$\max_{\theta} \sum_{x,y \in T_k} \log p(y \mid x; \theta) \text{ for } kin\{1, ..., n\} \tag{6.2}$$

Such a naïve continual fine-tuning may cause catastrophic forgetting, ultimately leading to a decline in overall performance on earlier tasks after the learning of new tasks [68].

In this part, we briefly introduce recent CL methods that aim to alleviate the forgetting phenomenon, which can be categorized into three groups: (1) Regularization, (2) Rehearsal, and (3) Modularization, following [69, 70].

Regularization. Regularization has been widely adopted to inhibit catastrophic forgetting by penalizing the model when it deviates significantly from its previous state. This is typically accomplished by determining the essential parameters for the prior task and domain data, and subsequently integrating a regularization term into the loss function, which encourages the model to preserve this knowledge while learning new tasks [69]. For example, [71] propose Elastic Weight Consolidation (EWC), which involves measuring important parameters using a Fisher information matrix, derived from the magnitude of the gradient update step corresponding to each parameter. Following this work, [72] introduce a pre-training simulation mechanism that enables the memorization of knowledge for pre-training tasks, eliminating the requirement for pre-training data access. [73] introduce a novel approach for continual learning in pre-trained models, which calibrates both the parameters and logits. This approach helps in retaining the acquired knowledge while also facilitating the learning of new concepts. As a concurrent work, [74] propose to selectively memorize important parameters from previous tasks through a recall optimization mechanism facilitated by regularization, which is evaluated on interactive dialogue generation datasets.

Rehearsal. A rehearsal-based approach to continual learning involves replaying previous task data or synthetic previous task data while learning new tasks. For example, [75] conduct preliminary experiments on interactive dialog models, assuming access to the pre-training corpus during fine-tuning. Throughout the training process, they mix random subsets of the pre-training corpus based on a mix-ratio that anneals towards the target task. ELLE [16] employs a memory replay mechanism, where several data from previous tasks is mixed with current task data for tuning. Moreover, the synthetic previous task data generated by the old model can be utilized to train the new model as memory replay [76].

Modularization. Modularization refers to separating model parameters into distinct modules or sub-networks, where each module is responsible for performing a specific task or set of tasks [69, 70, 77]. The modularization can be achieved by adapter-based methods, which introduce new parameters corresponding to new tasks; Alternatively, it can be based on partial tuning, which freezes or prunes previous task parameters as discussed in [70, 77] and Sect. 6.3. For example, [78] propose a plug-and-play approach for incorporating the target knowledge into new parameters by applying multiple large-scale updates on plug-in modules, thereby avoiding the risk of forgetting the previously acquired source knowledge in old parameters. In SupSup [79], a network is first initialized as a base network, and task-specific sub-networks are separately learned for different tasks and acquired with a network-level mask. During testing, the task identity can be provided or inferred using gradient-based optimization, allowing the appropriate sub-network to be retrieved.

Continual learning is crucial for interactive NLP (iNLP), enabling the model to adapt to the dynamic and diverse nature of user inputs, tasks, and environments. Despite considerable efforts to address the issue of catastrophic forgetting, the problem persists, especially

Fig. 6.3 Parameter-efficient fine-tuning

in large language models, as they impose substantial computational requirements. Additionally, various other questions, such as evaluation for CL and sample efficiency, remain unanswered. Researchers persistently strive to improve the efficiency and effectiveness of continual learning for NLP, and advancements in this field will significantly impact the future of iNLP systems (Fig. 6.3).

6.3 Parameter-Efficient Fine-Tuning

When the size of a pre-trained language model grows larger and larger, it becomes more and more difficult to fine-tune the model, especially when the size of interaction message data is limited. This is because very large models often require an impractical amount of GPU memory to fine-tune, and over-fitting easily happens when the training data is limited. Thus, to avoid these issues, various parameter-efficient tuning methods are proposed. Parameter-efficient fine-tuning, as its name suggests, only updates a small number of parameters compared to the number of parameters of the full model. By reducing the number of trainable parameters during fine-tuning, parameter-efficient fine-tuning methods consume much less GPU memory and substantially diminish the risk of overfitting. Therefore, parameter-efficient fine-tuning methods have become the de facto practice for fine-tuning LLMs.

Parameter-efficient fine-tuning can be formalized as:

$$\mathcal{L}_{\theta_0 \subsetneq h(\theta)}(LM(x; h(\theta))) \tag{6.3}$$

where θ_0 is the set of tunable parameters, $h(\theta)$ is the change of model parameters by introducing additional modules, pruning, and reparameterization [18].

Following Delta-Tuning [18], parameter-efficient fine-tuning methods for iNLP can be
divided into two main categories based on whether they introduce additional modules or
parameters. The first category includes **partial fine-tuning** techniques, which select a small
number of parameters of the pre-trained model during fine-tuning. In contrast, the second
category of parameter-efficient fine-tuning methods keeps the model frozen and adds extra
parameters that are updated during fine-tuning. These newly added and tuned modules are
called adapters, belonging to **adapter-based methods**.

Partial Fine-tuning. This fine-tuning method, also known as "specification-based tuning", updates only a strict subset of the model parameters while keeping the remaining parameters unchanged during fine-tuning [18]. This approach does not add any new parameters
to the model, meaning that the number of parameters remains the same (i.e., $|h(\theta)| = |\theta|$).
Instead, the method specifies explicitly or implicitly which parts of the model's parameters
should be optimized by indicating θ_0 and $|\theta_0| << |\theta|$ [18]. For example, we can choose a
few layers to freeze and only update the remaining layers. Howard and Ruder [80] initially
freeze all layers except for the last layer, which contains less general knowledge, and then
gradually unfreeze the remaining layers during the fine-tuning process. Moreover, BitFit [81]
optimizes only the bias terms of a pre-trained model, specifically the "query" and "middle-
of-MLP" bias terms, resulting in a significant reduction in the number of tunable parameters
while still maintaining high performance on several benchmarks. Diff pruning [82] involves
learning a delta vector to be applied to the initial pre-trained model parameters. It uses
a differentiable approximation of the *L0*-norm penalty which facilitates the delta vector
becoming more sparse, thereby resulting in a more parameter-efficient approach to fine-
tuning. Similarly, [83] encourage the model to prune less important attention heads through
regularization, effectively reducing the number of attention heads that need to be fine-tuned.
SupSup [79] selectively updates the critical weights of a PLM for specific tasks by learning a
sub-network called supermask. It learns a superposition of supermasks with gradient-based
optimization for an unseen task during inference.

Adapter-based Methods. Adapter tuning [84] inserts small modules called *adapters*
to a model: $\theta \subsetneq h(\theta)$ and $|h(\theta)| - |\theta| << |\theta|$ [18]. With adapters, we can completely
freeze the model and only optimize the newly introduced adapters, i.e., $\theta_0 = h(\theta) - \theta$.
Conventional adapters follow [21]'s practice of placing a two-layer feed-forward neural network with a bottleneck after each sub-layer within a Transformer layer, including both the
multi-head attention sub-layer and the feed-forward network sub-layer. In addition, recent
works have introduced many other parameterizations of adapters. The most representative
ones include Prefix Tuning [19], Prompt Tuning [20], LoRA [22], and Compacter [85]. As
pointed out by [86], these approaches can be unified into a single framework, wherein a
module is added as a residual to specific components of the original computational graph in
the Transformer architecture. As a result, they can all essentially be considered as different
variants of adapters. For example, Prefix Tuning [19] introduces learnable tokens to the

input, which are prepended to the keys and values of self-attention, with the model weights being frozen. LoRA [22] introduces a module of low-rank trainable parameters, while the pre-trained model's weights keep frozen. In addition to low-rank approximation, Compacter [85] also utilizes parameterized hypercomplex multiplications layers [87]. Compared to partial fine-tuning, adapter-based methods incorporate task-specific additional modules, offering increased flexibility, modularity, compositionality, and shareability [77, 88]. These features also allow adapters to function as a form of task representation. For example, recent works [89, 90] show that the additional parameters introduced in adapter-based approaches reveal inter-task similarities and transferability between tasks. Specifically, [89] show that intermediate task transferability can be effectively predicted by calculating the cosine similarity between adapter parameters for two tasks under the same backbone model.

Applications in iNLP. Parameter-efficient fine-tuning has been successfully applied in iNLP scenarios since it substantially reduces the parameters required for training and storage. For example, in the model-in-the-loop setting, BLIP-2 [23] uses prompt tuning-like method to implement vision model-language model interaction, where the interaction interface is learnable soft tokens which is mapped from the output of a vision model. In the KB-in-the-loop setting, K-Adapter [91] uses trainable adapters to infuse knowledge retrieved from knowledge bases into pre-trained language models while keeping the model parameter frozen. Liang et al. [92] also investigate adapter tuning techniques to train a Transformer-based policy model through imitation learning for robotic manipulation (environment-in-the-loop). Additionally, parameter-efficient fine-tuning can be applied to improve the modularity of models, leading to better out-of-distribution generalization, knowledge composition, prevention of catastrophic interference, and continual learning [16, 77, 88], which are beneficial for iNLP (Fig. 6.4).

6.4   Semi-Supervised Fine-Tuning

According to [93–95], semi-supervised learning aims to use both labeled data and unlabeled data to train a model. Semi-supervised learning can be used for interactive natural language processing in that interaction messages are without sufficient supervision in some cases such as lack of instructions [24, 43], misaligned image-text signals [96], etc. We can formulate semi-supervised fine-tuning as follows:

Let D_l be the set of labeled data points, D_u be the set of unlabeled data points, and $D = D_l \cup D_u$ be the full dataset. Denote $P(y \mid x; \theta)$ as the language model with initial parameters θ_0. The initial training of the model on the labeled data can be formalized as:

$$\theta_1 = \text{argmax}_\theta \frac{1}{|D_l|} \sum_{(x,y) \in D_l} \log P(y \mid x; \theta) \tag{6.4}$$

The model is then used to make predictions on the unlabeled data:

Fig. 6.4 Semi-supervised fine-tuning

$$y_u = \mathrm{argmax}_y \, P\left(y \mid x_u; \theta_1\right) \tag{6.5}$$

where y_u can refer to both generated labels (for sample matching) or output distribution (for distribution matching) [97]. The top s fraction of the most confident predictions (with high probability) is added to the labeled data set:

$$D_l' = \{(x_u, y_u) \mid P\left(y_u \mid x_u; \theta_1\right) > \text{ threshold }\} \tag{6.6}$$

which is finally used to train the model $P(y \mid x)$ in turn (**self-training**) or another model $P'(y \mid x)$ (**semi-supervised knowledge distillation**).

    Self-Training. Self-Training uses the model-generated data to train the model itself, which is also known as bootstrapping. For example, BLIP [96] employs a captioner to generate novel captions derived from the given image, and utilizes a filter to eliminate noisy generations. Subsequently, the models are fine-tuned using the refined data obtained after filtering. Huang et al. [27] generate multiple reasoning paths with the help of a language model, and then majority voting [98] is used to predict the most reliable answer. The reasoning paths with this answer are then used as synthetic training data for language model self-training. Zelikman et al. [25] aim to train the language model to perform CoT reasoning in a semi-supervised setting, where the rationales are not always available. They first generate an intermediate rationale, and if it induces an incorrect answer, the model attempts to produce an inversely rationalized explanation. Subsequently, the model is fine-tuned using the question, the newly generated rationale, and the answer. Alpaca [24] uses a set of 175 seed tasks, each comprising a single instruction and a single example, to enable the large language model to generate additional instructions and examples through In-Context Learning for self-training. This technique is known as Self-Instruct [43]. TALM [99] uses a self-play approach to improve performance. It generates a tool input based on the task input, and then incorporates the output of the tool and the tool input to produce the final task output. These synthetic data generated through self-play are employed to iteratively fine-tune the language model.

Semi-Supervised Knowledge Distillation. Similar to self-training, semi-supervised knowledge distillation also leverages a model to annotate unlabeled data for model tuning. However, instead of generating the synthetic data with the trainable model itself, semi-supervised knowledge distillation uses another teacher model for data generation, and the student model is trained with this new data. For example, [26] use large language models to generate chain-of-thought data, which is then used to fine-tune a smaller language model. Fu et al. [97] first use a LLM to generate responses for unlabeled questions and then use its output distribution to train the smaller student language model for specialization. Shridhar et al. [100] propose a method to disassemble a LLM into two smaller models, namely a problem decomposer and a problem solver, using a distillation approach.

References

1. Wei, J., Bosma, M., Zhao, V., Guu, K., Yu, A., Lester, B., Du, N., Dai, A.M., Le, Q.V.: Finetuned language models are zero-shot learners. International Conference On Learning Representations (2021)
2. Sanh, V., Webson, A., Raffel, C., Bach, S.H., Sutawika, L., Alyafeai, Z., Chaffin, A., Stiegler, A., Scao, T.L., Raja, A., Dey, M., Bari, M.S., Xu, C., Thakker, U., Sharma, S., Szczechla, E., Kim, T., Chhablani, G., Nayak, N.V., Datta, D., Chang, J., Jiang, M.T.J., Wang, H., Manica, M., Shen, S., Yong, Z.X., Pandey, H., Bawden, R., Wang, T., Neeraj, T., Rozen, J., Sharma, A., Santilli, A., Févry, T., Fries, J.A., Teehan, R., Biderman, S.R., Gao, L., Bers, T., Wolf, T., Rush, A.M.: Multitask prompted training enables zero-shot task generalization. International Conference On Learning Representations (2021)
3. Aribandi, V., Tay, Y., Schuster, T., Rao, J., Zheng, H., Mehta, S.V., Zhuang, H., Tran, V., Bahri, D., Ni, J., Gupta, J., Hui, K., Ruder, S., Metzler, D.: Ext5: Towards extreme multi-task scaling for transfer learning. International Conference On Learning Representations (2021)
4. Xu, H., Chen, Y., Du, Y., Shao, N., Wang, Y., Li, H., Yang, Z.: Zeroprompt: Scaling prompt-based pretraining to 1,000 tasks improves zero-shot generalization. arXiv preprint arXiv:2201.06910 (2022)
5. Ouyang, L., Wu, J., Jiang, X., Almeida, D., Wainwright, C.L., Mishkin, P., Zhang, C., Agarwal, S., Slama, K., Ray, A., et al.: Training language models to follow instructions with human feedback. arXiv preprint arXiv:2203.02155 (2022)

6. Fu Yao; Peng, H., Khot, T.: How does GPT obtain its ability? tracing emergent abilities of language models to their sources. Yao Fu's Notion (2022). https://yaofu.notion.site/How-does-GPT-Obtain-its-Ability-Tracing-Emergent-Abilities-of-Language-Models-to-their-Sources-b9a57ac0fcf74f30a1ab9e3e36fa1dc1

7. Xie, T., Wu, C.H., Shi, P., Zhong, R., Scholak, T., Yasunaga, M., Wu, C.S., Zhong, M., Yin, P., Wang, S.I., et al.: Unifiedskg: Unifying and multi-tasking structured knowledge grounding with text-to-text language models. arXiv preprint arXiv:2201.05966 (2022)

8. Sharma, P., Torralba, A., Andreas, J.: Skill induction and planning with latent language. Annual Meeting Of The Association For Computational Linguistics (2021). https://doi.org/10.18653/v1/2022.acl-long.120

9. Suglia, A., Gao, Q., Thomason, J., Thattai, G., Sukhatme, G.: Embodied bert: A transformer model for embodied, language-guided visual task completion. arXiv preprint arXiv: Arxiv-2108.04927 (2021)

10. Pashevich, A., Schmid, C., Sun, C.: Episodic transformer for vision-and-language navigation. In: 2021 IEEE/CVF International Conference on Computer Vision, ICCV 2021, Montreal, QC, Canada, October 10–17, 2021, pp. 15922–15932. IEEE (2021). https://doi.org/10.1109/ICCV48922.2021.01564

11. Gururangan, S., Marasović, A., Swayamdipta, S., Lo, K., Beltagy, I., Downey, D., Smith, N.A.: Don't stop pretraining: Adapt language models to domains and tasks. In: Proceedings of the 58th Annual Meeting of the Association for Computational Linguistics, pp. 8342–8360. Association for Computational Linguistics, Online (2020). https://doi.org/10.18653/v1/2020.acl-main.740. https://aclanthology.org/2020.acl-main.740

12. He, R., Liu, L., Ye, H., Tan, Q., Ding, B., Cheng, L., Low, J., Bing, L., Si, L.: On the effectiveness of adapter-based tuning for pretrained language model adaptation. In: Proceedings of the 59th Annual Meeting of the Association for Computational Linguistics and the 11th International Joint Conference on Natural Language Processing (Volume 1: Long Papers), pp. 2208–2222. Association for Computational Linguistics, Online (2021). https://doi.org/10.18653/v1/2021.acl-long.172. https://aclanthology.org/2021.acl-long.172

13. Dhingra, B., Cole, J.R., Eisenschlos, J.M., Gillick, D., Eisenstein, J., Cohen, W.W.: Time-Aware Language Models as Temporal Knowledge Bases. Transactions of the Association for Computational Linguistics 10, 257–273 (2022). https://doi.org/10.1162/tacl_a_00459

14. Jang, J., Ye, S., Yang, S., Shin, J., Han, J., Kim, G., Choi, S.J., Seo, M.: Towards continual knowledge learning of language models. International Conference On Learning Representations (2021)

15. Jin, X., Zhang, D., Zhu, H., Xiao, W., Li, S.W., Wei, X., Arnold, A.O., Ren, X.: Lifelong pretraining: Continually adapting language models to emerging corpora. BIGSCIENCE (2021). https://doi.org/10.18653/v1/2022.bigscience-1.1

16. Qin, Y., Zhang, J., Lin, Y., Liu, Z., Li, P., Sun, M., Zhou, J.: ELLE: Efficient lifelong pre-training for emerging data. In: Findings of the Association for Computational Linguistics: ACL 2022. Association for Computational Linguistics, Dublin, Ireland (2022). https://doi.org/10.18653/v1/2022.findings-acl.220. https://aclanthology.org/2022.findings-acl.220

17. Liu, P., Yuan, W., Fu, J., Jiang, Z., Hayashi, H., Neubig, G.: Pre-train, prompt, and predict: A systematic survey of prompting methods in natural language processing. arXiv preprint arXiv: Arxiv-2107.13586 (2021)

18. Ding, N., Qin, Y., Yang, G., Wei, F., Yang, Z., Su, Y., Hu, S., Chen, Y., Chan, C., Chen, W., Yi, J., Zhao, W., Wang, X., Liu, Z., Zheng, H., Chen, J., Liu, Y., Tang, J., Li, J., Sun, M.: Delta tuning: A comprehensive study of parameter efficient methods for pre-trained language models. CoRR **abs/2203.06904** (2022). https://doi.org/10.48550/arXiv.2203.06904

19. Li, X.L., Liang, P.: Prefix-tuning: Optimizing continuous prompts for generation. Annual Meeting Of The Association For Computational Linguistics (2021). https://doi.org/10.18653/v1/2021.acl-long.353
20. Lester, B., Al-Rfou, R., Constant, N.: The power of scale for parameter-efficient prompt tuning. In: Proceedings of the 2021 Conference on Empirical Methods in Natural Language Processing, pp. 3045–3059. Association for Computational Linguistics, Online and Punta Cana, Dominican Republic (2021). https://doi.org/10.18653/v1/2021.emnlp-main.243. https://aclanthology.org/2021.emnlp-main.243
21. Houlsby, N., Giurgiu, A., Jastrzebski, S., Morrone, B., De Laroussilhe, Q., Gesmundo, A., Attariyan, M., Gelly, S.: Parameter-efficient transfer learning for NLP. In: K. Chaudhuri, R. Salakhutdinov (eds.) Proceedings of the 36th International Conference on Machine Learning, *Proceedings of Machine Learning Research*, vol. 97, pp. 2790–2799. PMLR (2019). https://proceedings.mlr.press/v97/houlsby19a.html
22. Hu, E.J., yelong shen, Wallis, P., Allen-Zhu, Z., Li, Y., Wang, S., Wang, L., Chen, W.: LoRA: Low-rank adaptation of large language models. In: International Conference on Learning Representations (2022). https://openreview.net/forum?id=nZeVKeeFYf9
23. Li, J., Li, D., Savarese, S., Hoi, S.: Blip-2: Bootstrapping language-image pre-training with frozen image encoders and large language models. arXiv preprint arXiv: Arxiv-2301.12597 (2023)
24. Taori, R., Gulrajani, I., Zhang, T., Dubois, Y., Li, X., Guestrin, C., Liang, P., Hashimoto, T.B.: Stanford alpaca: An instruction-following llama model. https://github.com/tatsu-lab/stanford_alpaca (2023)
25. Zelikman, E., Mu, J., Goodman, N.D., Wu, Y.T.: Star: Self-taught reasoner bootstrapping reasoning with reasoning (2022)
26. Ho, N., Schmid, L., Yun, S.Y.: Large language models are reasoning teachers. arXiv preprint arXiv: Arxiv-2212.10071 (2022)
27. Huang, J., Gu, S.S., Hou, L., Wu, Y., Wang, X., Yu, H., Han, J.: Large language models can self-improve. arXiv preprint arXiv:2210.11610 (2022)
28. Raffel, C., Shazeer, N., Roberts, A., Lee, K., Narang, S., Matena, M., Zhou, Y., Li, W., Liu, P.J.: Exploring the limits of transfer learning with a unified text-to-text transformer. Journal of Machine Learning Research **21**(140), 1–67 (2020). http://jmlr.org/papers/v21/20-074.html
29. Li, Z., Liu, X., Wong, D.F., Chao, L.S., Zhang, M.: Consisttl: Modeling consistency in transfer learning for low-resource neural machine translation. arXiv preprint arXiv:2212.04262 (2022)
30. Weller, O., Seppi, K., Gardner, M.: When to use multi-task learning vs intermediate fine-tuning for pre-trained encoder transfer learning. arXiv preprint arXiv:2205.08124 (2022)
31. Chung, H.W., Hou, L., Longpre, S., Zoph, B., Tay, Y., Fedus, W., Li, E., Wang, X., Dehghani, M., Brahma, S., et al.: Scaling instruction-finetuned language models. arXiv preprint arXiv:2210.11416 (2022)
32. Longpre, S., Hou, L., Vu, T., Webson, A., Chung, H.W., Tay, Y., Zhou, D., Le, Q.V., Zoph, B., Wei, J., Roberts, A.: The flan collection: Designing data and methods for effective instruction tuning. ARXIV.ORG (2023). https://doi.org/10.48550/arXiv.2301.13688
33. Iyer, S., Lin, X.V., Pasunuru, R., Mihaylov, T., Simig, D., Yu, P., Shuster, K., Wang, T., Liu, Q., Koura, P.S., et al.: Opt-iml: Scaling language model instruction meta learning through the lens of generalization. arXiv preprint arXiv:2212.12017 (2022)
34. Glaese, A., McAleese, N., Trębacz, M., Aslanides, J., Firoiu, V., Ewalds, T., Rauh, M., Weidinger, L., Chadwick, M., Thacker, P., et al.: Improving alignment of dialogue agents via targeted human judgements. arXiv preprint arXiv:2209.14375 (2022)
35. Zhang, G., Shi, Y., Liu, R., Yuan, R., Li, Y., Dong, S., Shu, Y., Li, Z., Wang, Z., Lin, C., et al.: Chinese open instruction generalist: A preliminary release. arXiv preprint arXiv:2304.07987 (2023)

36. Wang, Y., Mishra, S., Alipoormolabashi, P., Kordi, Y., Mirzaei, A., Naik, A., Ashok, A., Dhanasekaran, A.S., Arunkumar, A., Stap, D., et al.: Super-naturalinstructions: Generalization via declarative instructions on 1600+ nlp tasks. In: Proceedings of the 2022 Conference on Empirical Methods in Natural Language Processing, pp. 5085–5109 (2022)

37. Muennighoff, N., Wang, T., Sutawika, L., Roberts, A., Biderman, S., Scao, T.L., Bari, M.S., Shen, S., Yong, Z.X., Schoelkopf, H., et al.: Crosslingual generalization through multitask finetuning. arXiv preprint arXiv:2211.01786 (2022)

38. Kaplan, J., McCandlish, S., Henighan, T., Brown, T.B., Chess, B., Child, R., Gray, S., Radford, A., Wu, J., Amodei, D.: Scaling laws for neural language models. arXiv preprint arXiv:2001.08361 (2020)

39. Wei, J., Tay, Y., Bommasani, R., Raffel, C., Zoph, B., Borgeaud, S., Yogatama, D., Bosma, M., Zhou, D., Metzler, D., Chi, E.H., Hashimoto, T., Vinyals, O., Liang, P., Dean, J., Fedus, W.: Emergent abilities of large language models. arXiv preprint arXiv: Arxiv-2206.07682 (2022)

40. Brown, T., Mann, B., Ryder, N., Subbiah, M., Kaplan, J.D., Dhariwal, P., Neelakantan, A., Shyam, P., Sastry, G., Askell, A., et al.: Language models are few-shot learners. Advances in neural information processing systems 33, 1877–1901 (2020)

41. Liu, G.K.M.: Perspectives on the social impacts of reinforcement learning with human feedback. arXiv preprint arXiv:2303.02891 (2023)

42. Chowdhery, A., Narang, S., Devlin, J., Bosma, M., Mishra, G., Roberts, A., Barham, P., Chung, H.W., Sutton, C., Gehrmann, S., et al.: Palm: Scaling language modeling with pathways. arXiv preprint arXiv:2204.02311 (2022)

43. Wang, Y., Kordi, Y., Mishra, S., Liu, A., Smith, N.A., Khashabi, D., Hajishirzi, H.: Self-instruct: Aligning language model with self generated instructions. arXiv preprint arXiv: Arxiv-2212.10560 (2022)

44. Xu, C., Guo, D., Duan, N., McAuley, J.: Baize: An open-source chat model with parameter-efficient tuning on self-chat data. arXiv preprint arXiv:2304.01196 (2023)

45. Peng, B., Li, C., He, P., Galley, M., Gao, J.: Instruction tuning with GPT-4. arXiv preprint arXiv:2304.03277 (2023)

46. Zhou, W., Jiang, Y.E., Wilcox, E., Cotterell, R., Sachan, M.: Controlled text generation with natural language instructions (2023)

47. Honovich, O., Scialom, T., Levy, O., Schick, T.: Unnatural instructions: Tuning language models with (almost) no human labor. arXiv preprint arXiv: Arxiv-2212.09689 (2022)

48. Ding, N., Chen, Y., Xu, B., Qin, Y., Zheng, Z., Hu, S., Liu, Z., Sun, M., Zhou, B.: Enhancing chat language models by scaling high-quality instructional conversations. arXiv preprint arXiv:2305.14233 (2023)

49. Li, G., Hammoud, H.A.A.K., Itani, H., Khizbullin, D., Ghanem, B.: Camel: Communicative agents for "mind" exploration of large scale language model society. arXiv preprint arXiv:2303.17760 (2023)

50. Yue, X., Qu, X., Zhang, G., Fu, Y., Huang, W., Sun, H., Su, Y., Chen, W.: Mammoth: Building math generalist models through hybrid instruction tuning. arXiv preprint arXiv: 2309.05653 (2023)

51. Chen, W., Ma, X., Wang, X., Cohen, W.W.: Program of thoughts prompting: Disentangling computation from reasoning for numerical reasoning tasks. arXiv preprint arXiv:2211.12588 (2022)

52. Wang, Z.M., Peng, Z., Que, H., Liu, J., Zhou, W., Wu, Y., Guo, H., Gan, R., Ni, Z., Zhang, M., Zhang, Z., Ouyang, W., Xu, K., Chen, W., Fu, J., Peng, J.: Rolellm: Benchmarking, eliciting, and enhancing role-playing abilities of large language models. arXiv preprint arXiv: 2310.00746 (2023)

53. Jiang, D., Li, Y., Zhang, G., Huang, W., Lin, B.Y., Chen, W.: Tigerscore: Towards building explainable metric for all text generation tasks. arXiv preprint arXiv:2310.00752 (2023)

54. Dong, H., Xiong, W., Goyal, D., Pan, R., Diao, S., Zhang, J., Shum, K., Zhang, T.: Raft: Reward ranked finetuning for generative foundation model alignment. ARXIV.ORG (2023). https://doi.org/10.48550/arXiv.2304.06767

55. Yuan, Z., Yuan, H., Tan, C., Wang, W., Huang, S., Huang, F.: Rrhf: Rank responses to align language models with human feedback without tears. ARXIV.ORG (2023). https://doi.org/10.48550/arXiv.2304.05302

56. Rafailov, R., Sharma, A., Mitchell, E., Ermon, S., Manning, C.D., Finn, C.: Direct preference optimization: Your language model is secretly a reward model. arXiv preprint arXiv: 2305.18290 (2023)

57. Sun, Z., Shen, Y., Zhou, Q., Zhang, H., Chen, Z., Cox, D.D., Yang, Y., Gan, C.: Principle-driven self-alignment of language models from scratch with minimal human supervision. ARXIV.ORG (2023). https://doi.org/10.48550/arXiv.2305.03047

58. Kang, J., Luo, H., Zhu, Y., Glass, J., Cox, D., Ritter, A., Feris, R., Karlinsky, L.: Self-specialization: Uncovering latent expertise within large language models. arXiv preprint arXiv: 2310.00160 (2023)

59. Song, F., Yu, B., Li, M., Yu, H., Huang, F., Li, Y., Wang, H.: Preference ranking optimization for human alignment. arXiv preprint arXiv: 2306.17492 (2023)

60. Liu, R., Yang, R., Jia, C., Zhang, G., Zhou, D., Dai, A.M., Yang, D., Vosoughi, S.: Training socially aligned language models in simulated human society (2023)

61. Liu, H., Sferrazza, C., Abbeel, P.: Chain of hindsight aligns language models with feedback. arXiv preprint arXiv: Arxiv-2302.02676 (2023)

62. Zhao, Y., Khalman, M., Joshi, R., Narayan, S., Saleh, M., Liu, P.J.: Calibrating sequence likelihood improves conditional language generation. International Conference on Learning Representations (2022). https://doi.org/10.48550/arXiv.2210.00045

63. Schulman, J., Wolski, F., Dhariwal, P., Radford, A., Klimov, O.: Proximal policy optimization algorithms. arXiv preprint arXiv:1707.06347 (2017)

64. Schick, T., Dwivedi-Yu, J., Dessì, R., Raileanu, R., Lomeli, M., Zettlemoyer, L., Cancedda, N., Scialom, T.: Toolformer: Language models can teach themselves to use tools. CoRR **abs/2302.04761** (2023). https://doi.org/10.48550/arXiv.2302.04761

65. Dalvi, B., Tafjord, O., Clark, P.: Towards teachable reasoning systems: Using a dynamic memory of user feedback for continual system improvement. Conference On Empirical Methods In Natural Language Processing (2022)

66. Robins, A.: Catastrophic forgetting, rehearsal and pseudorehearsal. Connection Science **7**(2), 123–146 (1995). https://doi.org/10.1080/09540099550039318

67. Wu, T., Caccia, M., Li, Z., Li, Y.F., Qi, G., Haffari, G.: Pretrained language model in continual learning: A comparative study. In: International Conference on Learning Representations (2022). https://openreview.net/forum?id=figzpGMrdD

68. McCloskey, M., Cohen, N.J.: Catastrophic interference in connectionist networks: The sequential learning problem. In: Psychology of learning and motivation, vol. 24, pp. 109–165. Elsevier (1989)

69. Delange, M., Aljundi, R., Masana, M., Parisot, S., Jia, X., Leonardis, A., Slabaugh, G., Tuytelaars, T.: A continual learning survey: Defying forgetting in classification tasks. IEEE Transactions on Software Engineering **PP** (2021). https://doi.org/10.1109/TPAMI.2021.3057446

70. Biesialska, M., Biesialska, K., Costa-jussà, M.R.: Continual lifelong learning in natural language processing: A survey. In: Proceedings of the 28th International Conference on Computational Linguistics, pp. 6523–6541. International Committee on Computational Linguistics, Barcelona, Spain (Online) (2020). https://doi.org/10.18653/v1/2020.coling-main.574. https://aclanthology.org/2020.coling-main.574

71. Kirkpatrick, J., Pascanu, R., Rabinowitz, N., Veness, J., Desjardins, G., Rusu, A.A., Milan, K., Quan, J., Ramalho, T., Grabska-Barwinska, A., et al.: Overcoming catastrophic forgetting in neural networks. Proceedings of the national academy of sciences **114**(13), 3521–3526 (2017)

72. Chen, S., Hou, Y., Cui, Y., Che, W., Liu, T., Yu, X.: Recall and learn: Fine-tuning deep pretrained language models with less forgetting. In: Proceedings of the 2020 Conference on Empirical Methods in Natural Language Processing (EMNLP), pp. 7870–7881 (2020)

73. Li, X., Wang, Z., Li, D., Khan, L., Thuraisingham, B.: Lpc: A logits and parameter calibration framework for continual learning. In: Findings of the Association for Computational Linguistics: EMNLP 2022, pp. 7142–7155 (2022)

74. Li, D., Chen, Z., Cho, E., Hao, J., Liu, X., Xing, F., Guo, C., Liu, Y.: Overcoming catastrophic forgetting during domain adaptation of seq2seq language generation. In: Proceedings of the 2022 Conference of the North American Chapter of the Association for Computational Linguistics: Human Language Technologies, pp. 5441–5454 (2022)

75. He, T., Liu, J., Cho, K., Ott, M., Liu, B., Glass, J., Peng, F.: Analyzing the forgetting problem in pretrain-finetuning of open-domain dialogue response models. In: Proceedings of the 16th Conference of the European Chapter of the Association for Computational Linguistics: Main Volume, pp. 1121–1133 (2021)

76. Cappellazzo, U., Falavigna, D., Brutti, A.: Exploring the joint use of rehearsal and knowledge distillation in continual learning for spoken language understanding. ARXIV.ORG (2022). https://doi.org/10.48550/arXiv.2211.08161

77. Pfeiffer, J., Ruder, S., Vulić, I., Ponti, E.M.: Modular deep learning. arXiv preprint arXiv: Arxiv-2302.11529 (2023)

78. Lee, K., Han, W., Hwang, S.w., Lee, H., Park, J., Lee, S.W.: Plug-and-play adaptation for continuously-updated qa. In: Findings of the Association for Computational Linguistics: ACL 2022, pp. 438–447 (2022)

79. Wortsman, M., Ramanujan, V., Liu, R., Kembhavi, A., Rastegari, M., Yosinski, J., Farhadi, A.: Supermasks in superposition. Neural Information Processing Systems (2020)

80. Howard, J., Ruder, S.: Universal language model fine-tuning for text classification. In: Proceedings of the 56th Annual Meeting of the Association for Computational Linguistics (Volume 1: Long Papers), pp. 328–339. Association for Computational Linguistics, Melbourne, Australia (2018). https://doi.org/10.18653/v1/P18-1031. https://aclanthology.org/P18-1031

81. Zaken, E.B., Goldberg, Y., Ravfogel, S.: Bitfit: Simple parameter-efficient fine-tuning for transformer-based masked language-models. In: S. Muresan, P. Nakov, A. Villavicencio (eds.) Proceedings of the 60th Annual Meeting of the Association for Computational Linguistics (Volume 2: Short Papers), ACL 2022, Dublin, Ireland, May 22–27, 2022, pp. 1–9. Association for Computational Linguistics (2022). https://doi.org/10.18653/v1/2022.acl-short.1

82. Guo, D., Rush, A.M., Kim, Y.: Parameter-efficient transfer learning with diff pruning. In: C. Zong, F. Xia, W. Li, R. Navigli (eds.) Proceedings of the 59th Annual Meeting of the Association for Computational Linguistics and the 11th International Joint Conference on Natural Language Processing, ACL/IJCNLP 2021, (Volume 1: Long Papers), Virtual Event, August 1–6, 2021, pp. 4884–4896. Association for Computational Linguistics (2021). https://doi.org/10.18653/v1/2021.acl-long.378

83. Voita, E., Talbot, D., Moiseev, F., Sennrich, R., Titov, I.: Analyzing multi-head self-attention: Specialized heads do the heavy lifting, the rest can be pruned. In: Proceedings of the 57th Annual Meeting of the Association for Computational Linguistics, pp. 5797–5808. Association for Computational Linguistics, Florence, Italy (2019). https://doi.org/10.18653/v1/P19-1580. https://aclanthology.org/P19-1580

84. Rebuffi, S.A., Bilen, H., Vedaldi, A.: Learning multiple visual domains with residual adapters. In: I. Guyon, U.V. Luxburg, S. Bengio, H. Wallach, R. Fergus, S. Vishwanathan, R. Garnett

(eds.) Advances in Neural Information Processing Systems, vol. 30. Curran Associates, Inc. (2017). https://proceedings.neurips.cc/paper/2017/file/e7b24b112a44fdd9ee93bdf998c6ca0e-Paper.pdf

85. Karimi Mahabadi, R., Henderson, J., Ruder, S.: Compacter: Efficient low-rank hypercomplex adapter layers. Advances in Neural Information Processing Systems **34**, 1022–1035 (2021)

86. He, J., Zhou, C., Ma, X., Berg-Kirkpatrick, T., Neubig, G.: Towards a unified view of parameter-efficient transfer learning. In: International Conference on Learning Representations (2022). https://openreview.net/forum?id=0RDcd5Axok

87. Zhang, A., Tay, Y., Zhang, S., Chan, A., Luu, A.T., Hui, S.C., Fu, J.: Beyond fully-connected layers with quaternions: Parameterization of hypercomplex multiplications with $1/n$ parameters. arXiv preprint arXiv:2102.08597 (2021)

88. Pfeiffer, J., Rücklé, A., Poth, C., Kamath, A., Vulić, I., Ruder, S., Cho, K., Gurevych, I.: Adapterhub: A framework for adapting transformers. In: Proceedings of the 2020 Conference on Empirical Methods in Natural Language Processing (EMNLP 2020): Systems Demonstrations, pp. 46–54. Association for Computational Linguistics, Online (2020). https://www.aclweb.org/anthology/2020.emnlp-demos.7

89. Zhou, W., Xu, C., McAuley, J.: Efficiently tuned parameters are task embeddings. In: Proceedings of the 2022 Conference on Empirical Methods in Natural Language Processing, pp. 5007–5014. Association for Computational Linguistics, Abu Dhabi, United Arab Emirates (2022). https://aclanthology.org/2022.emnlp-main.334

90. Vu, T., Lester, B., Constant, N., Al-Rfou', R., Cer, D.: SPoT: Better frozen model adaptation through soft prompt transfer. In: Proceedings of the 60th Annual Meeting of the Association for Computational Linguistics (Volume 1: Long Papers), pp. 5039–5059. Association for Computational Linguistics, Dublin, Ireland (2022). https://doi.org/10.18653/v1/2022.acl-long.346. https://aclanthology.org/2022.acl-long.346

91. Wang, R., Tang, D., Duan, N., Wei, Z., Huang, X., Ji, J., Cao, G., Jiang, D., Zhou, M.: K-adapter: Infusing knowledge into pre-trained models with adapters. In: C. Zong, F. Xia, W. Li, R. Navigli (eds.) Findings of the Association for Computational Linguistics: ACL/IJCNLP 2021, Online Event, August 1–6, 2021, *Findings of ACL*, vol. ACL/IJCNLP 2021, pp. 1405–1418. Association for Computational Linguistics (2021). https://doi.org/10.18653/v1/2021.findings-acl.121

92. Liang, A., Singh, I., Pertsch, K., Thomason, J.: Transformer adapters for robot learning. In: CoRL 2022 Workshop on Pre-training Robot Learning (2022). https://openreview.net/forum?id=H--wvRYBmF

93. Liang, P.: Semi-supervised learning for natural language (2005)

94. Chapelle, O., Schölkopf, B., Zien, A. (eds.): Semi-Supervised Learning. The MIT Press (2006). http://dblp.uni-trier.de/db/books/collections/CSZ2006.html

95. Zhou, Z.H.: A brief introduction to weakly supervised learning. National Science Review **5**(1), 44–53 (2017). https://doi.org/10.1093/nsr/nwx106.

96. Li, J., Li, D., Xiong, C., Hoi, S.: Blip: Bootstrapping language-image pre-training for unified vision-language understanding and generation. In: ICML (2022)

97. Fu, Y., Peng, H., Ou, L., Sabharwal, A., Khot, T.: Specializing smaller language models towards multi-step reasoning. arXiv preprint arXiv: Arxiv-2301.12726 (2023)

98. Wang, X., Wei, J., Schuurmans, D., Le, Q., Chi, E., Narang, S., Chowdhery, A., Zhou, D.: Self-consistency improves chain of thought reasoning in language models. arXiv preprint arXiv: Arxiv-2203.11171 (2022)

99. Parisi, A., Zhao, Y., Fiedel, N.: Talm: Tool augmented language models. arXiv preprint arXiv: Arxiv-2205.12255 (2022)

100. Shridhar, K., Stolfo, A., Sachan, M.: Distilling multi-step reasoning capabilities of large language models into smaller models via semantic decompositions. arXiv preprint arXiv: Arxiv-2212.00193 (2022)

Reinforcement Learning

7

Zekun Wang, Guangzheng Xiong and Jie Fu

With the advent of reinforcement learning from environment feedback (RLEF) [1, 2], human feedback (RLHF) [3, 4], and artificial intelligence feedback (RLAIF) [5, 6], there has been a surge in reinforcement learning for natural language processing. In the context of Reinforcement Learning (RL), generated text or tool-use can be treated as an "action", while both surrounding text and non-text inputs can be viewed as "observations". The "reward" for the language model mainly includes human preference feedback [3, 4] and affordance grounding feedback [1, 2]. Therefore, our survey focuses on examining the interaction of language models with environments and humans. We also refer the readers to [7] for more information. In the following part, we will provide a brief introduction to RL for Interactive NLP (iNLP), focusing on two aspects: (1) feedback loop, and (2) reward modeling. Additionally, we introduce a special scenario, reasoning, to explore how RL can enhance the reasoning capabilities of LLMs (Fig. 7.1).

Z. Wang (✉)
Beihang University, Beijing, China
e-mail: zenmoore@buaa.edu.cn

G. Xiong
Hong Kong University of Science and Technology, Hong Kong, China

J. Fu
Beijing, China
e-mail: jiefu@ust.hk

Fig. 7.1 Reinforcement learning

7.1 Feedback Loop

There are two main approaches to build RL frameworks with language models: online reinforcement learning and offline reinforcement learning. In these approaches, language models serve as RL agents or policies. As shown in Fig. 7.2, online reinforcement learning involves updating models with real-time synchronous rewards during training [8, 9], whereas offline reinforcement learning leverages rewards derived from a static data source [10]. The choice between online RL and offline RL depends primarily on practical scenarios. Generally, online RL is more suitable for LM-environment interaction as obtaining feedback from environments is a more automated process [7]. On the other hand, offline RL is more practical for LM-human interaction since human feedback may not always be readily available [11]. Consequently, while there are studies on online RL for LM-human interaction [12] and offline RL for LM-environment interaction [7], our survey mainly focuses on online RL for LM-environment interaction and offline RL for LM-human interaction.[1]

Online RL. Online RL empowers language models to tackle specific tasks by learning from real-time feedback provided by external entities such as the environment or more intricate reward models [8, 15, 16]. This feedback, similar to other RL models, is typically used as rewards for the online RL models. Standard online RL allows language models to learn from immediate rewards after each step. These dense rewards are often scalar values [8]

---

[1] For the counterpart, we refer readers to [12] and [7].

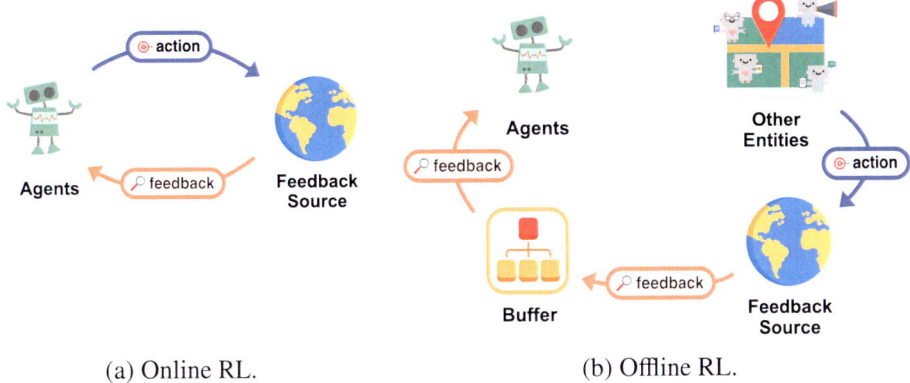

(a) Online RL.   (b) Offline RL.

Fig. 7.2 Online RL uses the actions of the training agent to generate feedback in real-time, or synchronously [13]. This means that the agent interacts with the environment, takes actions, and immediately learns from the consequences of those actions. On the other hand, Offline RL, also known as batch RL, uses feedback from actions that were generated by other entities, not the training agent itself [14]. This feedback is calculated and stored in a buffer, allowing the training agent to learn from it asynchronously

or Boolean values [16] derived from the agent's state, observation, and outputs. For example, in tasks that necessitate continuous movement and multiple actions, RL agents can learn from real-time feedback during the process. [17] sample multiple discrete time points during the agent's continuous movement and provides real-time binary feedback (positive or negative) based on the relative progress of the task. Moreover, in tasks with a long horizon, where each task involves multiple actions, sparse rewards can be employed to avoid the inefficiency of dense rewards. These rewards are given once per episode instead of after each step, as seen in studies like [1, 18, 19]. For large-scale, production-ready online services, language models can be trained with buffered feedback and re-deployed periodically to minimize learning costs [20].

Offline RL. According to recent research [4, 10], offline reinforcement learning agents can utilize PLMs as the policy and share the same reward mechanism as online reinforcement agents while using different learning procedures. Specifically, they asynchronously apply feedback by fine-tuning the PLM rather than updating it in real time. During such a procedure, offline reinforcement learning with language models can utilize several algorithms to align language models with human preference. These algorithms include Proximal Policy Optimization (PPO) [4, 21], Direct Preference Optimization (DPO) [22], Advantage Actor-Critic (A2C) [23, 24], Implicit Language Q-Learning (ILQL) [25], Trust Region Policy Optimization (TRPO) [26], Natural Language Policy Optimization (NLPO) [27], among others.[2] These RL methods usually introduce alignment costs, resulting in performance degradation for other tasks, known as alignment tax [4, 28]. To address this issue,

---

[2] Please note that these algorithms are not specific to offline RL.

[29] employ offline reinforcement learning during model pre-training, leading to better alignment while preserving superior performance. RAFT [30] demonstrates that employing early stopping can find a more favorable balance between text quality and preference reward. A significant shift from the point-wise reward algorithms like RAFT came with [22], which introduced Direct Preference Optimization (DPO), utilizing pair-wise preference learning. Building on this, [31] proposes the Statistical Rejection Sampling Optimization (RSO) to improve sampling from the optimal policy. Meng et al. [32] introduces SimPO, offering a more efficient approach by using average log probability as an implicit reward. Additionally, [33] developed Group Robust Preference Optimization (GRPO), focusing on aligning language models with individual group preferences robustly. Furthermore, a distinct branch of research explores pre-training transformer architectures on decision-making sequences [34], extending this approach to both pure textual and multimodal environments [35–38]. Another line of works focus on improving the optimization algorithms such as computational efficiency [30], robustness [39], and training stability [27].

7.2   Reward Modeling

Reinforcement learning agents are trained using rewards that are computed based on feedback from external entities, which primarily include interactive environments, and humans. Various feedback sources and mechanism contribute to the development of distinct reward models.

RL from Environment Feedback. As discussed in Sect. 2.4, reinforcement learning from environment feedback (RLEF) facilitates affordance grounding of language models [1]. For tasks with clear and easy-to-compare evaluation metrics, such as shopping and object arrangement tasks [9, 15, 18], reinforcement learning agents can be optimized using absolute reward evaluated by corresponding environments and other evaluation models. Typically, reinforcement learning agents can be trained with binary reward functions, receiving positive reward if they conducted correct actions, and negative reward otherwise [16, 40]. Further, [41] involve evaluators that map visual environment and natural language descriptions to scale reward. Additionally, reinforcement learning agents can also receive rewards that are more complex and specifically handcrafted. For example, [18] involve a carefully-designed reward function that considers various attributes of the chosen product and text-based instruction-product similarity, thereby reducing the need for human-in-the-loop evaluation. For tasks where it's hard to build such efficient and sensitive evaluation metrics that maps agents' states and actions to absolute scores and establish totally-ordered relations among all actions, such as text generation tasks, reinforcement learning agents can be optimized using comparative reward, such as the relative relationship between generated actions, where the policy is rewarded whether the generated action is better than the previous one [42]. In addition to considering the relative relationship between actions within the same state and input, reinforcement learning agents in long-horizon tasks can also be rewarded based on the relative progress made in reducing the distance between the current state and the target state [19].

RL from Human Feedback. Reinforcement Learning from Human Feedback (RLHF) is receiving increasing attention as a crucial post-training procedure for Large Language Models (LLMs) [4, 11, 24, 43]. For general-purpose text generation tasks, which is an open-ended task without deterministic answers, reward models are often preferred than handcrafted reward functions due to their higher sensitivity and better performance [10, 15, 44]. Specifically, the reward models for offline RL are trained with a set of human feedback that is cost-effective to gather, and subsequently are used to reward reinforcement learning agents [5, 45]. Similar to environment feedback, the reward mechanism of human feedback also varies among tasks. In embodied tasks, the reward model can observe the agent's history and provide binary feedback [17]. In natural language generation tasks, reward models simulate human annotators to label outputs and provide numeric scores [4, 20, 46]. In more complicated tasks like reasoning tasks, reward models learn to follow specific rules and provide more complex and structural feedback such as Salmon [47] and Sparrow [24]. Note that feedback isn't limited to simple binary or numeric forms. It can also be generated in natural languages. Language models can process this type of feedback, make adjustments, and ultimately produce corrected outputs [48–53]. Further, the use of a reward model trained on human preference data can also be considered as "RL from Model Feedback". For example, [5] propose "RL from AI Feedback" (RLAIF), where agents can self-adjust their outputs by providing self-feedback and self-prompts, resulting in higher-quality outputs. These AI feedback mechanisms essentially provide indirect human feedback as they are trained on data annotated by humans. The primary motivation behind AI feedback mechanisms is the significant cost of collecting real human preferences. As a result, there's a line of research focused on differentiating models' outputs to automatically synthesize preference-contrastive response pairs. For example, ALMoST [54] differentiates models' outputs based on the model size and the number of in-context examples. Responses from larger models with more demonstrations are considered better aligned than those from smaller models with fewer demonstrations. RLCD [55] creates contrasts in input prefixes, such as "Harmless" v.s. "Harmful", to guide the model in generating contrastive responses based on the keyword instruction. Contrastive Post-Training [56] employs better and worse LLMs to generate corresponding higher and lower quality responses, thereby creating preference-contrastive pairs. A significant advancement in the designing of reward modeling is the development of Process Reward Models (PRO), which aims to enhance the reasoning capabilities in language models. Initially proposed by [57] as process-supervised reward models (PRMs), this approach breaks down the rewards of the chaining [58] reasoning process, offering a more nuanced alternative to traditional outcome-supervised reward models (ORMs) only for the final results. Subsequent work by [59, 60] has expanded PRM's applicability to a wider range of tasks and demonstrated its potential in mathematical reasoning. Recent research has also explored PRM's effectiveness in code generation for reasoning [61, 62], and the automatic data curation for such a paradigm [63], showcasing the versatility and growing importance of this approach in improving LM performance across various domains.

7.3 RL for Reasoning

Reinforcement Learning offers a robust framework for enhancing the reasoning capabilities of Large Language Models by optimizing their decision-making processes within a formalized reasoning environment. By modeling reasoning as a Markov Decision Process (MDP), RL algorithms such as Proximal Policy Optimization (PPO) [21], Group-Relative Policy Optimization (GRPO) [64], and Step-level Direct Preference Optimization (Step-DPO) [65] can be applied to improve reasoning performance. These methods rely on verifiers to provide reward signals, evaluating either the correctness of each reasoning step or the full response. Verifiers, categorized as Outcome-based Reward Models (ORMs), Process-based Reward Models (PRMs), Generative Verifiers, and Oracle Verifiers, play a critical role in guiding the RL process.

7.3.1 Reasoning as an MDP

Reasoning in LLMs can be formalized as an MDP, defined by the tuple $\tau = ((s_0, a_0, r_0), \ldots, (s_T, a_T, r_T))$, where s_t represents the state (e.g., the current reasoning context or prompt), a_t denotes the action (e.g., generating a reasoning step or token), and r_t is the reward received at time step t. The state transition function $P(s_{t+1} \mid s_t, a_t)$ governs the progression of reasoning, either through the model's internal generation or interactions with external environments. The objective is to maximize the expected cumulative reward, $\max \mathbb{E}_{\tau \sim P(\tau \mid s_0, \pi)} \left[\sum_{t=1}^{T} r_t \right]$, where π is the policy (the LLM's reasoning strategy). This framework enables RL to optimize reasoning by refining the policy based on reward signals.

7.3.2 RL Methods for Reasoning Optimization

Several RL algorithms have been tailored to improve the reasoning capabilities of LLMs by refining their ability to generate accurate and coherent reasoning trajectories. These methods leverage feedback from verifiers to optimize decision-making, focusing on either step-wise improvements or overall solution quality. Techniques such as Proximal Policy Optimization (PPO) [21], Group-Relative Policy Optimization (GRPO) [64], and Step-level Direct Preference Optimization (Step-DPO) [65] employ distinct strategies to enhance reasoning performance. Each method relies on reward signals to guide the model toward better reasoning outcomes, making them critical for tasks like mathematical problem-solving [66].

Proximal Policy Optimization (PPO) [21]. PPO is a widely used RL algorithm that optimizes the policy of a LLM for reasoning tasks by balancing exploration and exploitation through constrained policy updates. It maximizes the expected cumulative reward while ensuring training stability using a clipped objective function. The PPO loss is defined as:

$$\mathcal{L}_{PPO}(\theta) = -\mathbb{E}\left[\min\left(\frac{\pi_\theta(a_t \mid s_t)}{\pi_{\theta_{ref}}(a_t \mid s_t)}\hat{A}_t, \text{clip}\left(\frac{\pi_\theta(a_t \mid s_t)}{\pi_{\theta_{ref}}(a_t \mid s_t)}, 1 - \epsilon, 1 + \epsilon\right)\hat{A}_t\right)\right],$$

where, $\hat{A}_t$ is the advantage function, and ϵ (typically 0.2) defines the clipping range. The PPO algorithm requires four distinct models to be loaded into GPU memory: a reference model $\pi_{\theta_{ref}}$, a policy model π_θ, a value model, and a reward model (for calculating the advantage).

Group-Relative Policy Optimization (GRPO) [64]. GRPO, introduced in DeepSeek-R1 [64], enhances LLM reasoning by eliminating the need for a separate value model and instead using group-based advantage estimation. It generates multiple reasoning trajectories (e.g., 8) for a given prompt, evaluates their rewards, and computes advantages relative to the group mean. The GRPO loss function is:

$$\mathcal{L}_{GRPO}(\theta) = -\frac{1}{G}\sum_{i=1}^{G}\frac{1}{|\tau_i|}\sum_{t=1}^{|\tau_i|}\Bigg[$$
$$\min\left(\frac{\pi_\theta(a_{i,t} \mid s_{i,t})}{\pi_{\theta_{ref}}(a_{i,t} \mid s_{i,t})}\hat{A}_{i,t}, \text{clip}\left(\frac{\pi_\theta(a_{i,t} \mid s_{i,t})}{\pi_{\theta_{ref}}(a_{i,t} \mid s_{i,t})}, 1 - \epsilon, 1 + \epsilon\right)\hat{A}_{i,t}\right)\Bigg]$$

where G is the number of trajectories, $|\tau_i|$ is the length of trajectory τ_i, and $\hat{A}_{i,t} = \frac{r_i - \text{mean}(\mathbf{r})}{\text{std}(\mathbf{r})}$ is the standardized advantage. GRPO reduces GPU memory usage by eliminating the value model, making it deployment-friendly.

Step-level Direct Preference Optimization (Step-DPO) [65]. Step-DPO optimizes reasoning by constructing preference pairs at the step level, focusing on contrasting actions within the same trajectory prefix. Contrasting signals can be derived using methods like Tree-of-Thought [67] to estimate outcome rewards or by employing advanced models (e.g., GPT-4o) to identify incorrect steps. Iterative DPO [68] further refines this process by repeatedly updating preferences, enhancing reasoning accuracy.

7.3.3 Verifiers for Reward Signals

The effectiveness of RL for reasoning hinges on verifiers, which provide reward signals to evaluate the correctness of reasoning steps or full responses. Four types of verifiers are commonly used:

Outcome-based Reward Models (ORMs). ORMs assess the correctness of the final response, assigning rewards based on outcomes. They can be trained using pointwise loss:

$$\mathcal{L}_{orm}(\theta) = \mathbb{E}_{x,y \sim \mathcal{D}}\left[c \log \sigma(r_\theta(x, y)) + (1 - c) \log(1 - \sigma(r_\theta(x, y)))\right]$$

where $c \in \{0, 1\}$ indicates correctness and $\sigma()$ indicates sigmoid function, or pairwise loss based on the Bradley-Terry model:

$$\mathcal{L}_{\text{orm}}(\theta) = -\mathbb{E}_{x, y_w, y_l \sim \mathcal{D}} \left[\log \sigma \left(r_\theta(x, y_w) - r_\theta(x, y_l) \right) \right]$$

ORMs are effective for tasks with clear ground-truth answers but may overlook intermediate reasoning quality.

Process-based Reward Models (PRMs). PRMs evaluate the correctness of intermediate reasoning steps, using a loss function such as:

$$\mathcal{L}_{\text{prm}}(\theta) = \mathbb{E}_{x, y \sim \mathcal{D}} \left[-\frac{1}{T} \sum_{t=1}^{T} \left(c_t \log \sigma(r_\theta(x, y_{\leq t})) + (1 - c_t) \log(1 - \sigma(r_\theta(x, y_{\leq t}))) \right) \right]$$

PRMs often rely on human annotations or Monte Carlo Tree Search (MCTS) [69] to estimate step correctness.

Generative Verifiers. These verifiers are based on LLM-as-a-judge [70], generating natural language feedback or evaluating responses based on protocols like pairwise comparison or rating. Typical generative verifiers are less precise than reward models but can be combined with them (e.g., Generative Reward Models [71]) to enhance feedback quality. Their flexibility makes them suitable for both rewarding and refinement.

Oracle Verifiers. Oracle verifiers rely on ground-truth-based evaluation, such as code execution or exact matching, requiring no training. They are highly reliable for structured environments (e.g., mathematical proofs in Lean or code in HumanEval [72]) but are limited to tasks with programmatically verifiable, predefined correctness criteria. Loong [73] is a representative project that leverages program-based verifiers for reasoning in diverse disciplines.

References

1. Ahn, M., Brohan, A., Brown, N., Chebotar, Y., Cortes, O., David, B., Finn, C., Gopalakrishnan, K., Hausman, K., Herzog, A., et al.: Do as i can, not as i say: Grounding language in robotic affordances. arXiv preprint arXiv:2204.01691 (2022)
2. Chen, B., Xia, F., Ichter, B., Rao, K., Gopalakrishnan, K., Ryoo, M.S., Stone, A., Kappler, D.: Open-vocabulary queryable scene representations for real world planning. arXiv preprint arXiv: Arxiv-2209.09874 (2022)
3. Christiano, P., Leike, J., Brown, T.B., Martic, M., Legg, S., Amodei, D.: Deep reinforcement learning from human preferences. arXiv preprint arXiv: Arxiv-1706.03741 (2017)
4. Ouyang, L., Wu, J., Jiang, X., Almeida, D., Wainwright, C.L., Mishkin, P., Zhang, C., Agarwal, S., Slama, K., Ray, A., et al.: Training language models to follow instructions with human feedback. arXiv preprint arXiv:2203.02155 (2022)
5. Bai, Y., Kadavath, S., Kundu, S., Askell, A., Kernion, J., Jones, A., Chen, A., Goldie, A., Mirhoseini, A., McKinnon, C., et al.: Constitutional ai: Harmlessness from ai feedback. arXiv preprint arXiv:2212.08073 (2022)
6. Liu, R., Zhang, G., Feng, X., Vosoughi, S.: Aligning generative language models with human values. In: Findings of the Association for Computational Linguistics: NAACL 2022, pp. 241–

252. Association for Computational Linguistics, Seattle, United States (2022). https://doi.org/ 10.18653/v1/2022.findings-naacl.18. https://aclanthology.org/2022.findings-naacl.18

7. Yang, S., Nachum, O., Du, Y., Wei, J., Abbeel, P., Schuurmans, D.: Foundation models for decision making: Problems, methods, and opportunities. arXiv preprint arXiv:2303.04129 (2023)

8. Carta, T., Romac, C., Wolf, T., Lamprier, S., Sigaud, O., Oudeyer, P.Y.: Grounding large language models in interactive environments with online reinforcement learning. arXiv preprint arXiv:2302.02662 (2023)

9. Yu, Y., Chung, J., Yun, H., Hessel, J., Park, J., Lu, X., Ammanabrolu, P., Zellers, R., Bras, R.L., Kim, G., et al.: Multimodal knowledge alignment with reinforcement learning. arXiv preprint arXiv:2205.12630 (2022)

10. Li, S., Puig, X., Paxton, C., Du, Y., Wang, C., Fan, L., Chen, T., Huang, D.A., Akyürek, E., Anandkumar, A., et al.: Pre-trained language models for interactive decision-making. Advances in Neural Information Processing Systems **35**, 31199–31212 (2022)

11. Fernandes, P., Madaan, A., Liu, E., Farinhas, A., Martins, P.H., Bertsch, A., de Souza, J.G.C., Zhou, S., Wu, T.S., Neubig, G., Martins, A.F.T.: Bridging the gap: A survey on integrating (human) feedback for natural language generation. ARXIV.ORG (2023). https://doi.org/10. 48550/arXiv.2305.00955

12. Wang, Z.J., Choi, D., Xu, S., Yang, D.: Putting humans in the natural language processing loop: A survey. arXiv preprint arXiv:2103.04044 (2021)

13. Prudencio, R.F., Maximo, M.R., Colombini, E.L.: A survey on offline reinforcement learning: Taxonomy, review, and open problems. IEEE Transactions on Neural Networks and Learning Systems (2023)

14. Levine, S., Kumar, A., Tucker, G., Fu, J.: Offline reinforcement learning: Tutorial, review, and perspectives on open problems. arXiv preprint arXiv:2005.01643 (2020)

15. Fan, L., Wang, G., Jiang, Y., Mandlekar, A., Yang, Y., Zhu, H., Tang, A., Huang, D.A., Zhu, Y., Anandkumar, A.: Minedojo: Building open-ended embodied agents with internet-scale knowledge. arXiv preprint arXiv:2206.08853 (2022)

16. Huang, W., Xia, F., Xiao, T., Chan, H., Liang, J., Florence, P., Zeng, A., Tompson, J., Mordatch, I., Chebotar, Y., et al.: Inner monologue: Embodied reasoning through planning with language models. arXiv preprint arXiv:2207.05608 (2022)

17. Abramson, J., Ahuja, A., Carnevale, F., Georgiev, P., Goldin, A., Hung, A., Landon, J., Lhotka, J., Lillicrap, T., Muldal, A., Powell, G., Santoro, A., Scully, G., Srivastava, S., von Glehn, T., Wayne, G., Wong, N., Yan, C., Zhu, R.: Improving multimodal interactive agents with reinforcement learning from human feedback. arXiv preprint arXiv: Arxiv-2211.11602 (2022)

18. Yao, S., Chen, H., Yang, J., Narasimhan, K.: Webshop: Towards scalable real-world web interaction with grounded language agents. arXiv preprint arXiv:2207.01206 (2022)

19. Yuan, H., Zhang, C., Wang, H., Xie, F., Cai, P., Dong, H., Lu, Z.: Plan4mc: Skill reinforcement learning and planning for open-world minecraft tasks. arXiv preprint arXiv:2303.16563 (2023)

20. Bai, Y., Jones, A., Ndousse, K., Askell, A., Chen, A., DasSarma, N., Drain, D., Fort, S., Ganguli, D., Henighan, T., Joseph, N., Kadavath, S., Kernion, J., Conerly, T., El-Showk, S., Elhage, N., Hatfield-Dodds, Z., Hernandez, D., Hume, T., Johnston, S., Kravec, S., Lovitt, L., Nanda, N., Olsson, C., Amodei, D., Brown, T., Clark, J., McCandlish, S., Olah, C., Mann, B., Kaplan, J.: Training a helpful and harmless assistant with reinforcement learning from human feedback. arXiv preprint arXiv: Arxiv-2204.05862 (2022)

21. Schulman, J., Wolski, F., Dhariwal, P., Radford, A., Klimov, O.: Proximal policy optimization algorithms. arXiv preprint arXiv:1707.06347 (2017)

22. Rafailov, R., Sharma, A., Mitchell, E., Ermon, S., Manning, C.D., Finn, C.: Direct preference optimization: Your language model is secretly a reward model. arXiv preprint arXiv: 2305.18290 (2023)

23. Mnih, V., Badia, A.P., Mirza, M., Graves, A., Lillicrap, T., Harley, T., Silver, D., Kavukcuoglu, K.: Asynchronous methods for deep reinforcement learning. International Conference On Machine Learning (2016)

24. Glaese, A., McAleese, N., Trębacz, M., Aslanides, J., Firoiu, V., Ewalds, T., Rauh, M., Weidinger, L., Chadwick, M., Thacker, P., et al.: Improving alignment of dialogue agents via targeted human judgements. arXiv preprint arXiv:2209.14375 (2022)

25. Snell, C., Kostrikov, I., Su, Y., Yang, M., Levine, S.: Offline rl for natural language generation with implicit language q learning. arXiv preprint arXiv:2206.11871 (2022)

26. Schulman, J., Levine, S., Abbeel, P., Jordan, M., Moritz, P.: Trust region policy optimization. In: International conference on machine learning, pp. 1889–1897. PMLR (2015)

27. Ramamurthy, R., Ammanabrolu, P., Brantley, K., Hessel, J., Sifa, R., Bauckhage, C., Hajishirzi, H., Choi, Y.: Is reinforcement learning (not) for natural language processing: Benchmarks, baselines, and building blocks for natural language policy optimization. In: The Eleventh International Conference on Learning Representations (2023). https://openreview.net/forum?id=8aHzds2uUyB

28. Askell, A., Bai, Y., Chen, A., Drain, D., Ganguli, D., Henighan, T., Jones, A., Joseph, N., Mann, B., DasSarma, N., Elhage, N., Hatfield-Dodds, Z., Hernandez, D., Kernion, J., Ndousse, K., Olsson, C., Amodei, D., Brown, T.B., Clark, J., McCandlish, S., Olah, C., Kaplan, J.: A general language assistant as a laboratory for alignment. ARXIV.ORG (2021)

29. Korbak, T., Shi, K., Chen, A., Bhalerao, R., Buckley, C.L., Phang, J., Bowman, S.R., Perez, E.: Pretraining language models with human preferences. arXiv preprint arXiv:2302.08582 (2023)

30. Dong, H., Xiong, W., Goyal, D., Pan, R., Diao, S., Zhang, J., Shum, K., Zhang, T.: Raft: Reward ranked finetuning for generative foundation model alignment. ARXIV.ORG (2023). https://doi.org/10.48550/arXiv.2304.06767

31. Liu, T., Zhao, Y., Joshi, R., Khalman, M., Saleh, M., Liu, P.J., Liu, J.: Statistical rejection sampling improves preference optimization. arXiv preprint arXiv: 2309.06657 (2023). https://arxiv.org/abs/2309.06657v1

32. Meng, Y., Xia, M., Chen, D.: Simpo: Simple preference optimization with a reference-free reward. arXiv preprint arXiv:2405.14734 (2024)

33. Ramesh, S.S., Hu, Y., Chaimalas, I., Mehta, V., Sessa, P.G., Ammar, H.B., Bogunovic, I.: Group robust preference optimization in reward-free rlhf. arXiv preprint arXiv:2405.20304 (2024)

34. Chen, L., Lu, K., Rajeswaran, A., Lee, K., Grover, A., Laskin, M., Abbeel, P., Srinivas, A., Mordatch, I.: Decision transformer: Reinforcement learning via sequence modeling. Advances in neural information processing systems **34**, 15084–15097 (2021)

35. McCallum, S., Taylor-Davies, M., Albrecht, S.V., Suglia, A.: Is feedback all you need? leveraging natural language feedback in goal-conditioned reinforcement learning. arXiv preprint arXiv:2312.04736 (2023)

36. Lifshitz, S., Paster, K., Chan, H., Ba, J., McIlraith, S.: Steve-1: A generative model for text-to-behavior in minecraft. Advances in Neural Information Processing Systems **36** (2024)

37. Jin, Y., Zhang, G., Zhao, H., Zheng, T., Guo, J., Xiang, L., Yue, S., Huang, S.W., Chen, W., He, Z., et al.: Read to play (r2-play): Decision transformer with multimodal game instruction. arXiv preprint arXiv:2402.04154 (2024)

38. Zheng, T., Zhang, G., Qu, X., Kuang, M., Huang, S.W., He, Z.: More-3s: Multimodal-based offline reinforcement learning with shared semantic spaces. arXiv preprint arXiv:2402.12845 (2024)

39. Yuan, Z., Yuan, H., Tan, C., Wang, W., Huang, S., Huang, F.: Rrhf: Rank responses to align language models with human feedback without tears. ARXIV.ORG (2023). https://doi.org/10.48550/arXiv.2304.05302

40. Huang, W., Xia, F., Shah, D., Driess, D., Zeng, A., Lu, Y., Florence, P., Mordatch, I., Levine, S., Hausman, K., et al.: Grounded decoding: Guiding text generation with grounded models for robot control. arXiv preprint arXiv:2303.00855 (2023)

41. Goyal, P., Niekum, S., Mooney, R.: Pixl2r: Guiding reinforcement learning using natural language by mapping pixels to rewards. In: Conference on Robot Learning, pp. 485–497. PMLR (2021)

42. Zhou, W., Ge, T., Xu, K., Wei, F., Zhou, M.: Self-adversarial learning with comparative discrimination for text generation. In: ICLR. OpenReview.net (2020)

43. Fu Yao; Peng, H., Khot, T.: How does GPT obtain its ability? tracing emergent abilities of language models to their sources. Yao Fu's Notion (2022). https://yaofu.notion.site/How-does-GPT-Obtain-its-Ability-Tracing-Emergent-Abilities-of-Language-Models-to-their-Sources-b9a57ac0fcf74f30a1ab9e3e36fa1dc1

44. Mialon, G., Dessì, R., Lomeli, M., Nalmpantis, C., Pasunuru, R., Raileanu, R., Roziére, B., Schick, T., Dwivedi-Yu, J., Celikyilmaz, A., Grave, E., LeCun, Y., Scialom, T.: Augmented language models: a survey. arXiv preprint arXiv: Arxiv-2302.07842 (2023)

45. Kiseleva, J., Skrynnik, A., Zholus, A., Mohanty, S., Arabzadeh, N., Côté, M.A., Aliannejadi, M., Teruel, M., Li, Z., Burtsev, M., ter Hoeve, M., Volovikova, Z., Panov, A., Sun, Y., Srinet, K., Szlam, A., Awadallah, A.: Iglu 2022: Interactive grounded language understanding in a collaborative environment at neurips 2022. arXiv preprint arXiv:2205.13771 (2022)

46. Stiennon, N., Ouyang, L., Wu, J., Ziegler, D.M., Lowe, R., Voss, C., Radford, A., Amodei, D., Christiano, P.: Learning to summarize from human feedback. arXiv preprint arXiv: Arxiv-2009.01325 (2020)

47. Sun, Z., Shen, Y., Zhang, H., Zhou, Q., Chen, Z., Cox, D., Yang, Y., Gan, C.: Salmon: Self-alignment with principle-following reward models. arXiv preprint arXiv: 2310.05910 (2023)

48. Chen, A., Scheurer, J., Korbak, T., Campos, J.A., Chan, J.S., Bowman, S.R., Cho, K., Perez, E.: Improving code generation by training with natural language feedback (2023)

49. Scheurer, J., Campos, J.A., Korbak, T., Chan, J.S., Chen, A., Cho, K., Perez, E.: Training language models with language feedback at scale (2023)

50. Li, Y., Wei, F., Zhao, J., Zhang, C., Zhang, H.: Rain: Your language models can align themselves without finetuning. arXiv preprint arXiv: 2309.07124 (2023)

51. Ye, S., Jo, Y., Kim, D., Kim, S., Hwang, H., Seo, M.: Selfee: Iterative self-revising LLM empowered by self-feedback generation. Blog post (2023). https://kaistai.github.io/SelFee/

52. Lu, J., Zhong, W., Huang, W., Wang, Y., Mi, F., Wang, B., Wang, W., Shang, L., Liu, Q.: Self: Language-driven self-evolution for large language model. arXiv preprint arXiv: 2310.00533 (2023)

53. Liu, R., Yang, R., Jia, C., Zhang, G., Zhou, D., Dai, A.M., Yang, D., Vosoughi, S.: Training socially aligned language models in simulated human society (2023)

54. Kim, S., Bae, S., Shin, J., Kang, S., Kwak, D., Yoo, K.M., Seo, M.: Aligning large language models through synthetic feedback. arXiv preprint arXiv: 2305.13735 (2023)

55. Yang, K., Klein, D., Celikyilmaz, A., Peng, N., Tian, Y.: Rlcd: Reinforcement learning from contrast distillation for language model alignment. arXiv preprint arXiv: 2307.12950 (2023)

56. Xu, C., Rosset, C., Corro, L.D., Mahajan, S., McAuley, J., Neville, J., Awadallah, A.H., Rao, N.: Contrastive post-training large language models on data curriculum. arXiv preprint arXiv: 2310.02263 (2023)

57. Uesato, J., Kushman, N., Kumar, R., Song, F., Siegel, N., Wang, L., Creswell, A., Irving, G., Higgins, I.: Solving math word problems with process-and outcome-based feedback. arXiv preprint arXiv:2211.14275 (2022)

58. Wei, J., Wang, X., Schuurmans, D., Bosma, M., Chi, E.H., Le, Q., Zhou, D.: Chain of thought prompting elicits reasoning in large language models. CoRR **abs/2201.11903** (2022). https://arxiv.org/abs/2201.11903

59. Lightman, H., Kosaraju, V., Burda, Y., Edwards, H., Baker, B., Lee, T., Leike, J., Schulman, J., Sutskever, I., Cobbe, K.: Let's verify step by step. arXiv preprint arXiv:2305.20050 (2023)
60. Luo, H., Sun, Q., Xu, C., Zhao, P., Lou, J., Tao, C., Geng, X., Lin, Q., Chen, S., Zhang, D.: Wizardmath: Empowering mathematical reasoning for large language models via reinforced evol-instruct. arXiv preprint arXiv:2308.09583 (2023)
61. Ma, Q., Zhou, H., Liu, T., Yuan, J., Liu, P., You, Y., Yang, H.: Let's reward step by step: Step-level reward model as the navigators for reasoning. arXiv preprint arXiv:2310.10080 (2023)
62. Dou, S., Liu, Y., Jia, H., Xiong, L., Zhou, E., Shan, J., Huang, C., Shen, W., Fan, X., Xi, Z., et al.: Stepcoder: Improve code generation with reinforcement learning from compiler feedback. arXiv preprint arXiv:2402.01391 (2024)
63. Wang, Z., Li, Y., Wu, Y., Luo, L., Hou, L., Yu, H., Shang, J.: Multi-step problem solving through a verifier: An empirical analysis on model-induced process supervision. arXiv preprint arXiv:2402.02658 (2024)
64. Guo, D., Yang, D., Zhang, H., Song, J., Zhang, R., et al.: Deepseek-r1: Incentivizing reasoning capability in LLMs via reinforcement learning. arXiv preprint arXiv: 2501.12948 (2025)
65. Lai, X., Tian, Z., Chen, Y., Yang, S., Peng, X., Jia, J.: Step-dpo: Step-wise preference optimization for long-chain reasoning of LLMs. arXiv preprint arXiv: 2406.18629 (2024)
66. Shao, Z., Wang, P., Zhu, Q., Xu, R., Song, J., Bi, X., Zhang, H., Zhang, M., Li, Y.K., Wu, Y., Guo, D.: Deepseekmath: Pushing the limits of mathematical reasoning in open language models. arXiv preprint arXiv: 2402.03300 (2024)
67. Yao, S., Yu, D., Zhao, J., Shafran, I., Griffiths, T.L., Cao, Y., Narasimhan, K.: Tree of thoughts: Deliberate problem solving with large language models. arXiv preprint arXiv: 2305.10601 (2023)
68. Pang, R.Y., Yuan, W., Cho, K., He, H., Sukhbaatar, S., Weston, J.: Iterative reasoning preference optimization. arXiv preprint arXiv: 2404.19733 (2024)
69. Zhang, D., Zhoubian, S., Hu, Z., Yue, Y., Dong, Y., Tang, J.: Rest-mcts*: LLM self-training via process reward guided tree search. arXiv preprint arXiv: 2406.03816 (2024)
70. Gu, J., Jiang, X., Shi, Z., Tan, H., Zhai, X., Xu, C., Li, W., Shen, Y., Ma, S., Liu, H., Wang, S., Zhang, K., Wang, Y., Gao, W., Ni, L., Guo, J.: A survey on LLM-as-a-judge. arXiv preprint arXiv: 2411.15594 (2024)
71. Mahan, D., Phung, D.V., Rafailov, R., Blagden, C., Lile, N., Castricato, L., Fränken, J.P., Finn, C., Albalak, A.: Generative reward models. arXiv preprint arXiv: 2410.12832 (2024)
72. Chen, M., Tworek, J., Jun, H., Yuan, Q., de Oliveira Pinto, H.P., et al.: Evaluating large language models trained on code. arXiv preprint arXiv: Arxiv-2107.03374 (2021)
73. CAMEL-AI.org: Loong: Synthesize long cots at scale through verifiers (2025). https://github.com/camel-ai/loong

Agents

8

Zekun Wang and Jie Fu

Since the release of ChatGPT, the development of large language model (LLM) agent architectures has garnered significant attention. These architectures are designed to enhance the capabilities of LLMs, enabling them to accomplish complex tasks more autonomously through the integration of components such as short-term memory (context), long-term memory (retrieval augmentation, c.f., Sect. 2.2), tool-use modules (c.f., Sect. 2.3), reflection modules, planning modules, and more. This chapter explores the evolution and components of these sophisticated LLM agent architectures, their applications, and the challenges they face.

The chapter begins with an in-depth examination of the **Single-Agent Architectures** (Sect. 8.1), detailing how individual agents utilize structured prompt chaining schemes and advanced modules to perform tasks with increasing complexity and efficiency. Next, we delve into the **Multi-Agent System** (Sect. 8.2), highlighting how multiple LLM agents collaborate, communicate, and coordinate their actions to achieve goals that a single agent cannot accomplish alone. Following this, the chapter presents some **Task-Specific Agents** (Sect. 8.3), focusing on architectures designed for specialized domains such as software development, gaming, and story generation. These agents are tailored to excel in their respective fields by leveraging domain-specific knowledge and capabilities. Subsequently, the chapter discusses **Agent Tuning** (Sect. 8.4), examining methods and strategies to optimize the performance of open-source LLMs to serve as agents, including efforts to enhance

Z. Wang (✉)
Beihang University, Haidian District, Beijing, China
e-mail: zenmoore@buaa.edu.cn

J. Fu (✉)
Hong Kong University of Science and Technology, Hong Kong, China
e-mail: jiefu@ust.hk

agent capabilities through techniques like GPT-generated trajectories. Lastly, we delve into the agentic workflows for LLM reasoning (Sect. 8.5).

Note that our survey, "Interactive Natural Language Processing", considers LLMs as agents, defined by their interactive capabilities with external tools, models, human users, and environments. When the first version of our survey was released, the concept of agents had not yet gained widespread attention. Subsequently, with the development of large language models such as ChatGPT, a significant amount of work related to LLM agents has emerged. The general structure and framework of these works fall under the scope described in our "Interactive Natural Language Processing". Therefore, our survey can be viewed broadly as a survey of LLM agents. This chapter aims to provide specific implementation examples of agents in the commonly accepted narrower sense. We recommend readers refer to our survey alongside [1] to gain a more comprehensive perspective and technical overview from our work, and a more targeted understanding about agents from [1].

8.1 Single-Agent Architectures

Single-agent architectures for large language models (LLMs) have evolved significantly, aiming to enhance the autonomy and efficiency of these systems. These architectures integrate various cognitive and functional components to enable LLMs to perform complex tasks with minimal human intervention. The evolution of single-agent architectures is marked by the incorporation of sophisticated systems that manage memory, plan tasks, and interact with external tools [1, 2].

In [1]'s breakdown, the agent architecture involves four components: (1) a profiling module that defines the characters and functions of the LLM agents, (2) a memory module that stores the agents' observations and memories, (3) a planning module that breaks down complex tasks into a series of simple steps, and (4) an action module that completes tasks using external tools or the internal knowledge of the LLMs (self-knowledge). In our breakdown, we treat LLMs as the core (i.e., the brain) and describe LLM agents from an interaction-based perspective. Typically, current LLM prompt templates consist of three parts: system, user, and assistant. Profiling can be simply implemented through system prompts, defining who the LLM agents are and what they can do (e.g., which tools they can use, how they should plan, etc.). Single-agent architectures focus on the interaction of LLMs with themselves for planning (cf. Sect. 5.3), with tools and environments through actions. The memory module can be seen as a special interaction interface (cf. Sect. 3). In this section, we focus on the factors that qualify an LLM as an agent: memory, tool-use, planning, and reflection.

Memory. There are two kinds of memories in terms of agent architectures: short-term memory and long-term memory. Short-term memory refers to the immediate context (i.e., the context window) within which the model operates, often including recent interactions and data points used to generate responses. Effective management of short-term memory ensures that the agent maintains coherence and relevance in ongoing interactions. In contrast, long-term memory involves storing and retrieving information over extended periods. Retrieval augmentation techniques are employed to access relevant information from external memory pools, enhancing the model's ability to generate informed and accurate responses.

Tool-Use. The tool-use module is another critical component of single-agent architectures. This module allows the agent to interact with external tools and APIs, performing tasks beyond mere text generation. For instance, the tool-use module can enable the agent to conduct data analysis, perform web searches, or execute code, significantly broadening the practical applications of LLM agents. We also refer the readers to Sect. 2.3.

Planning. Planning modules are integral to single-agent architectures, allowing LLM agents to break down complex tasks into manageable steps. This capability is essential for the autonomous execution of multi-stage tasks and for maintaining a logical flow of actions over extended interactions. The planning module enables an agent to analyze a user's request, outline the necessary steps to fulfill the request, and execute those steps in a coherent sequence. Techniques such as task decomposition and hierarchical planning are commonly employed to enhance the agent's ability to handle intricate tasks. By systematically organizing tasks, the planning module helps ensure that the agent can achieve its objectives efficiently and effectively.

Reflection. Reflection modules enhance the cognitive abilities of LLM agents by allowing them to assess and improve their performance over time. This involves evaluating past actions, learning from mistakes, and integrating new knowledge to refine future behavior. Reflection modules typically function by periodically reviewing interactions, identifying areas for improvement, and updating the agent's strategies and responses. This continuous feedback loop helps the agent to become more effective in its tasks, enabling it to handle increasingly complex scenarios with greater autonomy. The reflection process is crucial for developing a more adaptive and intelligent agent that can evolve based on its experiences, thus improving its overall functionality and reliability.

Examples. ReAct, or Reasoning and Acting [3], is a seminal example of a single-agent architecture that integrates chain-of-thought reasoning [4] with tool-using abilities. This architecture allows the agent to reason through complex problems and execute actions based on its reasoning, making it a versatile tool for various applications. Another notable single-agent architecture is the Generative Agent [5], which features comprehensive memory and planning modules. The Generative Agent architecture includes a memory stream, a retrieval module, a planning module, and a reflection module. This system records perceptions into the memory stream, assigning scores for recency, importance, and relevance. The retrieval module accesses these memories based on the assigned scores, while the planning module breaks down tasks into actionable steps. The reflection module periodically summarizes observations into higher-level knowledge, enhancing the agent's ability to adapt and learn from its experiences. Moreover, Reflexion [6] includes a trajectory (short-term memory), an evaluator for internal feedback through self-evaluation, a self-reflection module generating experience (long-term memory) based on internal and external feedback, and an actor interacting with the environment.

The strengths and weaknesses of different single-agent architectures can be analyzed through their unique capabilities and limitations. ReAct [3], for example, excels in integrating reasoning and action but may face challenges in handling highly dynamic environments. On the other hand, the Generative Agent [5]'s sophisticated memory and planning modules enable advanced task execution but may require significant computational resources. Evaluating these architectures involves metrics such as efficiency, accuracy, and adaptability, which help determine their performance in real-world scenarios.

8.2   Multi-agent System

While single-agent architectures have demonstrated notable capabilities, they encounter limitations in both effectiveness and efficiency when addressing complex tasks that require collective efforts, such as software engineering [7, 8]. Multi-agent systems (MAS) have been developed to address these limitations by leveraging the strengths of multiple LLM agents [9]. These systems enable agents to collaborate, communicate, and coordinate their actions to accomplish tasks that a single agent cannot perform alone. This section explores the components and functionalities of multi-agent systems, and the challenges they face.

8.2.1   Components and Functionalities

Multi-agent systems simulate the collaborative efforts seen in human teams. Each agent in a MAS can have distinct roles, responsibilities, and expertise, allowing the system to tackle a broader range of tasks more efficiently. Key components of multi-agent systems include:

Role Assignment and Specialization. Agents within a MAS can be assigned specific roles or personas based on their capabilities. This specialization allows each agent to focus on tasks that align with their strengths, enhancing the overall efficiency and performance of the system. Role assignment can be dynamic, adapting to the evolving needs of the task at hand. For example, AutoAgents [11] dynamically generates and coordinates multiple specialized agents to build an AI team according to different tasks. This framework adapts to task content and plans solutions based on the generated expert agents, demonstrating the significance of assigning different roles to different tasks.

Coordination and Communication. Effective communication protocols are essential for multi-agent systems. These protocols enable agents to share information, negotiate roles, and synchronize their actions. Coordination mechanisms ensure that agents work together harmoniously, avoiding conflicts and redundancy. For instance, in the Solo Performance Prompting (SPP) framework [12], agents collaborate, through multi-turn self-collaboration with multiple personas, dynamically identifying and simulating different personas based on task inputs. This approach unleashes cognitive synergy, enhancing problem-solving abilities in tasks such as Trivia Creative Writing, Codenames Collaborative, and Logic Grid Puzzles.

Conflict Resolution. In a multi-agent system, conflicts can arise when agents have competing objectives or when their actions interfere with each other. Conflict resolution mechanisms are implemented to detect and resolve such issues, ensuring that the agents can collaborate effectively. For example, in a debate scenario, [13] forces multiple debater agents to reach a consensus, while ChatEval [14] mitigates the impact of conflicts through majority voting or calculating averaged scores.

8.2.2 Examples

In this sub-section, we present several representative examples of MAS.

Solo Performance Prompting (SPP) [12] turns a single large language model (LLM) into a cognitive synergist through iterative self-collaboration across different personas. This method improves problem-solving by fostering cognitive synergy, minimizing factual errors, and preserving robust reasoning. SPP's efficacy is tested on tasks like Trivia Creative Writing, Codenames Collaborative, and Logic Grid Puzzles.

Agents [15] is an open-source platform designed for building and deploying autonomous language agents, featuring planning, memory, tool usage, and multi-agent communication. Its modular design allows non-specialists to easily create and customize agents, while also being extensible for advanced research purposes.

AutoGen [16] enables developers to construct LLM applications where multiple agents interact to complete tasks. These customizable agents can integrate LLMs, human inputs, and tools, supporting a variety of applications such as mathematics, coding, question answering, and operations research.

AutoAgents [11] dynamically generates and manages specialized agents to form AI teams tailored to specific tasks. This adaptive framework enhances task solution coherence and accuracy through collaborative efforts, highlighting the value of role-specific agent assignments in complex problem-solving.

ChatDev [7] is a software development framework that uses LLM-driven agents communicating via multi-turn dialogues. These agents assist in design, coding, and testing, showcasing how language-based interactions enhance multi-agent collaboration in software development.

SwiftSage [17] employs dual-process cognition, blending rapid intuitive responses with deliberate reasoning. This framework merges behavior cloning with LLM prompting to boost task performance, significantly outperforming other methods in complex interactive scenarios by balancing fast and slow thinking.

AgentVerse [18] utilizes multi-agent collaboration, exploring emergent behaviors within agent groups. Inspired by human group dynamics, it dynamically adjusts its composition for better task performance, demonstrating how positive social behaviors can enhance multi-agent collaboration.

OpenAgents [19] is an open platform designed to host language agents for everyday use. It features agents for data analysis, API interactions, and web browsing. OpenAgents offers an intuitive interface for users and a smooth deployment process for developers, enabling practical evaluations of language agents.

8.2.3   Challenges

The development and deployment of multi-agent systems present several challenges, including scalability, coordination complexity, and evaluation metrics. Scalability is a major concern, as the number of agents and the complexity of their interactions can grow exponentially. Effective coordination mechanisms are needed to manage these interactions and ensure seamless collaboration.

Moreover, establishing standard evaluation metrics for multi-agent systems is crucial for assessing their performance and effectiveness. These metrics should consider factors such as task completion time, resource utilization, the quality of the agents' interactions, and the utility [10].

We also refer readers to the survey [10] for more comprehensive information on Multi-Agent Systems.

8.3   Task-Specific Agents

Despite the strong abilities demonstrated by different agent architectures and multi-agent systems, they often need to be designed and optimized for specific tasks to achieve the best performance, which limits their general applicability. For instance, agent architectures tailored for software development are generally not suitable for gaming simulation. Although works like AutoAgents [11] claim to enhance task adaptability through automatic agent generation and task allocation, their effectiveness still lags behind agent architectures that have been manually optimized for specific tasks. This discrepancy is natural, as constructing complex workflows requires sophisticated capabilities that current foundational models are likely less capable of compared to humans.

In Table 8.1, we list some task-specific agents. Different tasks or application scenarios impose varying requirements on agent architecture. For example, agents designed for software engineering [7, 8, 20] need robust code-writing and debugging capabilities and must effectively understand user intent expressed in textual form to translate it into executable code. Agents for psychological assessment [24] need to ensure diagnostic accuracy, as any ambiguity in expression could lead to disastrous misdiagnoses. Agents for game simulation [21, 22] require a deep understanding of the game, such as the strong conspiracy capabilities needed for Avalon games.

Table 8.1 Examples of task-specific agents and their application scenarios

Application scenario	References
Software development	MetaGPT [8], ChatDev [7], SWE-Agent [20]
Game simulation	Avalon's Game of Thoughts [21, 22]
Story generation	RecurrentGPT [23]
Evaluation	ChatEval [14]
Psychological assessment	PsychoGAT [24]
Operating System	OmniACT [25], OSWorld [26], OS-Copilot [27]
Mobile phone manipulation	Mobile-Agent [28], AppAgent [29]
Reading comprehension	ReadAgents [30]
Daily Life	V-IRL [31]

We recommend readers refer to the respective articles and the relevant surveys [1, 10] for application scenarios suited to their needs.

8.4   Agent Tuning

Although the aforementioned agent architectures can address a wide array of tasks, they rely heavily on advanced, closed-source foundation models such as GPT-4 [32]. To empower open-source large language models (LLMs) with similar agent capabilities, various agent tuning approaches are employed. These methods aim to enhance the performance and capabilities of LLMs when they operate as agents. The process involves fine-tuning the models on specific tasks and trajectories, thereby improving their ability to plan, reason, and act autonomously. Several recent studies have investigated different approaches to agent tuning, striving to close the performance gap between open-source models and commercial models like ChatGPT and GPT-4 [32].

One notable approach is SwiftSage [17], which combines behavior cloning and large language model prompting to improve task performance. SwiftSage utilizes a dual-process cognitive framework inspired by human cognition, with two primary modules: The Swift module and the Sage module. The Swift module, fine-tuned on oracle agent (i.e., GPT-4) action trajectories, represents fast, intuitive thinking. In contrast, the Sage module, which employs LLMs like GPT-4, represents deliberate thought processes for subgoal planning and grounding. This dual-process approach has shown significant improvements in complex interactive tasks by harmoniously integrating fast and slow thinking processes.

Another approach is FireAct [33], which focuses on fine-tuning LLMs with trajectories from various tasks and prompting methods. FireAct demonstrates that fine-tuning LLMs, such as Llama2-7B [34], with agent trajectories generated by larger models like GPT-4 can significantly improve performance on tasks like HotpotQA [35]. This method emphasizes the benefits of diverse fine-tuning data and explores scaling effects, robustness, generalization, and efficiency, establishing the comprehensive advantages of fine-tuning LLMs for agent tasks.

AgentTuning [36] is another method that enhances LLMs' agent capabilities without compromising their general abilities. This approach involves constructing a lightweight instruction-tuning dataset called AgentInstruct, which contains high-quality interaction trajectories. By combining AgentInstruct with open-source instructions from general domains, AgentTuning instruction-tunes the Llama 2 series, resulting in models like AgentLM. Evaluations show that AgentLM achieves performance comparable to GPT-3.5-turbo on unseen agent tasks, demonstrating generalized agent capabilities while maintaining general LLM functionalities.

AutoAct [37] introduces an automated agent learning framework capable of synthesizing planning trajectories without the need for large-scale annotated data or synthetic trajectories from proprietary models. The framework follows a four-step process. Initially, it extends the task data using Self-Instruct [38], leveraging an open-source LLM referred to as the Meta-Agent. This Meta-Agent then autonomously selects the appropriate tools for various tasks. Subsequently, it generates trajectories based on the selected tools with a reward score for data filtering. These model-generated trajectories are ultimately used to fine-tune the Meta-Agent, resulting in specialized agents, such as the Plan-Agent, Tool-Agent, and Reflect-Agent, each of which is fine-tuned on a specific group of trajectories. These differentiated agents then collaborate to accomplish the designated tasks.

In summary, agent tuning, whether achieved through behavior cloning using trajectories generated by advanced models like GPT-4 or through self-training, has proven effective in enhancing the capabilities of LLM agents. However, despite these successes, fine-tuning models with knowledge distillation or self-training often requires access to high-quality, annotated data from powerful models (including the models themselves that undergo agent tuning), which may not be readily available to open-source communities.

8.5   Agentic Workflow for Reasoning

Agentic workflows can enhance LLM reasoning by enabling autonomous and interactive systems, either through single-agent or multi-agent workflows. Single-agent workflows focus on iterative reasoning processes, often guided by external tools or verifiers, while multi-agent workflows leverage collaborative interactions to achieve robust reasoning outcomes. These workflows not only facilitate reasoning at inference time but also generate valuable trajectories for LLM training (either through SFT or RL).

8.5.1   Single-Agent Workflows

Single-agent workflows combine input and output mechanisms to enhance reasoning. At the input level, agents can incorporate external resources like Tool-in-the-Loop (e.g., SymPy, Wolfram|Alpha) or Knowledge-Base-in-the-Loop (e.g., Chain-of-Knowledge [39], Think-on-Graph [40]) to augment reasoning, as discussed in prior sections (Sects. 2.2 and 2.3). Alternatively, task decomposition or elicitation of the LLM's chain of thought can be achieved by modifying the prompt, as highlighted in Sects. 5.2 and 5.3 . The primary focus of this subsection, however, lies in the output phase, where agents perform reasoning searches and refine responses under verifier guidance.

Reasoning Search with Verifiers. Single-agent workflows leverage verifiers to guide reasoning searches, ensuring accurate and efficient exploration of solution spaces. Notable approaches include (1) MCTS-based approaches [41], (2) rejection sampling-based approaches [42], and (3) look-ahead search-based approaches [43], among others.

(1) Monte Carlo Tree Search (MCTS) involves four steps: selection (often using Upper Confidence Bound applied to trees to select the best child nodes), expansion (appending all possible states from the leaf node), simulation (generating the full reasoning paths until the end), and backpropagation (updating the rewards of each node), enabling agents to explore multiple reasoning paths and assign process rewards to reasoning steps, balancing exploration and exploitation. Verifiers, such as outcome-based reward models, process-based reward models or LLM-as-a-judge, evaluate intermediate steps or the final answers, guiding the search toward high-quality solutions. MCTS is particularly effective for structured tasks like mathematical reasoning.

(2) Rejection Sampling (e.g., Best-of-N Sampling): In Best-of-N sampling, the agent generates multiple candidate responses, and a verifier selects the best based on reward signals. This approach ensures high-quality outputs but can be computationally expensive due to repeated sampling.

(3) Look-Ahead Search: To reduce computational costs, look-ahead search methods dynamically prioritize promising reasoning branches by estimating their future value [43]. Derived from traditional beam search, which evaluates candidates at each step based on the model's output probability, look-ahead search performs a k-step rollout from the current state, using a PRM to assess the value of the final step in the rollout. This PRM-derived value, instead of the model's output probability, serves as a proxy for the quality of the current step, enabling the model to focus on high-potential paths while discarding less promising ones.

Reflection and Refinement. Single-agent workflows can simulate reflection and refine-ment, mimicking special Chain-of-Thought (CoT) patterns to correct errors and improve responses like human thinking. For example, in Self-Reflection [44], agents iteratively eval-uate their own outputs, identifying errors without external verifiers. However, this approach often yields poor performance due to model's limited self-correction capabilities. Recursive Introspection [45] trains a correction model using RL to teach the LLMs to self-improve.

Step-level Critique [46] provides feedback for each reasoning step to help the LLM iteratively self-refine its response, in a generator-critic-refiner pattern.

8.5.2 Multi-agent Workflows

Multi-agent systems enhance reasoning through collaborative interactions, leveraging communication and coordination to tackle complex tasks, as illustrated in Sect. 8.2. These systems typically involve multiple LLMs working together, following predefined patterns like debates or role-playing. For example, Mixture-of-Agents [47] sequentially chains agents with distinct roles, combining their outputs to produce a consensus solution. Each agent contributes specialized reasoning, improving overall accuracy. EvoAgent [48] dynamically generates multiple specialized agents for a given task from an initial agent, selecting those with high-quality outputs for the final result. This dynamic planning enhances adaptability to diverse problems.

Multi-agent workflows often generate rich reasoning trajectories through interactions like debates [50] or centralized coordination [49]. These trajectories can also serve as high-quality data for training reasoners, enabling models to learn from diverse, collaborative reasoning patterns.

References

1. Wang, L., Ma, C., Feng, X., Zhang, Z., Yang, H., Zhang, J., Chen, Z., Tang, J., Chen, X., Lin, Y., Zhao, W.X., Wei, Z., Wen, J.R.: A survey on large language model based autonomous agents. arXiv preprint arXiv: 2308.11432 (2023). https://arxiv.org/abs/2308.11432v2
2. Weng, L.: LLM-powered autonomous agents. lilianweng.github.io (2023). https://lilianweng.github.io/posts/2023-06-23-agent/
3. Yao, S., Zhao, J., Yu, D., Du, N., Shafran, I., Narasimhan, K., Cao, Y.: React: Synergizing reasoning and acting in language models. arXiv preprint arXiv: Arxiv-2210.03629 (2022)
4. Wei, J., Wang, X., Schuurmans, D., Bosma, M., Chi, E.H., Le, Q., Zhou, D.: Chain of thought prompting elicits reasoning in large language models. CoRR **abs/2201.11903** (2022). https://arxiv.org/abs/2201.11903
5. Park, J.S., O'Brien, J.C., Cai, C.J., Morris, M.R., Liang, P., Bernstein, M.S.: Generative agents: Interactive simulacra of human behavior. arXiv preprint arXiv:2304.03442 (2023)
6. Shinn, N., Cassano, F., Labash, B., Gopinath, A., Narasimhan, K., Yao, S.: Reflexion: Language agents with verbal reinforcement learning. arXiv preprint arXiv: 2303.11366 (2023)
7. Qian, C., Cong, X., Liu, W., Yang, C., Chen, W., Su, Y., Dang, Y., Li, J., Xu, J., Li, D., Liu, Z., Sun, M.: Communicative agents for software development (2023)
8. Hong, S., Zheng, X., Chen, J., Cheng, Y., Wang, J., Zhang, C., Wang, Z., Yau, S.K.S., Lin, Z., Zhou, L., Ran, C., Xiao, L., Wu, C.: Metagpt: Meta programming for multi-agent collaborative framework. arXiv preprint arXiv: 2308.00352 (2023)
9. Li, J., Zhang, Q., Yu, Y., Fu, Q., Ye, D.: More agents is all you need. arXiv preprint arXiv: 2402.05120 (2024)

10. Xi, Z., Chen, W., Guo, X., He, W., Ding, Y., Hong, B., Zhang, M., Wang, J., Jin, S., Zhou, E., Zheng, R., Fan, X., Wang, X., Xiong, L., Zhou, Y., Wang, W., Jiang, C., Zou, Y., Liu, X., Yin, Z., Dou, S., Weng, R., Cheng, W., Zhang, Q., Qin, W., Zheng, Y., Qiu, X., Huang, X., Gui, T.: The rise and potential of large language model based agents: A survey. arXiv preprint arXiv: 2309.07864 (2023)

11. Chen, G., Dong, S., Shu, Y., Zhang, G., Sesay, J., Karlsson, B.F., Fu, J., Shi, Y.: Autoagents: A framework for automatic agent generation. arXiv preprint arXiv: 2309.17288 (2023). https://arxiv.org/abs/2309.17288v1

12. Wang, Z., Mao, S., Wu, W., Ge, T., Wei, F., Ji, H.: Unleashing cognitive synergy in large language models: A task-solving agent through multi-persona self-collaboration. arXiv preprint arXiv: 2307.05300 (2023)

13. Du, Y., Li, S., Torralba, A., Tenenbaum, J.B., Mordatch, I.: Improving factuality and reasoning in language models through multiagent debate. arXiv preprint arXiv: 2305.14325 (2023)

14. Chan, C.M., Chen, W., Su, Y., Yu, J., Xue, W., Zhang, S., Fu, J., Liu, Z.: Chateval: Towards better LLM-based evaluators through multi-agent debate. arXiv preprint arXiv: 2308.07201 (2023). https://arxiv.org/abs/2308.07201v1

15. Zhou, W., Jiang, Y.E., Li, L., Wu, J., Wang, T., Qiu, S., Zhang, J., Chen, J., Wu, R., Wang, S., Zhu, S., Chen, J., Zhang, W., Zhang, N., Chen, H., Cui, P., Sachan, M.: Agents: An open-source framework for autonomous language agents. arXiv preprint arXiv: 2309.07870 (2023)

16. Wu, Q., Bansal, G., Zhang, J., Wu, Y., Zhang, S., Zhu, E., Li, B., Jiang, L., Zhang, X., Wang, C.: Autogen: Enabling next-gen LLM applications via multi-agent conversation framework (2023)

17. Lin, B.Y., Fu, Y., Yang, K., Ammanabrolu, P., Brahman, F., Huang, S., Bhagavatula, C., Choi, Y., Ren, X.: Swiftsage: A generative agent with fast and slow thinking for complex interactive tasks. arXiv preprint arXiv: 2305.17390 (2023)

18. Chen, W., Su, Y., Zuo, J., Yang, C., Yuan, C., Chan, C.M., Yu, H., Lu, Y., Hung, Y.H., Qian, C., Qin, Y., Cong, X., Xie, R., Liu, Z., Sun, M., Zhou, J.: Agentverse: Facilitating multi-agent collaboration and exploring emergent behaviors. arXiv preprint arXiv: 2308.10848 (2023)

19. Xie, T., Zhou, F., Cheng, Z., Shi, P., Weng, L., Liu, Y., Hua, T.J., Zhao, J., Liu, Q., Liu, C., Liu, L.Z., Xu, Y., Su, H., Shin, D., Xiong, C., Yu, T.: Openagents: An open platform for language agents in the wild. arXiv preprint arXiv: 2310.10634 (2023)

20. Yang, J., Jimenez, C.E., Wettig, A., Lieret, K., Yao, S., Narasimhan, K., Press, O.: Swe-agent: Agent-computer interfaces enable automated software engineering. arXiv preprint arXiv: 2405.15793 (2024)

21. Wang, S., Liu, C., Zheng, Z., Qi, S., Chen, S., Yang, Q., Zhao, A., Wang, C., Song, S., Huang, G.: Avalon's game of thoughts: Battle against deception through recursive contemplation. arXiv preprint arXiv: 2310.01320 (2023)

22. Xu, Y., Wang, S., Li, P., Luo, F., Wang, X., Liu, W., Liu, Y.: Exploring large language models for communication games: An empirical study on werewolf. arXiv preprint arXiv: 2309.04658 (2023)

23. Zhou, W., Jiang, Y.E., Cui, P., Wang, T., Xiao, Z., Hou, Y., Cotterell, R., Sachan, M.: Recurrentgpt: Interactive generation of (arbitrarily) long text. arXiv preprint arXiv: 2305.13304 (2023)

24. Yang, Q., Wang, Z., Chen, H., Wang, S., Pu, Y., Gao, X., Huang, W., Song, S., Huang, G.: Psychogat: A novel psychological measurement paradigm through interactive fiction games with LLM agents (2024). https://openreview.net/forum?id=efvlIIRKeO&referrer=%5BAuthor%20Console%5D(%2Fgroup%3Fid%3Daclweb.org%2FACL%2F2024%2FARR_Commitment%2FAuthors%23your-submissions)

25. Kapoor, R., Butala, Y.P., Russak, M., Koh, J.Y., Kamble, K., Alshikh, W., Salakhutdinov, R.: Omniact: A dataset and benchmark for enabling multimodal generalist autonomous agents for desktop and web. arXiv preprint arXiv: 2402.17553 (2024). https://arxiv.org/abs/2402.17553v2

26. Xie, T., Zhang, D., Chen, J., Li, X., Zhao, S., Cao, R., Hua, T.J., Cheng, Z., Shin, D., Lei, F., Liu, Y., Xu, Y., Zhou, S., Savarese, S., Xiong, C., Zhong, V., Yu, T.: Osworld: Benchmarking multimodal agents for open-ended tasks in real computer environments. arXiv preprint arXiv: 2404.07972 (2024)
27. Wu, Z., Han, C., Ding, Z., Weng, Z., Liu, Z., Yao, S., Yu, T., Kong, L.: Os-copilot: Towards generalist computer agents with self-improvement. arXiv preprint arXiv: 2402.07456 (2024). https://arxiv.org/abs/2402.07456v2
28. Wang, J., Xu, H., Ye, J., Yan, M., Shen, W., Zhang, J., Huang, F., Sang, J.: Mobile-agent: Autonomous multi-modal mobile device agent with visual perception. arXiv preprint arXiv: 2401.16158 (2024). https://arxiv.org/abs/2401.16158v2
29. Zhang, C., Yang, Z., Liu, J., Han, Y., Chen, X., Huang, Z., Fu, B., Yu, G.: Appagent: Multimodal agents as smartphone users. arXiv preprint arXiv: 2312.13771 (2023). https://arxiv.org/abs/2312.13771v2
30. Lee, K.H., Chen, X., Furuta, H., Canny, J., Fischer, I.: A human-inspired reading agent with gist memory of very long contexts. arXiv preprint arXiv: 2402.09727 (2024). https://arxiv.org/abs/2402.09727v2
31. Yang, J., Ding, R., Brown, E., Qi, X., Xie, S.: V-irl: Grounding virtual intelligence in real life. arXiv preprint arXiv: 2402.03310 (2024). https://arxiv.org/abs/2402.03310v2
32. OpenAI: GPT-4 technical report. PREPRINT (2023)
33. Chen, B., Shu, C., Shareghi, E., Collier, N., Narasimhan, K., Yao, S.: Fireact: Toward language agent fine-tuning. arXiv preprint arXiv: 2310.05915 (2023)
34. Cheng, Z., Leng, S., Zhang, H., Xin, Y., Li, X., Chen, G., Zhu, Y., Zhang, W., Luo, Z., Zhao, D., Bing, L.: Videollama 2: Advancing spatial-temporal modeling and audio understanding in video-LLMs. arXiv preprint arXiv: 2406.07476 (2024)
35. Yang, Z., Qi, P., Zhang, S., Bengio, Y., Cohen, W.W., Salakhutdinov, R., Manning, C.D.: Hotpotqa: A dataset for diverse, explainable multi-hop question answering. arXiv preprint arXiv: Arxiv-1809.09600 (2018)
36. Zeng, A., Liu, M., Lu, R., Wang, B., Liu, X., Dong, Y., Tang, J.: Agenttuning: Enabling generalized agent abilities for LLMs. arXiv preprint arXiv: 2310.12823 (2023)
37. Qiao, S., Zhang, N., Fang, R., Luo, Y., Zhou, W., Jiang, Y.E., Lv, C., Chen, H.: Autoact: Automatic agent learning from scratch for qa via self-planning. arXiv preprint arXiv: 2401.05268 (2024). https://arxiv.org/abs/2401.05268v4
38. Wang, Y., Kordi, Y., Mishra, S., Liu, A., Smith, N.A., Khashabi, D., Hajishirzi, H.: Self-instruct: Aligning language model with self generated instructions. arXiv preprint arXiv: Arxiv-2212.10560 (2022)
39. Li, X., Zhao, R., Chia, Y.K., Ding, B., Joty, S., Poria, S., Bing, L.: Chain-of-knowledge: Grounding large language models via dynamic knowledge adapting over heterogeneous sources. arXiv preprint arXiv: 2305.13269 (2023)
40. Sun, J., Xu, C., Tang, L., Wang, S., Lin, C., Gong, Y., Ni, L.M., Shum, H.Y., Guo, J.: Think-on-graph: Deep and responsible reasoning of large language model on knowledge graph. arXiv preprint arXiv: 2307.07697 (2023)
41. Zhang, D., Zhoubian, S., Hu, Z., Yue, Y., Dong, Y., Tang, J.: Rest-mcts*: LLM self-training via process reward guided tree search. arXiv preprint arXiv: 2406.03816 (2024)
42. Sessa, P.G., Dadashi, R., Hussenot, L., Ferret, J., Vieillard, N., Ramé, A., Shariari, B., Perrin, S., Friesen, A., Cideron, G., Girgin, S., Stanczyk, P., Michi, A., Sinopalnikov, D., Ramos, S., Héliou, A., Severyn, A., Hoffman, M., Momchev, N., Bachem, O.: Bond: Aligning LLMs with best-of-n distillation. arXiv preprint arXiv: 2407.14622 (2024)
43. Snell, C., Lee, J., Xu, K., Kumar, A.: Scaling LLM test-time compute optimally can be more effective than scaling model parameters. arXiv preprint arXiv: 2408.03314 (2024)

44. Renze, M., Guven, E.: Self-reflection in LLM agents: Effects on problem-solving performance. arXiv preprint arXiv: 2405.06682 (2024)
45. Qu, Y., Zhang, T., Garg, N., Kumar, A.: Recursive introspection: Teaching language model agents how to self-improve. arXiv preprint arXiv: 2407.18219 (2024)
46. Xi, Z., Yang, D., Huang, J., Tang, J., Li, G., Ding, Y., He, W., Hong, B., Do, S., Zhan, W., Wang, X., Zheng, R., Ji, T., Shi, X., Zhai, Y., Weng, R., Wang, J., Cai, X., Gui, T., Wu, Z., Zhang, Q., Qiu, X., Huang, X., Jiang, Y.G.: Enhancing LLM reasoning via critique models with test-time and training-time supervision. arXiv preprint arXiv: 2411.16579 (2024)
47. Wang, J., Wang, J., Athiwaratkun, B., Zhang, C., Zou, J.: Mixture-of-agents enhances large language model capabilities. arXiv preprint arXiv: 2406.04692 (2024)
48. Yuan, S., Song, K., Chen, J., Tan, X., Li, D., Yang, D.: Evoagent: Towards automatic multi-agent generation via evolutionary algorithms. arXiv preprint arXiv: 2406.14228 (2024)
49. Matta, M., Cardarilli, G.C., Nunzio, L.D., Fazzolari, R., Giardino, D., Re, M., Silvestri, F., Spanò, S.: Q-rts: A real-time swarm intelligence based on multi-agent q-learning. Electronics Letters **55**(10), 589–591 (2019)
50. Estornell, A., Ton, J.F., Yao, Y., Liu, Y.: Acc-collab: An actor-critic approach to multi-agent LLM collaboration. arXiv preprint arXiv: 2411.00053 (2024)

Other Methods 9

Zekun Wang, Ge Zhang, Ning Shi and Jie Fu

9.1 Active Learning

Active learning (AL) is a machine learning approach where an algorithm selects a subset of unlabeled data points iteratively to be manually labeled by a human annotator or automatically labeled by an automated labeling system. The goal of this approach is twofold: (1) to obtain a larger and more desirable set of labeled data points that can be used to further train the model, and (2) to maximize the model's performance gain with minimal data expansion, thereby improving sample efficiency (Fig. 9.1).

The human annotator or automated labeling system is commonly referred to as the Oracle ($O(\cdot)$), which is an (interactive) object or a function that can label the data points [1]. While traditional AL research has primarily focused on finding an optimal query strategy ($Q(\cdot)$) with humans as the oracle [1, 2], recent work has shown that other interactive objects such as LMs can also serve as oracles [3].

Z. Wang (✉)
Beihang University, Haidian District, Beijing, China
e-mail: zenmoore@buaa.edu.cn

G. Zhang
University of Michigan, Haidian District, Beijing, China
e-mail: gezhang@umich.edu

N. Shi
University of Alberta, Edmonton, AB, Canada
e-mail: ning.shi@ualberta.ca

J. Fu
Hong Kong University of Science and Technology, Hong Kong, China
e-mail: jiefu@ust.hk

© The Author(s), under exclusive license to Springer Nature Switzerland AG 2026
Z. Wang et al. (eds.), *Interactive Natural Language Processing*, Synthesis Lectures on Human Language Technologies, https://doi.org/10.1007/978-3-032-06264-2_9

Fig. 9.1 Active learning

Furthermore, AL involves selecting data points from an unlabeled data pool, which can include sources such as the internet [4] and knowledge graph [5]. The presence of interactive objects in the loop highlights the strong relationship between active learning and Interactive NLP (iNLP).

Formally, let D_u be the unlabeled dataset. We aim to find the optimal subset $Q(D_u) \subseteq D_u$ of samples to be labeled. Subsequently, when we construct a newly labeled set $D_l = O(Q(D_u))$ or then add them to the existing labeled dataset, we aim to minimize the loss of the model $P(y|x;\theta)$ on the labeled dataset. This process is repeated iteratively. Thus, given an unlabeled data pool such as corpus, internet [4], and knowledge graph [5], the main effort on active learning is twofold: (1) find the optimal **query strategy** $Q(\cdot)$, and (2) choose a suitable **oracle** $O(\cdot)$.

Query Strategy. Zhang et al. [6] summarize two major concerns of AL query strategy design: informativeness and representativeness. Informativeness-based query strategies aim to identify unlabeled data points that can provide the maximum amount of additional information when being labeled, with the objective of maximizing the information gained at each iteration. Common informativeness-based query strategies include uncertainty sampling-based strategies [7], disagreement based-strategies [8], and performance based-strategies [9, 10]. Representativeness-based query strategies aim to account for correlations among samples to avoid sampling bias and excessive weighting of outliers. Common representativeness-based query strategies include density based-strategies [11] and batch diversity based-strategies [12]. We refer the readers to [6] for more details. Traditional AL query strategies rely on statistical metrics [13, 14] that may lack representation richness compared to neural representations, particularly those based on PLMs. As such, there is a growing interest in exploring how to effectively leverage PLMs and LLMs in the design of AL query strategies. For example, several useful empirical conclusions on the combination of BERT [15] and traditional AL query strategies are illustrated in [16]. Seo et al. [17] propose a query strategy that is based on a task-independent triplet loss, which leverages task-related features provided by task classifiers and PLM-based knowledge features to enhance batch sample diversity. ALLSH [18] designs a query strategy that is guided by local sensitivity and hardness utilizing a PLM-based representation, in order to improve the performance of prompt-based few-shot fine-tuning on various NLP tasks. However, adapting PLM-based representations to AL query strategies poses several delicate challenges. For example, [7] point out that adapting state-of-the-art LLMs for query strategies can lead to prohibitive costs, even outweighing expected savings. They also conduct preliminary experiments that explore the combination of transformer-based models with several traditional uncertainty-based AL strategies. Margatina et al. [19] suggest that a PLM-based representation that is fine-tuned using poor training strategies can be detrimental to AL performance. They highlight the significance of effectively adapting PLMs to the specific downstream tasks during the AL process. These challenges suggest that there is a vast research space for the NLP community to explore more efficient and effective AL query strategies. This effort may be in line with the roadmap of retriever methods (c.f., 2.2), incorporating metrics that are specific to active learning. For example, [20] situate the RL-based retrieval method mentioned in Sect. para:retrieval within the context of active learning (AL), with the "additional gain" upon acquiring a label for an example being the AL-specific metric.

Oracle. The oracle applied for labeling are usually humans in most previous NLP AL research [2, 6], including annotation of coreference resolution [21], word sense disambiguation [22], non-literal language identification [23], and especially small corpora construction with intensive professional knowledge or cost [24–27]. PLM, as an alternative, has shown considerable potential in playing a role as an oracle for AL. For example, [12] propose to actively annotate highly uncertain samples with pseudo-labels generated by PLMs and use low uncertainty data points for self-training. Dossou et al. [3] adopt PLM to generate new sentences to enrich the corpora of low-resourced African languages for pre-training a multilingual model. Wang et al. [28] formulate the data augmentation problem as a relabeling mechanism. They iteratively collect a batch of data points with removed instruction or input-output example for relabeling, and then leverage PLM and ICL to generate new instruction labels or input-output labels to train the model. Furthermore, to the best of our knowledge, despite the scarcity of related work, other interactive objects such as knowledge bases, tools, and environments may have the potential to serve as oracles for active learning. For example, as a special type of knowledge base, ontology may contain structured information that can facilitate labeling [29], making it a potential candidate to serve as an oracle. ReAct [30] and PoT [31] integrate web searching and code execution, respectively, into the answer generation process. During the AL process, we can intuitively leverage these tools to label data points. MineDojo [32] finds action guidance from an internet-scale knowledge base when faced with a difficult open-ended task (i.e., task without guidance label) in the

MineCraft environment. NLMap-SayCan [33] equips the LLM with an open-vocabulary and queryable scene representation which can actively propose involved objects and their locations in the environment to label incomplete and non-reifiable planning of the LLM. These works highlight the potential for using knowledge bases, tools, and environments for labeling.

9.2   Imitation Learning

Imitation learning enables an agent to learn a policy [34, 35, 37] or a reward function [38] by mimicking an expert behavior represented by given demonstrations. In contrast to reinforcement learning, a primary advantage of imitation learning is that it does not depend on a manually designed reward function or a learned reward model, but solely relies on behavior demonstrations.

Such an advantage becomes especially prominent when accessing an expert at a low cost is possible, leading to a scalable generalization to various fields such as autonomous driving [39], computer control [40], game playing [35, 41, 42], human-agent interaction [43], robotic learning [44–46], skill acquisition [47, 48], and even surgery [49].

It has also been applied widely to tasks in NLP, including paraphrase generation [50], textual adversarial attack [51], text editing [52, 53], and text generation [54] (Fig. 9.2).

In imitation learning, the goal is to learn a policy π_θ parameterized by θ that mimics the behavior of an expert policy π_E in a given task. The behavior of the expert policy is represented by a set of demonstrations $D = \{(s_1, a_1), (s_2, a_2), ..., (s_T, a_T)\}$, where s_t is the state at time step t and a_t is the action taken by the expert policy π_E in that state. The objective of imitation learning is typically formulated as:

$$\max_\theta \sum_{i=1}^{T} \log \pi_\theta (a_i \mid s_i) \tag{9.1}$$

Imitation learning for interactive natural language processing can be divided into offline and online imitation learning.

Offline Imitation Learning. Demonstrations can be collected and stored offline as pairs of state observations and corresponding expert actions. Directly training models on such datasets in a supervised learning manner frames the basic type of imitation learning, namely behavior cloning [34]. Numerous supervised text generation approaches can be classified into this group by reformatting the autoregressive decoding into a Markov decision process at either token [52, 55] or sequence [53, 56] level. The learning process from the local expert's demonstrations can be considered an offline interaction. The expert policy is encoded in the data as an offline source to acquire the correct action given the current state. The model learns through offline interaction with the expert, and can later perform the task or behavior independently.

Fig. 9.2 Imitation learning

Online Imitation Learning. For some cases where the model can consult an expert, imitation learning can be conducted through an online interaction for additional supervision and evaluation. When incorporating online interaction or online imitation learning [37], the trained model can be updated on-the-fly using feedback from the expert, thus improving its performance over time. In particular, the expert response can be simulated by a separate system that knows the goal state, which proves beneficial in scenarios where accessing a human expert is infeasible or impractical [56]. This approach also provides the advantage of reinforcing or discouraging specific behaviors to aligning models with human expectations by adjusting the expert simulator.

Imitation learning often suffers from exposure bias, leading to distribution shifts and error accumulation in sequential decision-making tasks [37], such as robotic control or text generation [36]. When the model is only exposed to the prior trajectories generated by the expert policy, it rarely experiences the state updated by the action of its own policy. This mismatch can lead to a distribution shift where the model may encounter states it has not seen during training, thus exacerbating error accumulation. To address exposure bias in imitation learning, researchers have proposed methods that alternate between the policy of the expert and that of the model being trained. Upon the model predicting actions, the expert provides feedback or corrections, allowing for fine-tuning [37]. These approaches share the same principle with interactive natural language processing which involves both offline and online interaction. How to eliminate exposure bias in language generation, especially when the online interaction with experts is limited, remains an area of active exploration [57]. Furthermore, imitation learning has other limitations, including its reliance on the quality of expert demonstrations and the need for a large amount of demonstration data, impeding its broader application in interactive natural language processing.

9.3 Model Surgery

Model surgery encompasses targeted interventions to modify the behavior or knowledge of LLMs post-training, addressing the need to adapt models to evolving requirements without the computational overhead of full retraining. This framework includes two primary techniques: Knowledge Editing (Sect. 9.3.1) and Machine Unlearning (Sect. 9.3.2), which respectively focus on updating or correcting specific knowledge and removing the influence of designated data. These methods are critical for Interactive NLP, enabling LLMs to dynamically incorporate user feedback, correct errors, or comply with ethical and legal standards. By leveraging precise modifications, model surgery enhances the adaptability, safety, and reliability of LLMs in interactive settings. This section explores the motivations, methodologies, advantages, and limitations of these approaches, supported by recent research and their applications in interactive NLP (Fig. 9.3).

9.3.1 Knowledge Editing

Knowledge editing enables targeted modifications to an LLM's knowledge base, allowing the correction of erroneous facts, the integration of new information, or the adjustment of model behavior without affecting unrelated knowledge. This technique is pivotal in interactive NLP, where users may demand real-time updates to reflect new information or correct model outputs, such as updating a model to reflect a recent geopolitical event or fixing factual inaccuracies.

Fig. 9.3 Model surgery

The primary motivation for knowledge editing stems from the dynamic nature of real-world knowledge and the limitations of static LLMs. Pre-trained models often encode outdated or incorrect facts, such as claiming "the current U.S. president is Barack Obama" in 2025, necessitating efficient correction mechanisms. Retraining entire models is computationally prohibitive, especially for billion-parameter LLMs, making knowledge editing a cost-effective alternative. This survey [58] highlights the importance of knowledge editing in maintaining model reliability and accuracy in dynamic environments.

According to [58], the knowledge editing of LLMs involves three phases: (1) recognition phase, which exposes novel knowledge to the LLM and identifies the knowledge to be edited in the model [59, 60]; (2) association phase, which connects the novel knowledge with the related existing knowledge in the model [61, 62]; (3) mastery phase, which modifies the model parameters to enable the model to fully acquire the novel knowledge [63–65].

Several approaches have been developed for knowledge editing, each with distinct mechanisms:

Memory-Based Methods: Techniques like SERAC (Semi-Parametric Editing with a Retrieval-Augmented Counterfactual Model) [59] and MemPrompt [66] store new or corrected knowledge in an external memory module, which the LLM queries during inference.

Parameter Editing Methods: Methods like ROME (Rank-One Model Editing) [63] and MEMIT (Mass-Editing Memory in a Transformer) [64] first locate the factual knowledge on some critical MLP layers of the language model via causal tracing, and then modify their weights to write target knowledge into the model, achieving precise edits with minimal collateral impact.

Meta-Learning Approaches: MEND (Model Editor Networks using Gradient Decomposition) [67] enables fast, local, and scalable editing of LLMs by training small auxiliary networks to transform fine-tuning gradients using low-rank decomposition, allowing efficient correction of model behavior with just a single input-output pair.

Constrained Fine-tuning: MELO [62] performs plug-in model editing for LLMs by clustering edits and employing neuron-indexed dynamic LoRA [68] blocks, activating different non-overlapping LoRA modules for each cluster via an internal vector database. By fine-tuning each cluster separately with dedicated LoRA blocks, MELO enables efficient, flexible, and precise updates while minimizing catastrophic forgetting and computational cost.

Knowledge editing offers significant advantages, including computational efficiency compared to full retraining, and flexibility for applications like real-time fact correction in chatbots. However, challenges include potential side effects, where edits may inadvertently affect related knowledge, leading to inconsistencies, as noted in [69]. Additionally, evaluating the long-term stability of edits remains complex, as edited knowledge may revert under certain conditions, and ensuring robustness across diverse inputs is an ongoing research focus. These limitations necessitate careful design of editing procedures to maintain model reliability in interactive settings.

9.3.2   Machine Unlearning

Machine unlearning focuses on removing the influence of specific training data from an LLM, ensuring the model behaves as if it had never been trained on that data. The primary motivation for machine unlearning is to address privacy concerns, such as those mandated by regulations like GDPR (European Union's privacy regulation), which grant users the "right to be forgotten". LLMs often retain traces of sensitive or copyrighted data from their training sets, posing risks of unintended disclosure. Additionally, unlearning is critical for removing harmful or biased content that may perpetuate stereotypes or misinformation in interactive dialogues.

To achieve the goal of removing specific data influences from LLMs, machine unlearning employs a range of strategies, each leveraging different mechanisms:

Gradient-Based Methods: These approaches use techniques like gradient ascent to reverse the learning process by fine-tuning model parameters to minimize the impact of target data. For example, [70] proposes optimizing a loss function to "erase" data influence, defined as: $\mathcal{L}(\theta) = -\mathbb{E}_{x \sim \mathcal{D}_{forget}}[-log p_\theta(x)]$. Further research optimizes this loss function by introducing regularization such as rehearsal [71] or KL divergence [72]. For example, LLM Surgery [73] performs efficient model unlearning and editing by optimizing a three-component objective: it uses reverse gradient to unlearn outdated/problematic knowledge, applies gradient descent to integrate new information, and minimizes KL divergence on retained data to preserve unchanged knowledge. Additionally, some works utilize simple SFT on relabeled data [74], while others employ reinforcement learning [75] or preference learning [76] to improve machine unlearning performance.

Parameter Editing Methods: Approaches inspired by knowledge editing (Sect. 9.3.1) [77–79]. For example, DEPN [78] detects and edits privacy neurons in pretrained language models by first locating neurons associated with private information using a privacy neuron detector, then mitigating data leakage by setting these neurons' activations to zero, and further applying a privacy neuron aggregator to dememorize private information in batch–significantly reducing privacy risks without harming model performance.

Auxiliary-Guided Methods: These methods leverage additional models to facilitate targeted unlearning in LLMs. Typically, an auxiliary or teacher model is specifically trained on the data to be forgotten, and its outputs are used to influence or guide the main model's unlearning process. For example, ULD [80] trains an auxiliary model to capture only the knowledge within the forget set, encouraging the main model to avoid generating such information during inference using Contrastive Decoding [81]. RKLD [82] performs model unlearning in LLMs by leveraging a reverse KL-divergence-based knowledge distillation algorithm.

Prompt Engineering Methods: These approaches achieve machine unlearning by manipulating the input prompts or their representations during inference, rather than adjusting model parameters. For example, In-Context Unlearning [83] performs machine unlearning by demonstrating targeted instances with flipped labels in the context during inference, removing their influence without updating model parameters. ECO Prompts [84] performs unlearning by employing a prompt classifier to detect prompts containing knowledge to be forgotten and then applying learned corruptions to their embeddings during inference.

Machine unlearning provides significant benefits, including enhanced privacy protection, compliance with legal standards, and the ability to mitigate harmful biases in real-time interactive systems. However, challenges include the difficulty of ensuring complete data removal, as residual influences may persist, and the potential for unlearning to disrupt related knowledge, as highlighted in [85]. Moreover, evaluating unlearning efficacy requires robust and generalizable metrics, which are still under research. We refer the readers to [86] for a more comprehensive survey on Machine Unlearning.

9.4   Interaction Message Fusion

In this subsection, we strive to provide a comprehensive and unified framing of the interaction message fusion methods, including all the methods presented in this section. Note that this framing also systematically categorizes the knowledge integration methods as mentioned in 2.2.

As illustrated in Fig. 9.4, the interaction message fusion methods can be divided into three dimensions, each having three categories. Thus, we have a total of $3 * 3 * 3 = 27$ clusters of ways to incorporate interaction messages into language models. The following will present the basic definition and examples for each dimension.

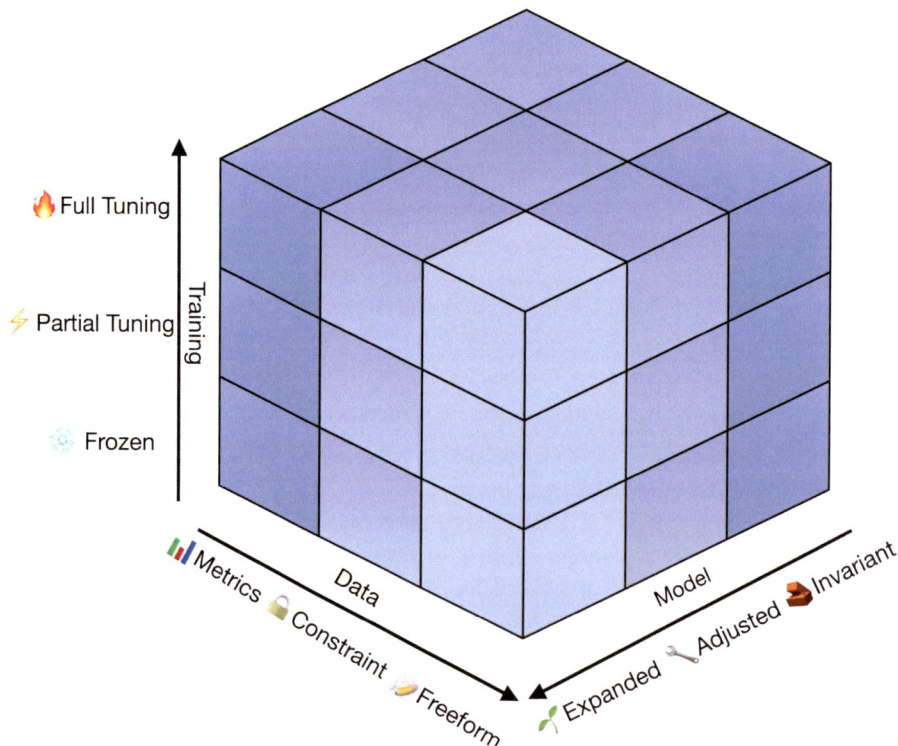

Fig. 9.4 The three dimensions of interaction message fusion methods

Along the data dimension. (1) "Metrics" refers to simple signals such as scalar rewards or ranking scores. (2) "Constraint" means that the interaction message is in form of keywords, templates, skeletons, constrained vocabularies and the like for message formatting, conditioning, refinement as well as data augmentation, etc. (3) "Freeform" involves providing an unconstrained and unstructured text as an interaction message. For example,

1. **Metrics**: Most reinforcement learning methods are related to this (6.4) as they rely on reward signals. For instance, RLHF [87, 88] utilizes a feedback mechanism wherein the language model outputs are scored to indicate their quality and safety. Several methods employ heuristics to filter interaction messages [89, 90], which can also be considered related, since the Boolean signals are leveraged. Sampling-based alignment[1] methods like Best-of-N (BoN) Sampling [91] and Rejection Sampling [92] also align with this approach.
2. **Constraint**: The constraint on the input end involves some conditional signals such as skeletons [93, 94], and outlines [95]. For instance, Re3 [95] generates an outline to condition long-text story generation. The constraint on the output end involves constrained decoding [96, 97], output refinement via a verbalizer which transforms the output [90], Best-of-N Sampling or Rejection Sametc.
3. **Freeform**: Most of the methods deal with freeform data, as defined by its unstructured nature.

Along the model dimension. (1) "Model-Invariant" approach does not modify any part of the model architecture. (2) "Model-Adjusted" approach makes changes to the existing modules of the model architecture. (3) "Model-Expanded" approach involves adding new modules to the model architecture. For example,

1. **Model-Invariant**: Prompting methods (Sect. prompting) belong to this line. Most of the tuning methods are also in this line of research. In most cases, the interaction message is concatenated or inserted to the input along sequence length dimension so that it is unnecessary to change the model architecture [98–101]. We can also only modify some critical mediating modules of the language model to incorporate knowledge [63, 64] with the model architecture unchanged. Since there is no change in the model architecture, it is crucial to effectively organize the inputs. For instance, Fusion in Decoder [102] suggests concatenating each retrieved passage with the query to encode each passage separately, and then concatenating these encoded passages for the decoder. While most other works [98, 99] concatenate all retrieved passages as a whole with the query.
2. **Model-Adjusted**: VaLM [103] replaces one layer of self-attention block with cross-attention block to fuse retrieved visual knowledge. K-BERT [104] uses a superposition of text embedding and knowledge embedding (i.e., token embedding+soft-position embedding+segment embedding) and an attention mask matrix to fan in the knowledge.

---

[1] C.f., 2.1.

3. **Model-Expanded**: K-Adapter [105] adds additional adapters to the model for knowledge enhancement as mentioned in Sect. paramspsefficient. Retro [106] adds cross-attention modules to the model for retrieval augmentation. RelationLM [107] uses an additional GRU function to fuse knowledge into the language model. KELM [108] uses additional hierarchical knowledge enhancement module to incorporate heterogeneous information from knowledge graphs and text into PLMs. This module employs a relational Graph Neural Network (GNN) to dynamically representing inputted knowledge, an attention mechanism to resolve knowledge ambiguity, and a self-attention mechanism for further interactions between knowledge-enriched tokens.

Along the training dimension. (1) "Frozen" means that all the model parameters are fixed and cannot be tuned. (2) "Partial Tuning" means that only a part of the model parameters are tuned. (3) "Full Tuning" refers to updating all model parameters. For example,

1. **Frozen**: This category mainly involves prompting methods (Sect. prompting). Besides, some constrained decoding-based methods [96, 97] also do not update model parameters. Another example is KPT [90], which refines the output verbalizer using metrics that consider the frequency and relevance of knowledgeable words.
2. **Partial Tuning**: For those which introduce additional modules or leverage a strict subset of model parameters to fan in external information [63, 64, 105], partial tuning is a trivial way as mentioned in Sect. paramspsefficient. For instance, K-Adapter [105] only tunes the adapter modules with the main body of the language model frozen. SpoT's generic variant [109] learns a source prompt on some source tasks with the language model frozen and then uses it to initialize another learnable soft prompt of the target task for knowledge transfer.
3. **Full Tuning**: Most methods require full tuning of the model parameters to integrate external information into the language model. For instance, ToolFormer [110] trains all the language model parameters on API call inserted corpus to enable the language model to use tools. Kepler [111] tunes all the model parameters with both knowledge embedding loss and masked language modeling loss. KnowBERT [112] embeds structured knowledge from multiple knowledge bases into PLMs. It uses a combined entity linker to obtain relevant entity embeddings and further enhances contextual word representations through word-to-entity attention, where the model parameters are also fully trained.

In summary, integrating external information into language models has emerged as a rapidly evolving research area in recent years, with numerous strategies available for its implementation. By breaking down the interaction message fusion methods into three dimensions along the data, model, and training dimensions, we provide a systematic categorization of the existing methods, which can help researchers and practitioners better understand and design new approaches. It is worth noting that these dimensions are not mutually exclusive, and different methods can combine different categories from each dimension. The selection of an appropriate category relies on the specific task, data, and resources, as well as the trade-off between performance and computational efficiency.

References

1. Ren, P., Xiao, Y., Chang, X., Huang, P.Y., Li, Z., Gupta, B.B., Chen, X., Wang, X.: A survey of deep active learning. ACM computing surveys (CSUR) **54**(9), 1–40 (2021)
2. Wang, Z.J., Choi, D., Xu, S., Yang, D.: Putting humans in the natural language processing loop: A survey. arXiv preprint arXiv:2103.04044 (2021)
3. Dossou, B.F.P., Tonja, A.L., Yousuf, O., Osei, S., Oppong, A., Shode, I., Awoyomi, O.O., Emezue, C.: AfroLM: A self-active learning-based multilingual pretrained language model for 23 African languages. In: Proceedings of The Third Workshop on Simple and Efficient Natural Language Processing (SustaiNLP), pp. 52–64. Association for Computational Linguistics, Abu Dhabi, United Arab Emirates (Hybrid) (2022). https://aclanthology.org/2022.sustainlp-1.11
4. Li, A.C., Brown, E., Efros, A.A., Pathak, D.: Internet explorer: Targeted representation learning on the open web. arXiv preprint arXiv:2302.14051 (2023)
5. Seo, S., Oh, B., Jo, E., Lee, S., Lee, D., Lee, K.H., Shin, D., Lee, Y.: Active learning for knowledge graph schema expansion. IEEE Transactions on Knowledge and Data Engineering **34**(12), 5610–5620 (2021)
6. Zhang, Z., Strubell, E., Hovy, E.: A survey of active learning for natural language processing. In: Proceedings of the 2022 Conference on Empirical Methods in Natural Language Processing, pp. 6166–6190. Association for Computational Linguistics, Abu Dhabi, United Arab Emirates (2022). https://aclanthology.org/2022.emnlp-main.414
7. Schröder, C., Niekler, A., Potthast, M.: Revisiting uncertainty-based query strategies for active learning with transformers. In: Findings of the Association for Computational Linguistics: ACL 2022, pp. 2194–2203. Association for Computational Linguistics, Dublin, Ireland (2022). https://doi.org/10.18653/v1/2022.findings-acl.172. https://aclanthology.org/2022.findings-acl.172
8. Shelmanov, A., Puzyrev, D., Kupriyanova, L., Belyakov, D., Larionov, D., Khromov, N., Kozlova, O., Artemova, E., Dylov, D.V., Panchenko, A.: Active learning for sequence tagging with deep pre-trained models and Bayesian uncertainty estimates. In: Proceedings of the 16th Conference of the European Chapter of the Association for Computational Linguistics: Main Volume, pp. 1698–1712. Association for Computational Linguistics, Online (2021). https://doi.org/10.18653/v1/2021.eacl-main.145. https://aclanthology.org/2021.eacl-main.145
9. Zhang, M., Plank, B.: Cartography active learning. In: Findings of the Association for Computational Linguistics: EMNLP 2021, pp. 395–406. Association for Computational Linguistics, Punta Cana, Dominican Republic (2021). https://doi.org/10.18653/v1/2021.findings-emnlp.36. https://aclanthology.org/2021.findings-emnlp.36
10. Shen, S., Li, Z., Qi, G.: Active learning for event extraction with memory-based loss prediction model. arXiv preprint arXiv:2112.03073 (2021)
11. Zhao, Y., Zhang, H., Zhou, S., Zhang, Z.: Active learning approaches to enhancing neural machine translation. In: Findings of the Association for Computational Linguistics: EMNLP 2020, pp. 1796–1806. Association for Computational Linguistics, Online (2020). https://doi.org/10.18653/v1/2020.findings-emnlp.162. https://aclanthology.org/2020.findings-emnlp.162
12. Yu, Y., Kong, L., Zhang, J., Zhang, R., Zhang, C.: Actune: Uncertainty-based active self-training for active fine-tuning of pretrained language models. In: Proceedings of the 2022 Conference

of the North American Chapter of the Association for Computational Linguistics: Human Language Technologies, pp. 1422–1436 (2022)

13. Tong, S.: Support vector machine active learning with applications to text classification. In: Proc. Seventeenth International Conference on Machine Learning, 2000 (2000)

14. Settles, B., Craven, M.: An analysis of active learning strategies for sequence labeling tasks. In: Proceedings of the 2008 Conference on Empirical Methods in Natural Language Processing, pp. 1070–1079. Association for Computational Linguistics, Honolulu, Hawaii (2008). https://aclanthology.org/D08-1112

15. Devlin, J., Chang, M.W., Lee, K., Toutanova, K.: Bert: Pre-training of deep bidirectional transformers for language understanding. arXiv preprint arXiv:1810.04805 (2018)

16. Ein-Dor, L., Halfon, A., Gera, A., Shnarch, E., Dankin, L., Choshen, L., Danilevsky, M., Aharonov, R., Katz, Y., Slonim, N.: Active Learning for BERT: An Empirical Study. In: Proceedings of the 2020 Conference on Empirical Methods in Natural Language Processing (EMNLP), pp. 7949–7962. Association for Computational Linguistics, Online (2020). https://doi.org/10.18653/v1/2020.emnlp-main.638. https://aclanthology.org/2020.emnlp-main.638

17. Seo, S., Kim, D., Ahn, Y., Lee, K.H.: Active learning on pre-trained language model with task-independent triplet loss. In: Proceedings of the AAAI Conference on Artificial Intelligence, vol. 36, pp. 11276–11284 (2022)

18. Zhang, S., Gong, C., Liu, X., He, P., Chen, W., Zhou, M.: Allsh: Active learning guided by local sensitivity and hardness. arXiv preprint arXiv:2205.04980 (2022)

19. Margatina, K., Barrault, L., Aletras, N.: On the importance of effectively adapting pretrained language models for active learning. In: Proceedings of the 60th Annual Meeting of the Association for Computational Linguistics (Volume 2: Short Papers), pp. 825–836. Association for Computational Linguistics, Dublin, Ireland (2022). https://doi.org/10.18653/v1/2022.acl-short.93. https://aclanthology.org/2022.acl-short.93

20. Zhang, Y., Feng, S., Tan, C.: Active example selection for in-context learning. arXiv preprint arXiv:2211.04486 (2022)

21. Li, B.Z., Stanovsky, G., Zettlemoyer, L.: Active learning for coreference resolution using discrete annotation. arXiv preprint arXiv:2004.13671 (2020)

22. Zhu, J., Hovy, E.: Active learning for word sense disambiguation with methods for addressing the class imbalance problem. In: Proceedings of the 2007 Joint Conference on Empirical Methods in Natural Language Processing and Computational Natural Language Learning (EMNLP-CoNLL), pp. 783–790. Association for Computational Linguistics, Prague, Czech Republic (2007). https://aclanthology.org/D07-1082

23. Birke, J., Sarkar, A.: Active learning for the identification of nonliteral language. In: Proceedings of the Workshop on Computational Approaches to Figurative Language, pp. 21–28. Association for Computational Linguistics, Rochester, New York (2007). https://aclanthology.org/W07-0104

24. Peshterliev, S., Kearney, J., Jagannatha, A., Kiss, I., Matsoukas, S.: Active learning for new domains in natural language understanding. arXiv preprint arXiv:1810.03450 (2018)

25. Quteineh, H., Samothrakis, S., Sutcliffe, R.: Textual data augmentation for efficient active learning on tiny datasets. In: Proceedings of the 2020 Conference on Empirical Methods in Natural Language Processing (EMNLP), pp. 7400–7410. Association for Computational Linguistics, Online (2020). https://doi.org/10.18653/v1/2020.emnlp-main.600. https://aclanthology.org/2020.emnlp-main.600

26. Grießhaber, D., Maucher, J., Vu, N.T.: Fine-tuning BERT for low-resource natural language understanding via active learning. In: Proceedings of the 28th International Conference on Computational Linguistics, pp. 1158–1171. International Committee on Computational Linguistics, Barcelona, Spain (Online) (2020). https://doi.org/10.18653/v1/2020.coling-main.100. https://aclanthology.org/2020.coling-main.100

27. Maekawa, S., Zhang, D., Kim, H., Rahman, S., Hruschka, E.: Low-resource interactive active labeling for fine-tuning language models. In: Findings of the Association for Computational Linguistics: EMNLP 2022, pp. 3230–3242 (2022)

28. Wang, Y., Kordi, Y., Mishra, S., Liu, A., Smith, N.A., Khashabi, D., Hajishirzi, H.: Self-instruct: Aligning language model with self generated instructions. arXiv preprint arXiv: Arxiv-2212.10560 (2022)

29. Ye, H., Zhang, N., Deng, S., Chen, X., Chen, H., Xiong, F., Chen, X., Chen, H.: Ontology-enhanced prompt-tuning for few-shot learning. arXiv preprint arXiv: Arxiv-2201.11332 (2022)

30. Yao, S., Zhao, J., Yu, D., Du, N., Shafran, I., Narasimhan, K., Cao, Y.: React: Synergizing reasoning and acting in language models. arXiv preprint arXiv: Arxiv-2210.03629 (2022)

31. Chen, W., Ma, X., Wang, X., Cohen, W.W.: Program of thoughts prompting: Disentangling computation from reasoning for numerical reasoning tasks. arXiv preprint arXiv:2211.12588 (2022)

32. Fan, L., Wang, G., Jiang, Y., Mandlekar, A., Yang, Y., Zhu, H., Tang, A., Huang, D.A., Zhu, Y., Anandkumar, A.: Minedojo: Building open-ended embodied agents with internet-scale knowledge. arXiv preprint arXiv:2206.08853 (2022)

33. Chen, B., Xia, F., Ichter, B., Rao, K., Gopalakrishnan, K., Ryoo, M.S., Stone, A., Kappler, D.: Open-vocabulary queryable scene representations for real world planning. arXiv preprint arXiv: Arxiv-2209.09874 (2022)

34. Pomerleau, D.A.: Alvinn: An autonomous land vehicle in a neural network. Advances in neural information processing systems **1** (1988)

35. Pomerleau, D.A.: Efficient Training of Artificial Neural Networks for Autonomous Navigation. Neural Computation **3**(1), 88–97 (1991). https://doi.org/10.1162/neco.1991.3.1.88. https://doi.org/10.1162/neco.1991.3.1.88

36. Williams, R.J., Zipser, D.: A learning algorithm for continually running fully recurrent neural networks. Neural Computation **1**(2), 270–280 (1989). https://doi.org/10.1162/neco.1989.1.2.270

37. Ross, S., Gordon, G., Bagnell, D.: A reduction of imitation learning and structured prediction to no-regret online learning. In: Proceedings of the fourteenth international conference on artificial intelligence and statistics, pp. 627–635. JMLR Workshop and Conference Proceedings (2011)

38. Ng, A.Y., Russell, S., et al.: Algorithms for inverse reinforcement learning. In: Icml, vol. 1, p. 2 (2000)

39. Bansal, M., Krizhevsky, A., Ogale, A.: Chauffeurnet: Learning to drive by imitating the best and synthesizing the worst. arXiv preprint arXiv:1812.03079 (2018)

40. Humphreys, P.C., Raposo, D., Pohlen, T., Thornton, G., Chhaparia, R., Muldal, A., Abramson, J., Georgiev, P., Santoro, A., Lillicrap, T.: A data-driven approach for learning to control computers. In: K. Chaudhuri, S. Jegelka, L. Song, C. Szepesvari, G. Niu, S. Sabato (eds.) Proceedings of the 39th International Conference on Machine Learning, *Proceedings of Machine Learning Research*, vol. 162, pp. 9466–9482. PMLR (2022). https://proceedings.mlr.press/v162/humphreys22a.html

41. Silver, D., Huang, A., Maddison, C.J., Guez, A., Sifre, L., Van Den Driessche, G., Schrittwieser, J., Antonoglou, I., Panneershelvam, V., Lanctot, M., et al.: Mastering the game of go with deep neural networks and tree search. nature **529**(7587), 484–489 (2016)

42. Baker, B., Akkaya, I., Zhokov, P., Huizinga, J., Tang, J., Ecoffet, A., Houghton, B., Sampedro, R., Clune, J.: Video pretraining (VPT): Learning to act by watching unlabeled online videos. In: A.H. Oh, A. Agarwal, D. Belgrave, K. Cho (eds.) Advances in Neural Information Processing Systems (2022). https://openreview.net/forum?id=AXDNM76T1nc

43. Team, D.I.A., Abramson, J., Ahuja, A., Brussee, A., Carnevale, F., Cassin, M., Fischer, F., Georgiev, P., Goldin, A., Gupta, M., et al.: Creating multimodal interactive agents with imitation and self-supervised learning. arXiv preprint arXiv:2112.03763 (2021)

44. Jang, E., Irpan, A., Khansari, M., Kappler, D., Ebert, F., Lynch, C., Levine, S., Finn, C.: Bc-z: Zero-shot task generalization with robotic imitation learning. In: A. Faust, D. Hsu, G. Neumann (eds.) Proceedings of the 5th Conference on Robot Learning, *Proceedings of Machine Learning Research*, vol. 164, pp. 991–1002. PMLR (2022). https://proceedings.mlr.press/v164/jang22a.html

45. Lynch, C., Wahid, A., Tompson, J., Ding, T., Betker, J., Baruch, R., Armstrong, T., Florence, P.: Interactive language: Talking to robots in real time (2022)

46. Karamcheti, S., Nair, S., Chen, A.S., Kollar, T., Finn, C., Sadigh, D., Liang, P.: Language-driven representation learning for robotics. arXiv preprint arXiv:2302.12766 (2023)

47. Zhang, T., McCarthy, Z., Jowl, O., Lee, D., Chen, X., Goldberg, K., Abbeel, P.: Deep imitation learning for complex manipulation tasks from virtual reality teleoperation. In: 2018 IEEE International Conference on Robotics and Automation (ICRA), p. 1–8. IEEE Press (2018). https://doi.org/10.1109/ICRA.2018.8461249. https://doi.org/10.1109/ICRA.2018.8461249

48. Peng, X.B., Coumans, E., Zhang, T., Lee, T.W.E., Tan, J., Levine, S.: Learning agile robotic locomotion skills by imitating animals. In: Robotics: Science and Systems (2020). https://doi.org/10.15607/RSS.2020.XVI.064

49. Tanwani, A.K., Sermanet, P., Yan, A., Anand, R., Phielipp, M., Goldberg, K.: Motion2vec: Semi-supervised representation learning from surgical videos. In: Proc. of IEEE Intl Conf. on Robotics and Automation (ICRA), pp. 1–8 (2020)

50. Du, W., Ji, Y.: An empirical comparison on imitation learning and reinforcement learning for paraphrase generation. In: Proceedings of the 2019 Conference on Empirical Methods in Natural Language Processing and the 9th International Joint Conference on Natural Language Processing (EMNLP-IJCNLP), pp. 6012–6018. Association for Computational Linguistics, Hong Kong, China (2019). https://doi.org/10.18653/v1/D19-1619. https://aclanthology.org/D19-1619

51. Chen, Y., Su, J., Wei, W.: Multi-granularity textual adversarial attack with behavior cloning. In: Proceedings of the 2021 Conference on Empirical Methods in Natural Language Processing, pp. 4511–4526. Association for Computational Linguistics, Online and Punta Cana, Dominican Republic (2021). https://doi.org/10.18653/v1/2021.emnlp-main.371. https://aclanthology.org/2021.emnlp-main.371

52. Agrawal, S., Carpuat, M.: An imitation learning curriculum for text editing with non-autoregressive models. In: Proceedings of the 60th Annual Meeting of the Association for Computational Linguistics (Volume 1: Long Papers), pp. 7550–7563. Association for Computational Linguistics, Dublin, Ireland (2022). https://doi.org/10.18653/v1/2022.acl-long.520. https://aclanthology.org/2022.acl-long.520

53. Shi, N., Tang, B., Yuan, B., Huang, L., Pu, Y., Fu, J., Lin, Z.: Text editing as imitation game. In: Findings of the Association for Computational Linguistics: EMNLP 2022, pp. 1583–1594. Association for Computational Linguistics, Abu Dhabi, United Arab Emirates (2022). https://aclanthology.org/2022.findings-emnlp.114

54. Hao, Y., Liu, Y., Mou, L.: Teacher forcing recovers reward functions for text generation. In: A.H. Oh, A. Agarwal, D. Belgrave, K. Cho (eds.) Advances in Neural Information Processing Systems (2022). https://openreview.net/forum?id=1_gypPuWUC3

55. Gu, J., Wang, C., Zhao, J.: Levenshtein transformer. In: H. Wallach, H. Larochelle, A. Beygelzimer, F. d'Alché-Buc, E. Fox, R. Garnett (eds.) Advances in Neural Information Processing Systems 32, pp. 11181–11191. Curran Associates, Inc. (2019)

56. Faltings, F., Galley, M., Peng, B., Brantley, K., Cai, W., Zhang, Y., Gao, J., Dolan, B.: Interactive text generation. arXiv preprint arXiv:2303.00908 (2023)

57. Arora, K., El Asri, L., Bahuleyan, H., Cheung, J.: Why exposure bias matters: An imitation learning perspective of error accumulation in language generation. In: Findings of the Association for Computational Linguistics: ACL 2022, pp. 700–710. Association for Computational

Linguistics, Dublin, Ireland (2022). https://doi.org/10.18653/v1/2022.findings-acl.58. https://aclanthology.org/2022.findings-acl.58

58. Zhang, N., Yao, Y., Tian, B., Wang, P., Deng, S., Wang, M., Xi, Z., Mao, S., Zhang, J., Ni, Y., Cheng, S., Xu, Z., Xu, X., Gu, J.C., Jiang, Y., Xie, P., Huang, F., Liang, L., Zhang, Z., Zhu, X., Zhou, J., Chen, H.: A comprehensive study of knowledge editing for large language models. arXiv preprint arXiv: 2401.01286 (2024)

59. Mitchell, E., Lin, C., Bosselut, A., Manning, C.D., Finn, C.: Memory-based model editing at scale. arXiv preprint arXiv: 2206.06520 (2022)

60. Zhong, Z., Wu, Z., Manning, C.D., Potts, C., Chen, D.: Mquake: Assessing knowledge editing in language models via multi-hop questions. arXiv preprint arXiv: 2305.14795 (2023)

61. Murty, S., Manning, C.D., Lundberg, S., Ribeiro, M.T.: Fixing model bugs with natural language patches. arXiv preprint arXiv: 2211.03318 (2022)

62. Yu, L., Chen, Q., Zhou, J., He, L.: Melo: Enhancing model editing with neuron-indexed dynamic lora. Proceedings of the AAAI Conference on Artificial Intelligence **38**(17), 19449–19457 (2024). https://doi.org/10.1609/aaai.v38i17.29916. https://ojs.aaai.org/index.php/AAAI/article/view/29916

63. Meng, K., Bau, D., Andonian, A., Belinkov, Y.: Locating and editing factual associations in GPT. Advances in Neural Information Processing Systems **36** (2022)

64. Meng, K., Sen Sharma, A., Andonian, A., Belinkov, Y., Bau, D.: Mass editing memory in a transformer. arXiv preprint arXiv:2210.07229 (2022)

65. Fang, J., Jiang, H., Wang, K., Ma, Y., Jie, S., Wang, X., He, X., seng Chua, T.: Alphaedit: Null-space constrained knowledge editing for language models. arXiv preprint arXiv: 2410.02355 (2024)

66. Madaan, A., Tandon, N., Clark, P., Yang, Y.: Memory-assisted prompt editing to improve GPT-3 after deployment. arXiv preprint arXiv:2201.06009 (2022)

67. Mitchell, E., Lin, C., Bosselut, A., Finn, C., Manning, C.D.: Fast model editing at scale. In: The Tenth International Conference on Learning Representations, ICLR 2022, Virtual Event, April 25-29, 2022. OpenReview.net (2022). https://openreview.net/forum?id=0DcZxeWfOPt

68. Hu, E.J., yelong shen, Wallis, P., Allen-Zhu, Z., Li, Y., Wang, S., Wang, L., Chen, W.: LoRA: Low-rank adaptation of large language models. In: International Conference on Learning Representations (2022). https://openreview.net/forum?id=nZeVKeeFYf9

69. Li, Z., Zhang, N., Yao, Y., Wang, M., Chen, X., Chen, H.: Unveiling the pitfalls of knowledge editing for large language models. arXiv preprint arXiv: 2310.02129 (2023)

70. Jang, J., Yoon, D., Yang, S., Cha, S., Lee, M., Logeswaran, L., Seo, M.: Knowledge unlearning for mitigating privacy risks in language models. arXiv preprint arXiv: 2210.01504 (2022)

71. Liu, B., Liu, Q., Stone, P.: Continual learning and private unlearning. arXiv preprint arXiv: 2203.12817 (2022)

72. Yao, J., Chien, E., Du, M., Niu, X., Wang, T., Cheng, Z., Yue, X.: Machine unlearning of pre-trained large language models. In: L.W. Ku, A. Martins, V. Srikumar (eds.) Proceedings of the 62nd Annual Meeting of the Association for Computational Linguistics (Volume 1: Long Papers), pp. 8403–8419. Association for Computational Linguistics, Bangkok, Thailand (2024). https://doi.org/10.18653/v1/2024.acl-long.457. https://aclanthology.org/2024.acl-long.457/

73. Veldanda, A.K., Zhang, S.X., Das, A., Chakraborty, S., Rawls, S., Sahu, S., Naphade, M.: LLM surgery: Efficient knowledge unlearning and editing in large language models. arXiv preprint arXiv: 2409.13054 (2024)

74. Eldan, R., Russinovich, M.: Who's harry potter? approximate unlearning for LLMs (2024). https://openreview.net/forum?id=PDct7vrcvT

75. Kassem, A., Mahmoud, O., Saad, S.: Preserving privacy through dememorization: An unlearning technique for mitigating memorization risks in language models. In: H. Bouamor,

J. Pino, K. Bali (eds.) Proceedings of the 2023 Conference on Empirical Methods in Natural Language Processing, pp. 4360–4379. Association for Computational Linguistics, Singapore (2023). https://doi.org/10.18653/v1/2023.emnlp-main.265. https://aclanthology.org/2023.emnlp-main.265/

76. Maini, P., Feng, Z., Schwarzschild, A., Lipton, Z.C., Kolter, J.Z.: Tofu: A task of fictitious unlearning for LLMs. arXiv preprint arXiv: 2401.06121 (2024)

77. Jia, J., Liu, J., Zhang, Y., Ram, P., Baracaldo, N., Liu, S.: WAGLE: strategic weight attribution for effective and modular unlearning in large language models. In: A. Globersons, L. Mackey, D. Belgrave, A. Fan, U. Paquet, J.M. Tomczak, C. Zhang (eds.) Advances in Neural Information Processing Systems 38: Annual Conference on Neural Information Processing Systems 2024, NeurIPS 2024, Vancouver, BC, Canada, December 10 - 15, 2024 (2024). http://papers.nips.cc/paper_files/paper/2024/hash/649ad92e7067b3553a0f15acac68806d-Abstract-Conference.html

78. Wu, X., Li, J., Xu, M., Dong, W., Wu, S., Bian, C., Xiong, D.: Depn: Detecting and editing privacy neurons in pretrained language models. Conference on Empirical Methods in Natural Language Processing (2023). https://doi.org/10.48550/arXiv.2310.20138

79. Hong, Y., Yu, L., Yang, H., Ravfogel, S., Geva, M.: Intrinsic evaluation of unlearning using parametric knowledge traces. arXiv preprint arXiv: 2406.11614 (2024)

80. Ji, J., Liu, Y., Zhang, Y., Liu, G., Kompella, R.R., Liu, S., Chang, S.: Reversing the forget-retain objectives: An efficient LLM unlearning framework from logit difference. In: A. Globerson, L. Mackey, D. Belgrave, A. Fan, U. Paquet, J. Tomczak, C. Zhang (eds.) Advances in Neural Information Processing Systems, vol. 37, pp. 12581–12611. Curran Associates, Inc. (2024). https://proceedings.neurips.cc/paper_files/paper/2024/file/171291d8fed723c6dfc76330aa827ff8-Paper-Conference.pdf

81. Li, X.L., Holtzman, A., Fried, D., Liang, P., Eisner, J., Hashimoto, T., Zettlemoyer, L., Lewis, M.: Contrastive decoding: Open-ended text generation as optimization. arXiv preprint arXiv: 2210.15097 (2022)

82. Wang, B., Zi, Y., Sun, Y., Zhao, Y., Qin, B.: Rkld: Reverse kl-divergence-based knowledge distillation for unlearning personal information in large language models. arXiv preprint arXiv: 2406.01983 (2024)

83. Pawelczyk, M., Neel, S., Lakkaraju, H.: In-context unlearning: Language models as few shot unlearners. International Conference on Machine Learning (2023). https://doi.org/10.48550/arXiv.2310.07579

84. Liu, C.Y., Wang, Y., Flanigan, J., Liu, Y.: Large language model unlearning via embedding-corrupted prompts. arXiv preprint arXiv: 2406.07933 (2024)

85. Liu, S., Yao, Y., Jia, J., Casper, S., Baracaldo, N., Hase, P., Yao, Y., Liu, C.Y., Xu, X., Li, H., Varshney, K.R., Bansal, M., Koyejo, S., Liu, Y.: Rethinking machine unlearning for large language models. arXiv preprint arXiv: 2402.08787 (2024)

86. Geng, J., Li, Q., Woisetschlaeger, H., Chen, Z., Cai, F., Wang, Y., Nakov, P., Jacobsen, H.A., Karray, F.: A comprehensive survey of machine unlearning techniques for large language models. arXiv preprint arXiv: 2503.01854 (2025). https://arxiv.org/abs/2503.01854

87. Christiano, P., Leike, J., Brown, T.B., Martic, M., Legg, S., Amodei, D.: Deep reinforcement learning from human preferences. arXiv preprint arXiv: Arxiv-1706.03741 (2017)

88. Ouyang, L., Wu, J., Jiang, X., Almeida, D., Wainwright, C.L., Mishkin, P., Zhang, C., Agarwal, S., Slama, K., Ray, A., et al.: Training language models to follow instructions with human feedback. arXiv preprint arXiv:2203.02155 (2022)

89. Li, J., Li, D., Xiong, C., Hoi, S.: Blip: Bootstrapping language-image pre-training for unified vision-language understanding and generation. In: ICML (2022)

90. Hu, S., Ding, N., Wang, H., Liu, Z., Li, J.Z., Sun, M.: Knowledgeable prompt-tuning: Incorporating knowledge into prompt verbalizer for text classification. Annual Meeting Of The Association For Computational Linguistics (2021). https://doi.org/10.18653/v1/2022.acl-long.158

91. Hilton, J., Gao, L.: Measuring goodhart's law. https://openai.com/research/measuring-goodharts-law (2022). Accessed: April 13, 2022

92. Liu, T., Zhao, Y., Joshi, R., Khalman, M., Saleh, M., Liu, P.J., Liu, J.: Statistical rejection sampling improves preference optimization. arXiv preprint arXiv: 2309.06657 (2023). https://arxiv.org/abs/2309.06657v1

93. Wu, Y., Wei, F., Huang, S., Li, Z., Zhou, M.: Response generation by context-aware prototype editing. Aaai Conference On Artificial Intelligence (2018). https://doi.org/10.1609/aaai.v33i01.33017281

94. Cai, D., Wang, Y., Bi, W., Tu, Z., Liu, X., Lam, W., Shi, S.: Skeleton-to-response: Dialogue generation guided by retrieval memory. In: Proceedings of the 2019 Conference of the North American Chapter of the Association for Computational Linguistics: Human Language Technologies, Volume 1 (Long and Short Papers), pp. 1219–1228. Association for Computational Linguistics, Minneapolis, Minnesota (2019). https://doi.org/10.18653/v1/N19-1124. https://aclanthology.org/N19-1124

95. Yang, K., Peng, N., Tian, Y., Klein, D.: Re3: Generating longer stories with recursive reprompting and revision. arXiv preprint arXiv:2210.06774 (2022)

96. Shin, R., Lin, C., Thomson, S., Chen, C., Roy, S., Platanios, E.A., Pauls, A., Klein, D., Eisner, J., Van Durme, B.: Constrained language models yield few-shot semantic parsers. In: Proceedings of the 2021 Conference on Empirical Methods in Natural Language Processing, pp. 7699–7715. Association for Computational Linguistics, Online and Punta Cana, Dominican Republic (2021). https://doi.org/10.18653/v1/2021.emnlp-main.608. https://aclanthology.org/2021.emnlp-main.608

97. Hokamp, C., Liu, Q.: Lexically constrained decoding for sequence generation using grid beam search. In: Proceedings of the 55th Annual Meeting of the Association for Computational Linguistics (Volume 1: Long Papers), pp. 1535–1546. Association for Computational Linguistics, Vancouver, Canada (2017). https://doi.org/10.18653/v1/P17-1141. https://aclanthology.org/P17-1141

98. Guu, K., Lee, K., Tung, Z., Pasupat, P., Chang, M.W.: Realm: Retrieval-augmented language model pre-training. arXiv preprint arXiv: Arxiv-2002.08909 (2020)

99. Izacard, G., Lewis, P., Lomeli, M., Hosseini, L., Petroni, F., Schick, T., Dwivedi-Yu, J., Joulin, A., Riedel, S., Grave, E.: Atlas: Few-shot learning with retrieval augmented language models. arXiv preprint arXiv: Arxiv-2208.03299 (2022)

100. Paranjape, B., Lundberg, S., Singh, S., Hajishirzi, H., Zettlemoyer, L., Ribeiro, M.T.: Art: Automatic multi-step reasoning and tool-use for large language models. arXiv preprint arXiv: Arxiv-2303.09014 (2023)

101. Xie, T., Wu, C.H., Shi, P., Zhong, R., Scholak, T., Yasunaga, M., Wu, C.S., Zhong, M., Yin, P., Wang, S.I., et al.: Unifiedskg: Unifying and multi-tasking structured knowledge grounding with text-to-text language models. arXiv preprint arXiv:2201.05966 (2022)

102. Izacard, G., Grave, E.: Leveraging passage retrieval with generative models for open domain question answering. Conference of the European Chapter of the Association for Computational Linguistics (2020). https://doi.org/10.18653/v1/2021.eacl-main.74. https://arxiv.org/abs/2007.01282v2

103. Wang, W., Dong, L., Cheng, H., Song, H., Liu, X., Yan, X., Gao, J., Wei, F.: Visually-augmented language modeling. arXiv preprint arXiv:2205.10178 (2022)

104. Liu, W., Zhou, P., Zhao, Z., Wang, Z., Ju, Q., Deng, H., Wang, P.: K-bert: Enabling language representation with knowledge graph. Aaai Conference On Artificial Intelligence (2019). https://doi.org/10.1609/AAAI.V34I03.5681

105. Wang, R., Tang, D., Duan, N., Wei, Z., Huang, X., Ji, J., Cao, G., Jiang, D., Zhou, M.: K-adapter: Infusing knowledge into pre-trained models with adapters. In: C. Zong, F. Xia, W. Li, R. Navigli (eds.) Findings of the Association for Computational Linguistics: ACL/IJCNLP 2021, Online Event, August 1-6, 2021, *Findings of ACL*, vol. ACL/IJCNLP 2021, pp. 1405–1418. Association for Computational Linguistics (2021). https://doi.org/10.18653/v1/2021.findings-acl.121. https://doi.org/10.18653/v1/2021.findings-acl.121

106. Borgeaud, S., Mensch, A., Hoffmann, J., Cai, T., Rutherford, E., Millican, K., van den Driessche, G., Lespiau, J., Damoc, B., Clark, A., de Las Casas, D., Guy, A., Menick, J., Ring, R., Hennigan, T., Huang, S., Maggiore, L., Jones, C., Cassirer, A., Brock, A., Paganini, M., Irving, G., Vinyals, O., Osindero, S., Simonyan, K., Rae, J.W., Elsen, E., Sifre, L.: Improving language models by retrieving from trillions of tokens. International Conference On Machine Learning (2021)

107. Liu, Q., Yogatama, D., Blunsom, P.: Relational memory augmented language models. arXiv preprint arXiv: Arxiv-2201.09680 (2022)

108. Lu, Y., Lu, H., Fu, G., Liu, Q.: Kelm: Knowledge enhanced pre-trained language representations with message passing on hierarchical relational graphs. arXiv preprint arXiv: Arxiv-2109.04223 (2021)

109. Vu, T., Lester, B., Constant, N., Al-Rfou', R., Cer, D.: SPoT: Better frozen model adaptation through soft prompt transfer. In: Proceedings of the 60th Annual Meeting of the Association for Computational Linguistics (Volume 1: Long Papers), pp. 5039–5059. Association for Computational Linguistics, Dublin, Ireland (2022). https://doi.org/10.18653/v1/2022.acl-long.346. https://aclanthology.org/2022.acl-long.346

110. Schick, T., Dwivedi-Yu, J., Dessì, R., Raileanu, R., Lomeli, M., Zettlemoyer, L., Cancedda, N., Scialom, T.: Toolformer: Language models can teach themselves to use tools. CoRR **abs/2302.04761** (2023). https://doi.org/10.48550/arXiv.2302.04761. https://doi.org/10.48550/arXiv.2302.04761

111. Wang, X., Gao, T., Zhu, Z., Liu, Z., Li, J.Z., Tang, J.: Kepler: A unified model for knowledge embedding and pre-trained language representation. Transactions Of The Association For Computational Linguistics (2019). https://doi.org/10.1162/tacl_a_00360

112. Peters, M.E., Neumann, M., Logan, R., Schwartz, R., Joshi, V., Singh, S., Smith, N.A.: Knowledge enhanced contextual word representations. In: Proceedings of the 2019 Conference on Empirical Methods in Natural Language Processing and the 9th International Joint Conference on Natural Language Processing (EMNLP-IJCNLP), pp. 43–54. Association for Computational Linguistics, Hong Kong, China (2019https://doi.org/10.18653/v1/D19-1005. https://aclanthology.org/D19-1005

After introducing the fundamental frameworks and methodologies of Interactive Natural Language Processing (iNLP), this part presents a comprehensive examination of its downstream applications and evaluation methodologies.

Unlike traditional evaluation methods, which primarily focus on assessing the standalone capabilities of language models, this part emphasizes how to evaluate the interactive abilities of large language models. This requires designing distinct metrics and evaluation systems tailored to different interactive objects. Specifically, for human-LM interaction, we explore general metrics for understanding and responding to human instructions, as well as task-specific evaluation methods. For other interactive entities, such as knowledge bases, tools, and environments, we discuss evaluation standards for knowledge acquisition and enhanced generation, tool-use proficiency, navigation and manipulation within environments, among others. By measuring these different aspects of interaction, we gain a detailed understanding of the interactive performance of Large Language Models (LLMs) and can monitor and improve their real-world effectiveness.

In addition to evaluation, this part also highlights several key applications of iNLP, including controllable text generation, writing assistance, embodied AI, text games, and others. These examples illustrate how iNLP can be applied to create more personalized, responsive, and context-aware systems that can effectively operate in dynamic and interactive environments.

Evaluation

Kexin Yang, Dayiheng Liu and Zekun Wang

Evaluation is undoubtedly important in tracking the NLP progress [1], and even helps to outline future directions of NLP, e.g., Turing test [2, 3]. Existing surveys have comparatively elaborated on both automatic evaluation and human evaluation methods from general [1] to task-specific metrics, e.g., evaluating controllable text generation [4]. However, iNLP is slightly different from the general NLP tasks, as it devotes greater attention to the quality and effectiveness of model interactions with humans, environments, etc. In the interest of brevity and simplicity, this section will shed light on the relationship between iNLP and evaluation, yet does not retrace nor depict the historical evolution of general NLP evaluation. For further information, please refer to related NLP evaluation surveys [1, 5]. Please note that our survey primarily focuses on generative natural language processing, i.e., natural language generation (NLG).

Interactive scenarios occur naturally in various general natural language generation (NLG) tasks, such as dialogue [6–8] and question answering [9, 10]. However, a large number of existing NLG metrics mainly evaluate the non-interactive performance of an NLG model [11], which ignores evaluating the interactive quality and effectiveness happening in the model inference stage. Specifically, these metrics focus on comparing the difference between the model completion with the pre-specified human reference to deter-

K. Yang
Chengdu, China

D. Liu
Sichuan University, Chengdu, China

Z. Wang (✉)
Beihang University, Haidian District, Beijing, China
e-mail: zenmoore@buaa.edu.cn

© The Author(s), under exclusive license to Springer Nature Switzerland AG 2026
Z. Wang et al. (eds.), *Interactive Natural Language Processing*, Synthesis Lectures
on Human Language Technologies, https://doi.org/10.1007/978-3-032-06264-2_10

mine whether the performance is beyond previous models, such as Meteor [12], Rouge [13], and CIDEr [14]. While they are convenient and time-saving, the interaction quality may not be improved if the model only resorts to feedback from such evaluation metrics, resulting in it being hardly fit for usage in the real world. For example, BLEU [15] is one of the most commonly used metrics for NLG systems. Instead of directly telling the generation quality of the system, it focuses on comparing the differences between the model output and reference by n-gram matching. In this case, a higher lexical similarity leads to a higher score. Thus, a sentence whose expression is not similar to the reference will get a low score, even if it meets the user's given preferences [16]. Although interaction evaluation is still in its fledgling stage, researchers are aware of such insufficiency in current NLG evaluation and are diverting their attention to the interaction performance of a NLG model.

10.1   Evaluating Human-in-the-Loop Interaction

Evaluating human-in-the-loop aims to evaluate the performance of human-system interaction, which can be divided into two technical routes: general and task-specific metrics.

General Metrics. These metrics are agnostic to specific tasks and primarily aim to evaluate the general performance of PLMs. The key challenge is to evaluate the alignment between the language model and humans, that is, whether the NLG model satisfies certain human preferences. To solve this problem, [17] gives a specific interpretation of this nebulous concept–a model is aligned if it is helpful (i.e., helps to solve the task), honest (i.e., ensures the authenticity of information), and harmless (i.e., conforms to ethics). In detail, they evaluate the helpfulness property by preference rating from human labelers, while the honesty property is automatically evaluated by a hallucination dataset and the TruthfulQA dataset [18]. As for the harmlessness property, they use both Real Toxicity Prompts [19] and CrowS-Pairs datasets [20] to evaluate the bias and toxicity of the NLG model. Another important metric is the Instruction Following Score (IFS [21]), which measures the models' ability to follow users' instructions. To further expand the dimensions of interaction evaluation, HALIE [11] proposes a human-language model interaction evaluation framework, which includes a user interface to facilitate the interaction between humans and language models. Based on this system, HALIE introduces metrics that expand the non-interaction evaluation in three dimensions: (1) **Targets**, evaluating extra interactive processes except model completions, such as user edits; (2) **Perspectives**, adding human evaluation involving users who come into direct interaction with PLMs; (3) **Criteria**, highlighting human preference when evaluating the model completions, e.g., enjoyment, and helpfulness. Comprehensive experiments in HALIE demonstrate a significant disparity between human evaluations conducted by third-party annotators and those provided by interactive system users. This disparity underscores the crucial need to investigate interaction-based evaluation methodologies.

Task-specific Metrics. Task-specific metrics are concerned with designing a customized interaction evaluation method for generation tasks, including task-specific feature measurement when evaluating human-model interaction. Such task-specific metrics are most commonly found in dialogue system [22], question answering [23], creative writing [24], etc. For example, dialogue system is interested in assessing the quality of human-model interaction in specific conversation scenarios [22, 25] and evaluating the performance gain of adding human feedback in the model inference stage [26]. In particular, the metrics of task-oriented dialogue interaction evaluation largely consist of three items: (1) the success rate of tasks [27], such as successfully booking flights or movies; (2) dialogue turn size [28], fewer numbers indicate that the dialogue agent is better at completing a task with less time cost; (3) the accuracy in the dialog state track (DST) [29], i.e., determining if the key information is accurate throughout the conversation. Additionally, human experience and knowledge have also been employed in the evaluation of question-answering systems. For instance, [23] involve human authors in the process to enhance the generator's ability to create a diverse set of adversarial questions. This approach enables a more comprehensive evaluation of the robustness of question-answering systems. As for creative writing, CoAuthor [24] introduces a dataset capturing interactions between writers and Large Language Model (LLM), and demonstrates its usefulness in exploring LLM's language, ideation, and collaboration capabilities for creative and argumentative writing.

Despite the advances made in Human-LM interaction evaluation metrics, the development of a comprehensive general benchmark remains largely unexplored [30, 31]. A unified benchmark and metric play a crucial role in evaluating models from an interactive perspective; however, designing and collecting such a benchmark pose significant challenges. One of the primary difficulties stems from the specificity of interaction evaluation, which necessitates detailed tracking and recording of the interactive process rather than merely collecting model completions. In addition, the interactive nature of human-in-the-loop evaluation can lead to data inconsistencies due to inherent individual diversity. This further complicates the task of constructing a general benchmark. Building such a benchmark necessitates careful consideration in terms of the user interface, means, and interaction process.

10.2 Evaluating KB-in-the-Loop Interaction

Evaluation of KB-in-the-loop interaction naturally arises in knowledge-augmented NLG models as a means of assessing their ability to acquire knowledge for enhanced generation.

Knowledge Acquisition. As discussed in Sect. 2.2, knowledge acquisition can be achieved through retrieving from external knowledge sources (e.g., knowledge graph, database, and even web browsing). Therefore, these KB-in-the-loop methods based on knowledge retrieval are interested in evaluating the effectiveness and efficiency of the knowledge retrieval process. For example, REALM [32] conducts ablation experiments comparing different retrievers used for knowledge acquisition, using "top-5 recall" metrics. Atlas [33] analyzes how the frequency of the text in the correct answer option appearing in retrieved passages is influenced by the number of retrieved passages. Such an evaluation is useful for analyzing and improving the intermediate component of KB-in-the-loop NLG systems.

Knowledge-Enhanced Generation. Another line of research focuses on evaluating the models' capabilities for knowledge-enhanced generation. The most common and important way lies in hallucination detection, which utilizes extra knowledge graphs or databases to detect factual errors [34]. For example, Knowledge F1 [35] measures the word overlap between the model's completion and the fact knowledge used to ground the dialogue. Several fact-checking benchmarks have been presented in recent years to facilitate this area. For example, BEGIN [36] introduces three types of knowledge-grounded dialogue responses based on the evidence from Wikipedia. In this case, factual knowledge is in the form of paragraph-level text. In contrast, DialFact [37] utilizes manually annotated instance-level knowledge snippets as evidence. The Attributable to Identified Sources (AIS) framework [38] assesses whether the statements produced by an NLG system can be attributed to a specific underlying source. Another type of evaluation method focuses on the comparison of the differences between humans and models when choosing to add additional knowledge. For example, WebGPT [39] evaluates model completions by comparing them with human-written answers based on the same web-browsing environment to determine whether the model's retained knowledge is different from that of users. GopherCite [40] computes the ratio at which the model produces plausible and supported claims (i.e., evidence) in human evaluation.

10.3   Evaluating Model/Tool-in-the-Loop Interaction

As discussed in Sect. 2.3, model/tool-in-the-loop encompasses three types of operations: (1) thinking, (2) acting, and (3) collaborating. Thus, evaluating model/tool-in-the-loop interaction involves assessing the following aspects: (1) (multi-stage) chain-of-thought capability, (2) tool-use ability, and (3) collaborative behavior analysis.

Chain-of-Thought Capability. Chain-of-Thought (CoT) [41] is regarded as one of the most crucial emergent abilities, which tends to become more pronounced as model size scales up [42, 43]. It is frequently triggered by elicitive prompts (Sect. 5.2) and can be instantiated through prompt chaining (Sect. 5.3). Typically, CoT aims at improving models' multi-step reasoning abilities [41, 44]. Thus, some reasoning-intensive benchmarks such as GSM8K [45], DROP [46], and ScienceQA [47], can be used to evaluate CoT capability. We refer the readers to [44] for a more comprehensive survey on the reasoning tasks. In our survey, we introduce additional perspectives focused on intermediate steps or reasoning trajectories for evaluating the CoT abilities. For example, we can assess the error propagation between reasoning steps. [48]'s successive prompting technique encounters three primary error sources: incorrect answer predictions, incorrect question decomposition, and out-of-scope reasoning types. By understanding and addressing these errors happened in the intermediate reasoning steps, we can develop more robust models that mitigate the impact of errors throughout the multi-step reasoning process. Another approach is to analyze the effects of varying the number of reasoning steps or the number of reasoning branches. For example, [49] conduct experiments to investigate the influences of different numbers of reasoning steps. [50] compare the influence of different numbers of sampled reasoning paths on performance.

Tool-Use Ability. Following [51, 52], evaluating the tool-use ability of language models involves examining: (1) **tool-use triggering**: whether the LM can determine when to use tools, (2) **tool-use accuracy**: whether the LM selects the correct tools to perform a specific task, (3) **tool-use proficiency**: whether the LM can effectively and efficiently use the tools for reasoning and decision making [53]. For example, Toolformer [54] evaluates tool-use triggering by calculating the API invoking rate, that is, how often the model requires the corresponding API to help with some text generation tasks such as machine translation, question answering, and mathematical calculation. ART [55] compares various strategies employed by language models to select prompts from a task library for correct tool-use prompting, which can also be viewed as a measurement of tool-use accuracy. WebShop [56] examines the success rate of language agents in utilizing web tools for shopping tasks, which serves as an indicator of the tool-use proficiency of language models. Since tool-use ability has been shown to be capable of enhancing the reasoning capability of language models in Tool-Augmented Learning settings [57], certain metrics used to evaluate CoT ability (Sect. 10.3) can also be employed to assess tool-use proficiency [45, 53, 58].

Collaborative Behavior Analysis. The analysis of collaborative behaviors of language model agents can be divided into **result-oriented analysis** and **process-oriented analysis**. Result-oriented analysis mainly focuses on the output or the final result of the collaboration. For example, MindCraft [59] measures whether a task can be solved through collaboration between two language model agents. Socratic Models [60] evaluates the success rate of tasks that are solved through closed-loop interaction involving a large language model, a vision-language model, and an audio-language model. BIG-bench [3] has incorporated several tasks for evaluating Theory of Mind (ToM) abilities. Furthermore, [61] demonstrates that ChatGPT has achieved a performance level comparable to that of nine-year-old children on certain ToM tasks. Process-oriented analysis is more concerned with the dynamics of the collaboration itself. For example, MindCraft [59] also assesses the communication efficiency by measuring the number of dialogue exchanges required to accomplish a task.

10.4 Evaluating Environment-in-the-Loop Interaction

Since the objective of environment-in-the-loop NLP primarily focuses on utilizing language model agents for embodied tasks, which significantly differ from typical text generation tasks, the evaluation of LM-Env interaction may require a distinct paradigm. Generally, the research in this field primarily focuses on two main aspects: (1) constructing embodied task platforms, and (2) determining the appropriate evaluation metrics. We also refer the readers to Sects. 11.3 and 11.4 for more information.

Embodied Task Platforms. To name a few, *EvalAI* [62] provides an open-source platform for evaluating models and agents, which involves tasks of visual question answering, visual dialog, and embodied question answering [63]. *SAPIEN* [64] introduces a simulated household environment specifically designed to simulate daily objects and their interactions within a household setting. Similarly, *Alexa Arena* [65] introduces a user-centric simulation platform for Embodied AI models and presents an instruction-following benchmark based on human-agent dialogue. Shridhar et al. [66] build a benchmark, *ALFRED*, to evaluate the ability of agents to transform visual observations and textual instructions into action sequences for everyday tasks. Building upon this setup, [67] introduce a new test set called *ALFRED-L*. This test set is created by modifying instructions from examples in the *ALFRED* validation splits. The modifications involve eliminating the need for actions manipulating objects and incorporating directives to approach more objects. Furthermore, [68] extend both *TextWorld* [69] and *ALFRED* platforms to create *ALFWorld*. This novel platform establishes a connection between textual descriptions and commands, and the physical simulation of embodied robots, offering an enhanced alignment between them. Lastly, [59] explore collaborative situations where agents are placed in interactive scenarios. They aim to model the mental states of participants in these collaborative interactions. To facilitate this research, they introduce *MindCraft*, a fine-grained dataset that includes the beliefs of partners when collaborating on tasks within the virtual world of *Minecraft*, a 3D environment consisting of blocks. We refer the readers to [51] for more examples.

Evaluation Metrics. To name a few, the mission success rate and the average number of robot actions are both metrics that can be utilized to assess the model's capabilities in terms of interacting with the environment [65]. The mission success rate measures the model's effectiveness by computing the average proportion of missions in which all the goal conditions are met, indicating successful completion of the mission. While the average action number examines the model's efficiency by recording the average number of actions performed by the robot for each task. Ahn et al. [70] employ the plan success rate and the execution success rate to evaluate the performance and effectiveness of the system. The plan success rate measures the system's ability to generate appropriate plans for given instructions, while the execution success rate assesses the system's capability to successfully execute those plans and accomplish the specified tasks. Huang et al. [71] report the text instruction generation speed in the inference stage (referred to as "token count" in the paper) to evaluate the efficiency of the involved large language model.

References

1. Sai, A.B., Mohankumar, A.K., Khapra, M.M.: A survey of evaluation metrics used for NLG systems. ACM Comput. Surv. **55**(2), 26:1–26:39 (2023). https://doi.org/10.1145/3485766. https://doi.org/10.1145/3485766
2. Elkins, K., Chun, J.: Can GPT-3 pass a writer's turing test? Journal of Cultural Analytics **5**(2) (2020)

3. Srivastava, A., Rastogi, A., Rao, A., Shoeb, A.A.M., Abid, A., Fisch, A., Brown, A.R., Santoro, A., Gupta, A., Garriga-Alonso, A., Kluska, A., Lewkowycz, A., Agarwal, A., Power, A., Ray, A., Warstadt, A., Kocurek, A.W., Safaya, A., Tazarv, A., Xiang, A., Parrish, A., Nie, A., Hussain, A., Askell, A., Dsouza, A., Slone, A., Rahane, A., Iyer, A.S., Andreassen, A., Madotto, A., Santilli, A., Stuhlmüller, A., Dai, A., La, A., Lampinen, A., Zou, A., Jiang, A., Chen, A., Vuong, A., Gupta, A., Gottardi, A., Norelli, A., Venkatesh, A., Gholamidavoodi, A., Tabassum, A., Menezes, A., Kirubarajan, A., Mullokandov, A., Sabharwal, A., Herrick, A., Efrat, A., Erdem, A., Karakaş, A., Roberts, B.R., Loe, B.S., Zoph, B., Bojanowski, B., Özyurt, B., Hedayatnia, B., Neyshabur, B., Inden, B., Stein, B., Ekmekci, B., Lin, B.Y., Howald, B., Diao, C., Dour, C., Stinson, C., Argueta, C., Ramírez, C.F., Singh, C., Rathkopf, C., Meng, C., Baral, C., Wu, C., Callison-Burch, C., Waites, C., Voigt, C., Manning, C.D., Potts, C., Ramirez, C., Rivera, C.E., Siro, C., Raffel, C., Ashcraft, C., Garbacea, C., Sileo, D., Garrette, D., Hendrycks, D., Kilman, D., Roth, D., Freeman, D., Khashabi, D., Levy, D., Gonz lez, D.M., Perszyk, D., Hernandez, D., Chen, D., Ippolito, D., Gilboa, D., Dohan, D., Drakard, D., Jurgens, D., Datta, D., Ganguli, D., Emelin, D., Kleyko, D., Yuret, D., Chen, D., Tam, D., Hupkes, D., Misra, D., Buzan, D., Mollo, D.C., Yang, D., Lee, D.H., Shutova, E., Cubuk, E.D., Segal, E., Hagerman, E., Barnes, E., Donoway, E., Pavlick, E., Rodola, E., Lam, E., Chu, E., Tang, E., Erdem, E., Chang, E., Chi, E.A., Dyer, E., Jerzak, E., Kim, E., Manyasi, E.E., Zheltonozhskii, E., Xia, F., Siar, F., Mart nez-Plumed, F., Happé, F., Chollet, F., Rong, F., Mishra, G., Winata, G.I., de Melo, G., Kruszewski, G., Parascandolo, G., Mariani, G., Wang, G., Jaimovitch-López, G., Betz, G., Gur-Ari, G., Galijasevic, H., Kim, H., Rashkin, H., Hajishirzi, H., Mehta, H., Bogar, H., Shevlin, H., Schütze, H., Yakura, H., Zhang, H., Wong, H.M., Ng, I., Noble, I., Jumelet, J., Geissinger, J., Kernion, J., Hilton, J., Lee, J., Fisac, J.F., Simon, J.B., Koppel, J., Zheng, J., Zou, J., Kocoń, J., Thompson, J., Kaplan, J., Radom, J., Sohl-Dickstein, J., Phang, J., Wei, J., Yosinski, J., Novikova, J., Bosscher, J., Marsh, J., Kim, J., Taal, J., Engel, J., Alabi, J., Xu, J., Song, J., Tang, J., Waweru, J., Burden, J., Miller, J., Balis, J.U., Berant, J., Frohberg, J., Rozen, J., Hernandez-Orallo, J., Boudeman, J., Jones, J., Tenenbaum, J.B., Rule, J.S., Chua, J., Kanclerz, K., Livescu, K., Krauth, K., Gopalakrishnan, K., Ignatyeva, K., Markert, K., Dhole, K.D., Gimpel, K., Omondi, K., Mathewson, K., Chiafullo, K., Shkaruta, K., Shridhar, K., McDonell, K., Richardson, K., Reynolds, L., Gao, L., Zhang, L., Dugan, L., Qin, L., Contreras-Ochando, L., Morency, L.P., Moschella, L., Lam, L., Noble, L., Schmidt, L., He, L., Colón, L.O., Metz, L., Şenel, L.K., Bosma, M., Sap, M., ter Hoeve, M., Farooqi, M., Faruqui, M., Mazeika, M., Baturan, M., Marelli, M., Maru, M., Quintana, M.J.R., Tolkiehn, M., Giulianelli, M., Lewis, M., Potthast, M., Leavitt, M.L., Hagen, M., Schubert, M., Baitemirova, M.O., Arnaud, M., McElrath, M., Yee, M.A., Cohen, M., Gu, M., Ivanitskiy, M., Starritt, M., Strube, M., Swędrowski, M., Bevilacqua, M., Yasunaga, M., Kale, M., Cain, M., Xu, M., Suzgun, M., Tiwari, M., Bansal, M., Aminnaseri, M., Geva, M., Gheini, M., T, M.V., Peng, N., Chi, N., Lee, N., Krakover, N.G.A., Cameron, N., Roberts, N., Doiron, N., Nangia, N., Deckers, N., Muennighoff, N., Keskar, N.S., Iyer, N.S., Constant, N., Fiedel, N., Wen, N., Zhang, O., Agha, O., Elbaghdadi, O., Levy, O., Evans, O., Casares, P.A.M., Doshi, P., Fung, P., Liang, P.P., Vicol, P., Alipoormolabashi, P., Liao, P., Liang, P., Chang, P., Eckersley, P., Htut, P.M., Hwang, P., Miłkowski, P., Patil, P., Pezeshkpour, P., Oli, P., Mei, Q., Lyu, Q., Chen, Q., Banjade, R., Rudolph, R.E., Gabriel, R., Habacker, R., Delgado, R.R., Millière, R., Garg, R., Barnes, R., Saurous, R.A., Arakawa, R., Raymaekers, R., Frank, R., Sikand, R., Novak, R., Sitelew, R., LeBras, R., Liu, R., Jacobs, R., Zhang, R., Salakhutdinov, R., Chi, R., Lee, R., Stovall, R., Teehan, R., Yang, R., Singh, S., Mohammad, S.M., Anand, S., Dillavou, S., Shleifer, S., Wiseman, S., Gruetter, S., Bowman, S.R., Schoenholz, S.S., Han, S., Kwatra, S., Rous, S.A., Ghazarian, S., Ghosh, S., Casey, S., Bischoff, S., Gehrmann, S., Schuster, S., Sadeghi, S., Hamdan, S., Zhou, S., Srivastava, S., Shi, S., Singh, S., Asaadi, S., Gu, S.S., Pachchigar, S., Toshniwal, S., Upadhyay, S., Shyamolima,

Debnath, Shakeri, S., Thormeyer, S., Melzi, S., Reddy, S., Makini, S.P., Lee, S.H., Torene, S., Hatwar, S., Dehaene, S., Divic, S., Ermon, S., Biderman, S., Lin, S., Prasad, S., Piantadosi, S.T., Shieber, S.M., Misherghi, S., Kiritchenko, S., Mishra, S., Linzen, T., Schuster, T., Li, T., Yu, T., Ali, T., Hashimoto, T., Wu, T.L., Desbordes, T., Rothschil: Beyond the imitation game: Quantifying and extrapolating the capabilities of language models. arXiv preprint arXiv: Arxiv-2206.04615 (2022)

4. Zhang, H., Song, H., Li, S., Zhou, M., Song, D.: A survey of controllable text generation using transformer-based pre-trained language models. CoRR **abs/2201.05337** (2022). https://arxiv. org/abs/2201.05337

5. Khapra, M.M., Sai, A.B.: A tutorial on evaluation metrics used in natural language generation. In: Proceedings of the 2021 Conference of the North American Chapter of the Association for Computational Linguistics: Human Language Technologies: Tutorials, pp. 15–19. Association for Computational Linguistics, Online (2021). https://doi.org/10.18653/v1/2021.naacl-tutorials.4. https://aclanthology.org/2021.naacl-tutorials.4

6. Chen, P.: AI alignment dialogues: An interactive approach to AI alignment in support agents. In: V. Conitzer, J. Tasioulas, M. Scheutz, R. Calo, M. Mara, A. Zimmermann (eds.) AIES '22: AAAI/ACM Conference on AI, Ethics, and Society, Oxford, United Kingdom, May 19 - 21, 2021, p. 894. ACM (2022). https://doi.org/10.1145/3514094.3539531. https://doi.org/10.1145/3514094.3539531

7. Stoyanchev, S., Pandey, S., Keizer, S., Braunschweiler, N., Doddipatla, R.S.: Combining structured and unstructured knowledge in an interactive search dialogue system. In: O. Lemon, D. Hakkani-Tür, J.J. Li, A. Ashrafzadeh, D.H. García, M. Alikhani, D. Vandyke, O. Dusek (eds.) Proceedings of the 23rd Annual Meeting of the Special Interest Group on Discourse and Dialogue, SIGDIAL 2022, Edinburgh, UK, 07-09 September 2022, pp. 531–540. Association for Computational Linguistics (2022). https://aclanthology.org/2022.sigdial-1.50

8. Chen, J., Zhang, R., Mao, Y., Xu, J.: Parallel interactive networks for multi-domain dialogue state generation. In: B. Webber, T. Cohn, Y. He, Y. Liu (eds.) Proceedings of the 2020 Conference on Empirical Methods in Natural Language Processing, EMNLP 2020, Online, November 16-20, 2020, pp. 1921–1931. Association for Computational Linguistics (2020). https://doi.org/10.18653/v1/2020.emnlp-main.151. https://doi.org/10.18653/v1/2020.emnlp-main.151

9. Gordon, D., Kembhavi, A., Rastegari, M., Redmon, J., Fox, D., Farhadi, A.: IQA: visual question answering in interactive environments. In: 2018 IEEE Conference on Computer Vision and Pattern Recognition, CVPR 2018, Salt Lake City, UT, USA, June 18-22, 2018, pp. 4089–4098. Computer Vision Foundation / IEEE Computer Society (2018). https://doi.org/10.1109/CVPR. 2018.00430

10. Yuan, X., Côté, M.A., Fu, J., Lin, Z., Pal, C., Bengio, Y., Trischler, A.: Interactive language learning by question answering. In: Proceedings of the 2019 Conference on Empirical Methods in Natural Language Processing and the 9th International Joint Conference on Natural Language Processing (EMNLP-IJCNLP), pp. 2796–2813. Association for Computational Linguistics, Hong Kong, China (2019). https://doi.org/10.18653/v1/D19-1280. https://aclanthology.org/D19-1280

11. Lee, M., Srivastava, M., Hardy, A., Thickstun, J., Durmus, E., Paranjape, A., Gerard-Ursin, I., Li, X.L., Ladhak, F., Rong, F., Wang, R.E., Kwon, M., Park, J.S., Cao, H., Lee, T., Bommasani, R., Bernstein, M.S., Liang, P.: Evaluating human-language model interaction. CoRR **abs/2212.09746** (2022). https://doi.org/10.48550/arXiv.2212.09746. https://doi.org/10.48550/arXiv.2212.09746

12. Banerjee, S., Lavie, A.: METEOR: An automatic metric for MT evaluation with improved correlation with human judgments. In: Proceedings of the ACL Workshop on Intrinsic and Extrinsic Evaluation Measures for Machine Translation and/or Summarization, pp. 65–72.

Association for Computational Linguistics, Ann Arbor, Michigan (2005). https://aclanthology.org/W05-0909

13. Lin, C.Y.: Rouge: A package for automatic evaluation of summaries. In: Text summarization branches out, pp. 74–81 (2004). https://aclanthology.org/W04-1013.pdf

14. Vedantam, R., Lawrence Zitnick, C., Parikh, D.: Cider: Consensus-based image description evaluation. In: Proceedings of the IEEE conference on computer vision and pattern recognition, pp. 4566–4575 (2015)

15. Papineni, K., Roukos, S., Ward, T., Zhu, W.: Bleu: a method for automatic evaluation of machine translation. In: Proceedings of the 40th Annual Meeting of the Association for Computational Linguistics, July 6-12, 2002, Philadelphia, PA, USA, pp. 311–318. ACL (2002). https://doi.org/10.3115/1073083.1073135. https://aclanthology.org/P02-1040/

16. Sellam, T., Das, D., Parikh, A.P.: BLEURT: learning robust metrics for text generation. In: D. Jurafsky, J. Chai, N. Schluter, J.R. Tetreault (eds.) Proceedings of the 58th Annual Meeting of the Association for Computational Linguistics, ACL 2020, Online, July 5-10, 2020, pp. 7881–7892. Association for Computational Linguistics (2020). https://doi.org/10.18653/v1/2020.acl-main.704. https://doi.org/10.18653/v1/2020.acl-main.704

17. Ouyang, L., Wu, J., Jiang, X., Almeida, D., Wainwright, C.L., Mishkin, P., Zhang, C., Agarwal, S., Slama, K., Ray, A., et al.: Training language models to follow instructions with human feedback. arXiv preprint arXiv:2203.02155 (2022)

18. Lin, S.C., Hilton, J., Evans, O.: Truthfulqa: Measuring how models mimic human falsehoods. Annual Meeting Of The Association For Computational Linguistics (2021). https://doi.org/10.18653/v1/2022.acl-long.229

19. Gehman, S., Gururangan, S., Sap, M., Choi, Y., Smith, N.A.: Realtoxicityprompts: Evaluating neural toxic degeneration in language models. In: T. Cohn, Y. He, Y. Liu (eds.) Findings of the Association for Computational Linguistics: EMNLP 2020, Online Event, 16-20 November 2020, *Findings of ACL*, vol. EMNLP 2020, pp. 3356–3369. Association for Computational Linguistics (2020). https://doi.org/10.18653/v1/2020.findings-emnlp.301. https://doi.org/10.18653/v1/2020.findings-emnlp.301

20. Nangia, N., Vania, C., Bhalerao, R., Bowman, S.R.: CrowS-pairs: A challenge dataset for measuring social biases in masked language models. In: Proceedings of the 2020 Conference on Empirical Methods in Natural Language Processing (EMNLP), pp. 1953–1967. Association for Computational Linguistics, Online (2020). https://doi.org/10.18653/v1/2020.emnlp-main.154. https://aclanthology.org/2020.emnlp-main.154

21. AlShikh, W., Daaboul, M., Goddard, K., Imel, B., Kamble, K., Kulkarni, P., Russak, M.: Becoming self-instruct: introducing early stopping criteria for minimal instruct tuning. arXiv preprint arXiv: 2307.03692 (2023)

22. Liu, B., Tür, G., Hakkani-Tür, D., Shah, P., Heck, L.P.: Dialogue learning with human teaching and feedback in end-to-end trainable task-oriented dialogue systems. In: M.A. Walker, H. Ji, A. Stent (eds.) Proceedings of the 2018 Conference of the North American Chapter of the Association for Computational Linguistics: Human Language Technologies, NAACL-HLT 2018, New Orleans, Louisiana, USA, June 1-6, 2018, Volume 1 (Long Papers), pp. 2060–2069. Association for Computational Linguistics (2018). https://doi.org/10.18653/v1/n18-1187. https://doi.org/10.18653/v1/n18-1187

23. Wallace, E., Rodriguez, P., Feng, S., Yamada, I., Boyd-Graber, J.: Trick Me If You Can: Human-in-the-Loop Generation of Adversarial Examples for Question Answering. Transactions of the Association for Computational Linguistics **7**, 387–401 (2019). https://doi.org/10.1162/tacl_a_00279. https://doi.org/10.1162/tacl_a_00279

24. Lee, M., Liang, P., Yang, Q.: Coauthor: Designing a human-ai collaborative writing dataset for exploring language model capabilities. In: Proceedings of the 2022 CHI Conference on Human Factors in Computing Systems, pp. 1–19 (2022)

25. Lu, Y., Srivastava, M., Kramer, J., Elfardy, H., Kahn, A., Wang, S., Bhardwaj, V.: Goal-oriented end-to-end conversational models with profile features in a real-world setting. In: A. Loukina, M. Morales, R. Kumar (eds.) Proceedings of the 2019 Conference of the North American Chapter of the Association for Computational Linguistics: Human Language Technologies, NAACL-HLT 2019, Minneapolis, MN, USA, June 2-7, 2019, Volume 2 (Industry Papers), pp. 48–55. Association for Computational Linguistics (2019). https://doi.org/10.18653/v1/n19-2007. https://doi.org/10.18653/v1/n19-2007
26. Li, J., Miller, A.H., Chopra, S., Ranzato, M., Weston, J.: Dialogue learning with human-in-the-loop. In: 5th International Conference on Learning Representations, ICLR 2017, Toulon, France, April 24-26, 2017, Conference Track Proceedings. OpenReview.net (2017). https://openreview.net/forum?id=HJgXCV9xx
27. Testoni, A., Bernardi, R.: The interplay of task success and dialogue quality: An in-depth evaluation in task-oriented visual dialogues. In: Proceedings of the 16th Conference of the European Chapter of the Association for Computational Linguistics: Main Volume, pp. 2071–2082. Association for Computational Linguistics, Online (2021). https://doi.org/10.18653/v1/2021.eacl-main.178. https://aclanthology.org/2021.eacl-main.178
28. Lu, W., Xu, Y., Li, E.: Efficient evaluation of task oriented dialogue systems. In: NeurIPS 2020 Workshop on Human in the Loop Dialogue Systems (2020). https://www.amazon.science/publications/efficient-evaluation-of-task-oriented-dialogue-systems
29. Henderson, M., Thomson, B., Williams, J.D.: The second dialog state tracking challenge. In: Proceedings of the 15th Annual Meeting of the Special Interest Group on Discourse and Dialogue (SIGDIAL), pp. 263–272. Association for Computational Linguistics, Philadelphia, PA, U.S.A. (2014). https://doi.org/10.3115/v1/W14-4337. https://aclanthology.org/W14-4337
30. Wang, Z.J., Choi, D., Xu, S., Yang, D.: Putting humans in the natural language processing loop: A survey. arXiv preprint arXiv:2103.04044 (2021)
31. Wu, X., Xiao, L., Sun, Y., Zhang, J., Ma, T., He, L.: A survey of human-in-the-loop for machine learning. Future Gener. Comput. Syst. **135**, 364–381 (2022). https://doi.org/10.1016/j.future.2022.05.014. https://doi.org/10.1016/j.future.2022.05.014
32. Guu, K., Lee, K., Tung, Z., Pasupat, P., Chang, M.W.: Realm: Retrieval-augmented language model pre-training. arXiv preprint arXiv: Arxiv-2002.08909 (2020)
33. Izacard, G., Lewis, P., Lomeli, M., Hosseini, L., Petroni, F., Schick, T., Dwivedi-Yu, J., Joulin, A., Riedel, S., Grave, E.: Atlas: Few-shot learning with retrieval augmented language models. arXiv preprint arXiv: Arxiv-2208.03299 (2022)
34. Ji, Z., Lee, N., Frieske, R., Yu, T., Su, D., Xu, Y., Ishii, E., Bang, Y., Madotto, A., Fung, P.: Survey of hallucination in natural language generation. CoRR **abs/2202.03629** (2022). https://arxiv.org/abs/2202.03629
35. Shuster, K., Poff, S., Chen, M., Kiela, D., Weston, J.: Retrieval augmentation reduces hallucination in conversation. In: M. Moens, X. Huang, L. Specia, S.W. Yih (eds.) Findings of the Association for Computational Linguistics: EMNLP 2021, Virtual Event / Punta Cana, Dominican Republic, 16-20 November, 2021, pp. 3784–3803. Association for Computational Linguistics (2021). https://doi.org/10.18653/v1/2021.findings-emnlp.320. https://doi.org/10.18653/v1/2021.findings-emnlp.320
36. Dziri, N., Rashkin, H., Linzen, T., Reitter, D.: Evaluating attribution in dialogue systems: The BEGIN benchmark. Trans. Assoc. Comput. Linguistics **10**, 1066–1083 (2022). https://transacl.org/ojs/index.php/tacl/article/view/3977
37. Gupta, P., Wu, C.S., Liu, W., Xiong, C.: DialFact: A benchmark for fact-checking in dialogue. In: Proceedings of the 60th Annual Meeting of the Association for Computational Linguistics (Volume 1: Long Papers), pp. 3785–3801. Association for Computational Linguistics, Dublin, Ireland (2022). https://doi.org/10.18653/v1/2022.acl-long.263. https://aclanthology.org/2022.acl-long.263

38. Rashkin, H., Nikolaev, V., Lamm, M., Collins, M., Das, D., Petrov, S., Tomar, G.S., Turc, I., Reitter, D.: Measuring attribution in natural language generation models. CoRR **abs/2112.12870** (2021). https://arxiv.org/abs/2112.12870

39. Nakano, R., Hilton, J., Balaji, S., Wu, J., Ouyang, L., Kim, C., Hesse, C., Jain, S., Kosaraju, V., Saunders, W., et al.: Webgpt: Browser-assisted question-answering with human feedback. arXiv preprint arXiv:2112.09332 (2021)

40. Menick, J., Trebacz, M., Mikulik, V., Aslanides, J., Song, F., Chadwick, M., Glaese, M., Young, S., Campbell-Gillingham, L., Irving, G., et al.: Teaching language models to support answers with verified quotes. arXiv preprint arXiv:2203.11147 (2022)

41. Wei, J., Wang, X., Schuurmans, D., Bosma, M., Chi, E.H., Le, Q., Zhou, D.: Chain of thought prompting elicits reasoning in large language models. CoRR **abs/2201.11903** (2022). https://arxiv.org/abs/2201.11903

42. Kaplan, J., McCandlish, S., Henighan, T., Brown, T.B., Chess, B., Child, R., Gray, S., Radford, A., Wu, J., Amodei, D.: Scaling laws for neural language models. arXiv preprint arXiv:2001.08361 (2020)

43. Wei, J., Tay, Y., Bommasani, R., Raffel, C., Zoph, B., Borgeaud, S., Yogatama, D., Bosma, M., Zhou, D., Metzler, D., Chi, E.H., Hashimoto, T., Vinyals, O., Liang, P., Dean, J., Fedus, W.: Emergent abilities of large language models. arXiv preprint arXiv: Arxiv-2206.07682 (2022)

44. Qiao, S., Ou, Y., Zhang, N., Chen, X., Yao, Y., Deng, S., Tan, C., Huang, F., Chen, H.: Reasoning with language model prompting: A survey. arXiv preprint arXiv: Arxiv-2212.09597 (2022)

45. Cobbe, K., Kosaraju, V., Bavarian, M., Chen, M., Jun, H., Kaiser, L., Plappert, M., Tworek, J., Hilton, J., Nakano, R., Hesse, C., Schulman, J.: Training verifiers to solve math word problems. arXiv preprint arXiv: Arxiv-2110.14168 (2021)

46. Dua, D., Wang, Y., Dasigi, P., Stanovsky, G., Singh, S., Gardner, M.: DROP: A reading comprehension benchmark requiring discrete reasoning over paragraphs. In: Proceedings of the 2019 Conference of the North American Chapter of the Association for Computational Linguistics: Human Language Technologies, Volume 1 (Long and Short Papers), pp. 2368–2378. Association for Computational Linguistics, Minneapolis, Minnesota (2019). https://doi.org/10.18653/v1/N19-1246. https://aclanthology.org/N19-1246

47. Lu, P., Mishra, S., Xia, T., Qiu, L., Chang, K.W., Zhu, S.C., Tafjord, O., Clark, P., Kalyan, A.: Learn to explain: Multimodal reasoning via thought chains for science question answering. In: The 36th Conference on Neural Information Processing Systems (NeurIPS) (2022)

48. Dua, D., Gupta, S., Singh, S., Gardner, M.: Successive prompting for decomposing complex questions. In: Y. Goldberg, Z. Kozareva, Y. Zhang (eds.) Proceedings of the 2022 Conference on Empirical Methods in Natural Language Processing, EMNLP 2022, Abu Dhabi, United Arab Emirates, December 7-11, 2022, pp. 1251–1265. Association for Computational Linguistics (2022). https://aclanthology.org/2022.emnlp-main.81

49. Zhou, D., Schärli, N., Hou, L., Wei, J., Scales, N., Wang, X., Schuurmans, D., Bousquet, O., Le, Q., Chi, E.: Least-to-most prompting enables complex reasoning in large language models. arXiv preprint arXiv:2205.10625 (2022)

50. Wang, X., Wei, J., Schuurmans, D., Le, Q., Chi, E., Narang, S., Chowdhery, A., Zhou, D.: Self-consistency improves chain of thought reasoning in language models. arXiv preprint arXiv: Arxiv-2203.11171 (2022)

51. Yang, S., Nachum, O., Du, Y., Wei, J., Abbeel, P., Schuurmans, D.: Foundation models for decision making: Problems, methods, and opportunities. arXiv preprint arXiv:2303.04129 (2023)

52. Li, M., Song, F., Yu, B., Yu, H., Li, Z., Huang, F., Li, Y.: Api-bank: A benchmark for tool-augmented LLMs (2023)

53. Yao, S., Zhao, J., Yu, D., Du, N., Shafran, I., Narasimhan, K., Cao, Y.: React: Synergizing reasoning and acting in language models. arXiv preprint arXiv: Arxiv-2210.03629 (2022)

54. Schick, T., Dwivedi-Yu, J., Dessì, R., Raileanu, R., Lomeli, M., Zettlemoyer, L., Cancedda, N., Scialom, T.: Toolformer: Language models can teach themselves to use tools. CoRR **abs/2302.04761** (2023). https://doi.org/10.48550/arXiv.2302.04761. https://doi.org/10.48550/arXiv.2302.04761

55. Paranjape, B., Lundberg, S., Singh, S., Hajishirzi, H., Zettlemoyer, L., Ribeiro, M.T.: Art: Automatic multi-step reasoning and tool-use for large language models. arXiv preprint arXiv: Arxiv-2303.09014 (2023)

56. Yao, S., Chen, H., Yang, J., Narasimhan, K.: Webshop: Towards scalable real-world web interaction with grounded language agents. arXiv preprint arXiv:2207.01206 (2022)

57. Qin, Y., Hu, S., Lin, Y., Chen, W., Ding, N., Cui, G., Zeng, Z., Huang, Y., Xiao, C., Han, C., Fung, Y., Su, Y., Wang, H., Qian, C., Tian, R., Zhu, K., Liang, S., Shen, X., Xu, B., Zhang, Z., Ye, Y., Li, B., Tang, Z., Yi, J., Zhu, Y., Dai, Z., Yan, L., Cong, X., Lu, Y.T., Zhao, W., Huang, Y., Yan, J.H., Han, X., Sun, X., Li, D., Phang, J., Yang, C., Wu, T., Ji, H., Liu, Z., Sun, M.: Tool learning with foundation models. ARXIV.ORG (2023). https://doi.org/10.48550/arXiv.2304.08354

58. Chen, W., Ma, X., Wang, X., Cohen, W.W.: Program of thoughts prompting: Disentangling computation from reasoning for numerical reasoning tasks. arXiv preprint arXiv:2211.12588 (2022)

59. Bara, C.P., CH-Wang, S., Chai, J.: MindCraft: Theory of mind modeling for situated dialogue in collaborative tasks. In: Proceedings of the 2021 Conference on Empirical Methods in Natural Language Processing, pp. 1112–1125. Association for Computational Linguistics, Online and Punta Cana, Dominican Republic (2021). https://doi.org/10.18653/v1/2021.emnlp-main.85. https://aclanthology.org/2021.emnlp-main.85

60. Zeng, A., Attarian, M., Ichter, B., Choromanski, K., Wong, A., Welker, S., Tombari, F., Purohit, A., Ryoo, M., Sindhwani, V., Lee, J., Vanhoucke, V., Florence, P.: Socratic models: Composing zero-shot multimodal reasoning with language. arXiv preprint arXiv: Arxiv-2204.00598 (2022)

61. Kosinski, M.: Theory of mind may have spontaneously emerged in large language models. arXiv preprint arXiv: Arxiv-2302.02083 (2023)

62. Yadav, D., Jain, R., Agrawal, H., Chattopadhyay, P., Singh, T., Jain, A., Singh, S., Lee, S., Batra, D.: Evalai: Towards better evaluation systems for AI agents. CoRR **abs/1902.03570** (2019). http://arxiv.org/abs/1902.03570

63. Das, A., Datta, S., Gkioxari, G., Lee, S., Parikh, D., Batra, D.: Embodied question answering. In: 2018 IEEE Conference on Computer Vision and Pattern Recognition, CVPR 2018, Salt Lake City, UT, USA, June 18-22, 2018, pp. 1–10. Computer Vision Foundation / IEEE Computer Society (2018). https://doi.org/10.1109/CVPR.2018.00008

64. Xiang, F., Qin, Y., Mo, K., Xia, Y., Zhu, H., Liu, F., Liu, M., Jiang, H., Yuan, Y., Wang, H., Yi, L., Chang, A.X., Guibas, L.J., Su, H.: SAPIEN: A simulated part-based interactive environment. In: 2020 IEEE/CVF Conference on Computer Vision and Pattern Recognition, CVPR 2020, Seattle, WA, USA, June 13-19, 2020, pp. 11094–11104. Computer Vision Foundation / IEEE (2020). https://doi.org/10.1109/CVPR42600.2020.01111. https://openaccess.thecvf.com/content_CVPR_2020/html/Xiang_SAPIEN_A_SimulAted_Part-Based_Interactive_ENvironment_CVPR_2020_paper.html

65. Gao, Q., Thattai, G., Gao, X., Shakiah, S., Pansare, S., Sharma, V., Sukhatme, G., Shi, H., Yang, B., Zheng, D., et al.: Alexa arena: A user-centric interactive platform for embodied ai. arXiv preprint arXiv:2303.01586 (2023)

66. Shridhar, M., Thomason, J., Gordon, D., Bisk, Y., Han, W., Mottaghi, R., Zettlemoyer, L., Fox, D.: ALFRED: A benchmark for interpreting grounded instructions for everyday tasks. In: 2020 IEEE/CVF Conference on Computer Vision and Pattern Recognition, CVPR 2020, Seattle, WA, USA, June 13-19, 2020, pp. 10737–10746. Computer Vision Foundation / IEEE (2020). https://doi.org/10.1109/CVPR42600.2020.01075

67. Akula, A.R., Gella, S., Padmakumar, A., Namazifar, M., BANSAL, M., Thomason, J., Hakkani-Tür, D.: Alfred-l: Investigating the role of language for action learning in interactive visual environments. In: EMNLP 2022 (2022)

68. Shridhar, M., Yuan, X., Côté, M., Bisk, Y., Trischler, A., Hausknecht, M.J.: Alfworld: Aligning text and embodied environments for interactive learning. In: 9th International Conference on Learning Representations, ICLR 2021, Virtual Event, Austria, May 3-7, 2021. OpenReview.net (2021). https://openreview.net/forum?id=0IOX0YcCdTn

69. Côté, M.A., Kádár, A., Yuan, X., Kybartas, B., Barnes, T., Fine, E., Moore, J., Hausknecht, M., El Asri, L., Adada, M., et al.: Textworld: A learning environment for text-based games. In: Computer Games: 7th Workshop, CGW 2018, Held in Conjunction with the 27th International Conference on Artificial Intelligence, IJCAI 2018, Stockholm, Sweden, July 13, 2018, Revised Selected Papers 7, pp. 41–75. Springer (2019)

70. Ahn, M., Brohan, A., Brown, N., Chebotar, Y., Cortes, O., David, B., Finn, C., Gopalakrishnan, K., Hausman, K., Herzog, A., et al.: Do as i can, not as i say: Grounding language in robotic affordances. arXiv preprint arXiv:2204.01691 (2022)

71. Huang, W., Xia, F., Shah, D., Driess, D., Zeng, A., Lu, Y., Florence, P., Mordatch, I., Levine, S., Hausman, K., et al.: Grounded decoding: Guiding text generation with grounded models for robot control. arXiv preprint arXiv:2303.00855 (2023)

Application

11

Zekun Wang, Ge Zhang, Xiuying Chen, Shaochun Hao, Yizhi Li, Guangzheng Xiong, Zhenzhu Yang, Kexin Yang, Mong Yuan Sim and Ke Xu

11.1 Controllable Text Generation

The Controllable Text Generation (CTG) technique is an NLP approach that empowers language models to generate text that is not only coherent and meaningful but also allows users to control particular aspects of the output. The need for CTG arises from the desire to customize generated text according to user-defined constraints, such as length constraint [1, 2], inclusion of particular keywords [3, 4], and adherence to a specific sentiment or style [5, 6]. Conventional CTG methods typically involve training models with explicit objectives that optimize for the control attributes [1, 2, 7–10], constrained decoding [3, 11–16], and

Z. Wang (✉)
Beihang University, Haidian District, Beijing, China
e-mail: zenmoore@buaa.edu.cn

G. Zhang
University of Michigan, Haidian District, Beijing, China
e-mail: gezhang@umich.edu

X. Chen
King Abdullah University of Science and Technology, Thuwal, Saudi Arabia
e-mail: xiuying.chen@kaust.edu.sa

S. Hao
Xi'an Jiaotong University, Hangzhou City, China
e-mail: haoshaochun@stu.xjtu.edu.cn

Y. Li
University of Manchester, Manchester, UK

G. Xiong
Haidian District, Beijing, China

prompting [17]. We refer the readers to [18, 19] for more information. In this section, we briefly discuss the potential applications of iNLP in CTG.

Interacting with humans can enhance controllability in CTG by enabling users to directly provide their preferences and constraints during the generation or training process. AI Chains [20], for example, allows users to chain together Large Language Model (LLM) steps and modify them in a modular way, improving transparency, controllability, and collaboration. Lee et al. [21] suggests the importance of controlling the intermediate generation process rather than just the final output, and highlight the need to consider more control attributes related to first-person subjective experience and user preferences. References [9, 22–24] imply the use of RL or RLHF for controlled text generation, which enables not only control over helpfulness, harmlessness, and honesty, but also allows potential optimization to meet length constraints and other criteria.

Interacting with knowledge bases can potentially enhance the robustness and factuality of CTG, as suggested by [25], who propose Knowledge-Aware Fine-Tuning (KAFT) to improve the controllability of language models while maintaining their robustness. KAFT fine-tunes LMs using a combination of the vanilla supervised dataset and augmented data, which includes instances with counterfactual contexts (i.e., contexts that contradict the model's memorized knowledge) and irrelevant contexts (i.e., contexts that are unrelated to the task).

Interacting with models and tools may have the potential for more complex and fine-grained control over text generation. For example, classifier-guided CTG approaches put a classifier in the loop to provide control signals or feedback [1, 5]. Similar to Diffusion-LM [1] which iteratively denoises the text with the control feedback from a classifier at each iteration, Self-Refine [26], lets a LLM generate an output and then provide multi-aspect feedback on it. This feedback is used to iteratively refine the output until it reaches a satisfactory quality or a specific criteria. Notably, typical classifier-guided CTG relies on external classifiers, while Self-Refine employs the LLM itself as a classifier through self-interaction.

Interacting with environments inherently requires great controllability due to the essential need for affordance grounding, as discussed in Sect. 2.4. SayCan [27], as a representative

Z. Yang
Haidian District, Beijing, China

K. Yang
Chengdu, China

M. Y. Sim
University of Adelaide, Adelaide, Australia
e-mail: mongyuan.sim@student.adelaide.edu.au

K. Xu
Beihang University, Beijing, China
e-mail: kexu@nlsde.buaa.edu.cn

example, leverages a scoring mechanism over action candidates to achieve such controllability.

Overall, various interactive objects may offer different avenues for optimizing CTG systems. By harnessing the power of interaction, we can achieve more user-oriented, robust, fine-grained, complex, and even reality-oriented control over text generation.

11.2 Writing Assistant

Intelligent and interactive writing assistants constitute a rapidly growing area of research that explores the potential of AI-powered tools to modify, enrich, and even co-create content with humans. These assistants can be broadly categorized into four types based on their level of involvement in the content generation process: (1) Content Supporting, (2) Content Checking and Polishing, (3) Content Enrichment, and (4) Content Co-creation.

Content Support. Content supporting writing assistants do not generate content for use, but just provide functional assistance for writers such as on-the-fly summarization [28] and real-time visualization [29]. For example, [30] proposes an interaction scheme where human writers are provided with questions for inspiration instead of content snippets for use. Dang et al. [28] proposes a writing assistant that continuously updates summaries, keywords, and central sentences of existing content for user reference, rather than generating content directly. Singh et al. [29] designs a writing assistant offering visual and aural suggestions as writing supports. Although content supporting writing assistants provide minimal aids for human writers, they benefit from avoiding the dominance over the writing process in some cases, which is one of the main challenges of PLM-based writing assistants [30, 31]. Moreover, writing support may reduce the manual effort to trigger or manipulate the writing assistant such as heavy prompt engineering [32]. Specifically, [32] indicate that manual effort is required for non-diegetic prompting, and thereby humans tend to prefer selecting suggestions from writing assistants over controlling automatic content generation through non-diegetic prompts. Jakesch et al. [31] point out that human writers' content creation can even be affected by opinionated PLM-powered writing assistants. These challenges can be mitigated by reducing the level of involvement of writing assistants to content supporting.

Content Checking and Polishing. The processing procedure for content checking and polishing typically involves taking manually written sentences as input and producing output that has been grammar-checked and rephrased, allowing users to interactively and iteratively improve their writing. Famous real-world products include QuillBot,[1] Effidit [33],[2] Pitaya,[3] Grammarly,[4] and Xiezuocat.[5] There is also growing interest in incorporating iterative editing

---

[1] https://quillbot.com/.

[2] https://effidit.qq.com/en.

[3] https://www.facebook.com/Mypitaya/.

[4] https://www.grammarly.com/.

[5] https://xiezuocat.com/.

operations into writing assistants [34–36], which can be considered an application of editing-based iNLP (Sect. 3.3). For example, [34] suggest incorporating transfer learning from other text editing tasks to improve the quality of iterative text revision by linking editing actions to content quality. Du et al. [35] raise a novel human-in-the-loop iterative text revision system that combines model-generated revisions with human judgments and specifically fine-tuning a PEGASUS model [37] as a revision generation model with which a revised sentence is generated based on a given sentence and an edit intention.

Content Enrichment. Content enrichment, unlike content checking and polishing, involves more creative content generation but still relies on manually provided context or configuration. Classic content enrichment features include text completion (**AutoCompletion**) [38–41], and keywords-to-sentence (**K2S**) [42–45]. Note that both AutoCompletion and K2S simply supplement manual input, rather than co-creating new content from scratch through manual collaboration or guidance. AutoCompletion is an interactive writing assistant feature that involves humans in the content generation process by providing suggestions to complete their prompts, thereby enhancing their overall writing experience. For example, [39] propose an Intent-Guided Authoring (IGA) Assistant, which follows fine-grained author specifications to process the input text for AutoCompletion. The scheme proposed in IGA is similar to the recent trend of instruction tuning (Sect. 6.1), which suggests that more complex and controllable user preferences in writing assistants can be formatted as instructions to further activate instruction-tuned LLMs. AutoCompletion can also be adapted to various NLP downstream tasks, including medical text simplification [38], human-computer collaborative translation [40], and interactive word completion of morphologically complex low-resource language [46]. Moreover, K2S is highly in line with controllable text generation (c.f. Sect. 11.1), but places a greater emphasis on controllable interactivity. Practical K2S applications typically allow users to customize the fine-grained control attributes according to their specific needs and preferences in an interactive manner. For example, CueBot [47] proposes a conversational assistant capable of generating responses that can be controlled by users using cues/keywords. It suggests responses for users to choose from and incorporates a keyword loss during training to generate lexically constrained outputs.

Content Co-creation. Content co-creation refers to the collaborative process between humans and AI systems to generate new content from scratch, rather than simply improving existing content. Content co-creation is widely explored in interactive fiction writing [48, 49], screenplays and theatre scripts writing [50], academic writing [51], and poem writing [52–54]. For example, [49] models story generation as simulating role-play games and tracing player interaction sequences. Yang et al. [55] develops *DOC*, which includes a detailed outline generator and a detailed controller, significantly improves the coherence of long story generation. Chakrabarty et al. [56] proposes *CoPoet*, an interactive poem writing assistant powered by instruction prompts and LLM, and verify that co-created poems are usually preferred compared to those written without *CoPoet* involved. Dramatron [50] adapts hierarchically controlled PLMs to allow expert writers to control style tags, logic lines, character descriptions, and environment descriptions. This enables writers to easily generate the necessary material for various use cases. Writing Path proposed in [57] also utilizes a title-outline-content generation pipeline given content guideline from users, as well as augments the writing outline by cooperating relative documents and keywords from RAG system. The Weaver series models [58] involve the creative writing knowledge into the full LLM development cycle including pre-training, instruction tuning and preference learning. The Weaver further engages RAG and function calling in the system to achieve a specialized writing that could surpass GPT-4 in corresponding evaluation. However, while practical writing assistants for content checking, polishing, and enriching have become increasingly mature, content co-creation writing assistants still face various challenges that need to be addressed. For example, [59] points out that current NLG techniques for story generation often exhibits poor performance in maintaining the author's voice and the coherence of the storyline. References [20, 60, 61] demonstrates that there is often a trade-off between controllability and creativity in the generated content. Moreover, the evaluation of content co-creation-based writing assistants can also be particularly challenging due to the subjective nature of creative writing. Despite various efforts to construct reliable benchmarks for evaluation [50, 59, 62–65] suggest that professional writers have become increasingly important for evaluation compared to crowd-sourcing annotators due to the ever-improved quality of artificial intelligence generated content (AIGC).

11.3 Embodied AI

Embodied AI enables language models to impact the real-world and virtual environments through which agents observe and update states of themselves and their surroundings. One method for bridging language models with the physical world is interaction with grounded language as mentioned in Sect. 2.4, which allows language models to see, listen, and control external objects.

Observation and Manipulation are fundamental to many embodied tasks, where agents acquire external states and perform actions to update those states. Thanks to text descriptions and textual controlling interfaces, language models usually observe their surrounding environment through input text and operate on objects by sending textual commands. For instance, a visual perception mapper converts visual input to text in natural language [66]. Additionally, human intervention can be part of agent observation, so that agents can be guided by real-time human feedback [67]. Typical observation and operation tasks include object rearrangements [68], tool usage [69], item creation and modification [70, 71], and other robotic controlling tasks.

Navigation and Exploration enable agents to move around and study their surrounding environment by using dynamic observation and manipulation. That is, unlike observation and manipulation tasks, navigation and exploration tasks allow agents to move within the environment to adjust their observation and manipulation. These agents not only plan routes and actions, but also combine observations collected from different locations to make decisions, answer questions and reason, allowing them to accomplish complicated tasks that require multi-location multi-object observation and long-horizon manipulation. Text commands in both natural languages and programming languages [72] bridge the gap between language model agents and available actions and tools. During such processes, these agents also combine different data sources, including cameras, microphones, other sensors [73], and textual commands from human controllers [74, 75]. Moreover, agents can also work as assistants to guide human operations. For instance, an interactive driving assistant can continuously observe the driving environment and guide human drivers to handle various situations [76].

Multi-agent Tasks require agents to cooperate and compete with humans and other agents to reach specific goals. Unlike agents with multiple skills, agents with social capabilities usually observe others' behaviors and communicate through textual messages, including messages in natural languages and data in more structured styles. Typical social tasks include multi-player gaming [77, 78], human-AI collaboration [79, 80], multi-agent collaboration [81], and other communication tasks, such as interview [82], negotiation [83], recruitment [84], and opinion gathering [85]. In text-based gaming tasks, agents learn from human behaviors and play as human players [86]. In multi-agent environments, agents coordinate with each other to accomplish complex tasks that cannot be accomplished by any single agent [87]. Agents also act as human delegates and communicate with others to complete day-to-day tasks, such as restaurant reservations and appointment scheduling [88]. MetaAI's Cicero [89] enables language model agents to play in an online Diplomacy league.

11.4   Text Game

Text games, also referred to as interactive fiction games [90], are capable of understanding player commands, simulating player states, and updating the current status of game environments [90]. Language models have shown great potential in these game-playing scenarios [89, 91–93], which are a specific type of Embodied AI (c.f., Sect. 11.3). Specifically, language models can be used to play or power text games through text-based interfaces, such as state descriptions [94], commands [95–97], situated dialogue [87], and multi-party dialogue [98]. Thus, text games are intrinsic applications of iNLP, in which the environment or other agents are involved in the game-playing loop. We can divide text games into two distinct categories: (1) text-only games which rely solely on text, and (2) text-aided games which use text as a supplement to other forms of media, such as graphics or audio. In this subsection, we will begin by discussing interactive text game platforms. We will then provide a brief overview of how language models are utilized to play text-only games and to power text-aided games.

Interactive Text Game Platforms. Interactive Text Game Platforms provide a framework and engine for building and running text-based games, often including features such as game state tracking, parser-based natural language understanding, and scripted events. Some examples of such platforms are:

(1) **Text Adventure Games** are games that allow players to interact with adventurous worlds solely through textual descriptions and actions [99]. Osborne et al. [90] summarize two major text adventure game platforms: *TextWorld* [100] and *Jericho* [101]. Additionally, they define seven major challenges that need to be addressed in developing solutions for Text Adventure Games, including partial observability, large state space, and long-term credit assignment, among others.

(2) **Social Deduction Games** are games where players attempt to discover each other's hidden role or team allegiance through strategic conversations, logical deduction, and deceitful actions.[6] For example, classic examples of social deduction games include *Werewolf,*[7] *Mush*[8] *SS13,*[9] and *Among Us.*[10] Specifically, [78] propose a multimodal dataset containing text and visual signals to model persuasion behaviors in *Werewolf.* Lin et al. [102] is another *Werewolf*-based corpus with self-revealing and role-estimation behavior annotation. Tuin and Rooijackers [103] construct a corpus aimed at player role detection, based on the game *Among Us*, and verify that it is a challenging yet learnable task.

(3) **Strategic Games** are games that heavily rely on player decision-making skills and situational awareness to determine the outcome.[11] For example, *Diplomacy*[12] is a strategic board game that involves multiple players who each assume control of the armed forces of a European power. The objective of the game is to move one's units skillfully and defeat those of opponents in order to gain possession of a majority of strategically important cities and provinces referred to as "supply centers." The contested nature of this gameplay often requires players to engage in extensive and complex interactions and diplomacy with each other in order to achieve their goals. *Diplomacy* is gaining increasing attention and is widely regarded as a benchmark for autonomous agents' ability to communicate and adjust strategies like humans, which is one of the essential elements for the success of human civilization [91]. Cicero [89] proposes an impressive autonomous agent that combines PLM with RL and achieves human-level performance in *Diplomacy*. Moreover, [91] make preliminary investigations on how negotiation algorithms and the inclination to punish traitors can enable autonomous agents to communicate like humans and cooperate more effectively in *Diplomacy*. Apart from Diplomacy, there are numerous classic strategic games that serve as potential resources for interactive text game platforms and related NLP research, such as *Eurogame,*[13] *Warhammer Fantacy,*[14] and *Paths of Glory.*[15]

---

[6] https://en.wikipedia.org/wiki/Social_deduction_game.

[7] https://en.wikipedia.org/wiki/Mafia_(party_game).

[8] https://en.wikipedia.org/wiki/Mush_(video_game).

[9] https://en.wikipedia.org/wiki/Space_Station_13.

[10] https://en.wikipedia.org/wiki/Among_Us.

[11] https://en.wikipedia.org/wiki/Strategy_game.

[12] https://en.wikipedia.org/wiki/Diplomacy_(game).

[13] https://en.wikipedia.org/wiki/Eurogame.

[14] https://en.wikipedia.org/wiki/Warhammer_(game).

[15] https://en.wikipedia.org/wiki/Paths_of_Glory_(board_game).

(4) **Tabletop role-playing games (TRPGs)**,[16] such as *Dungeons and Dragons (DND)*[17] and *Call of Cthulhu (COC)*,[18] as well as works of fiction, such as the *Harry Potter series* [104], have provided a rich source of situated and multi-party dialogue data that can be used to build challenging text game platforms [63, 105–107]. However, the raw dialogue data documenting the game process of TRPGs is usually a mixture of in-character action descriptions and out-of-character strategy explanations [105], frequently accompanied by lengthy world-building documents [108], which can differ from one game to another. The problem of extracting the golden-standard game states and game commands remains a challenging yet fascinating question [108]. For example, [106] provide 34,243 summary dialogue fragment pairs from raw dialogue data documenting the *DND* game process. The summaries in these summary-dialogue chunk pairs contain text descriptions of game states, which can serve as a good benchmark for abstractive game-state summarization of interactive text games. Callison-Burch et al. [105] frame *DND* as a dialogue system challenge, comprising both deterministic elements like dice rolls and imprecise descriptions of the game-play as partial state information. Zhou et al. [63] introduce a novel and highly interactive task, *G4C* (Goal-driven Guidance Generation in Grounded Communication), based on *DND*. They train an autonomous agent acting as a game host, also known as a *Dungeon Master (DM)*, using the theory of mind and RL. This approach significantly enhances the players' capacity to achieve their objectives.

(5) **Life Simulation Games** are games that enable players to control one or more virtual characters.[19] Classic life simulation games include *Virtual Pet*,[20] *Black and White*,[21] *MineCraft*,[22] and *GTA Series*.[23] For example, [87, 92, 93] explore how autonomous agents based on PLMs can learn to collaborate, communicate, and generalize across a range of tasks and objectives using *MineCraft* as the interactive game platform. Furthermore, **Social Simulation Games** are a sub-genre of life simulation games that simulates social interactions and relationships between multiple artificial characters or lives in a virtual world.[24] For example, *the Sims Series*[25] is one of the classic social simulation game. Mehta et al. [109] enhance the capability of AI agents to identify when they require additional information to enable more human-AI interactions and improve social simulations. As noted by [98, 110, 111], the role-playing agents and open sandbox worlds can be easily adapted as factors in social simulation games. Park et al. [98] configure PLM-based autonomous agents with

---

[16] https://en.wikipedia.org/wiki/Tabletop_role-playing_game.

[17] https://en.wikipedia.org/wiki/Dungeons_%26_Dragons.

[18] https://en.wikipedia.org/wiki/Call_of_Cthulhu_(role-playing_game).

[19] https://en.wikipedia.org/wiki/Life_simulation_game.

[20] https://en.wikipedia.org/wiki/Virtual_pet.

[21] https://en.wikipedia.org/wiki/Black_26_White_(video_game).

[22] https://en.wikipedia.org/wiki/Minecraft.

[23] https://en.wikipedia.org/wiki/Grand_Theft_Auto.

[24] https://en.wikipedia.org/wiki/Social_simulation_game.

[25] https://en.wikipedia.org/wiki/The_Sims.

social identity settings and conduct social simulations accordingly. Their experiment design is highly in line with the gameplay of *the Sims Series*.

Playing Text-Only Games. Early work on autonomous agents of text-only games mainly relies on handcrafted reward functions [112] or other well-formatted data structures, such as knowledge graphs, to preserve and retrieve past information and game states [96]. Although some exploratory methods before the emergence of PLMs also adapt trivial neural representations of text to help detect actions and states from text-only games, these methods focus on constrained hand-crafted template-based state and action space and are unable to understand complex and highly unstructured texts in many text-only games. Yao et al. [113] point out that these methods, based on constrained hand-crafted template-based state and action space, isolate autonomous agents from understanding the meanings of words or semantics by verifying that agents without understanding semantics can achieve similar performance on these text-only games by adopting similar methods but without understanding semantics. For designing better text-only games as testbeds for autonomous agents' ability of language understanding, the motivation and strategy implicitly expressed in words should not be detected through hand-crafted templates without understanding the semantics [113, 114]. The following work turns to autonomous agents based on PLM-powered neural representation which rely less on manual effort. For example, [115] adapt sentence-level semantics representation-based clustering and deep Q learning [116, 117] for playing text adventure games. Xu et al. [118] propose a lightweight transformer-based representation learning framework for text-only games and outperform previous SOTA methods. Recently, the success of LLMs has enabled text-only games to explore handling any user input [114, 119]. Understanding user inputs that are complex and ambiguous in their meaning requires careful attention to the actions and states explicitly described in the text.

Powering Text-Aided Games. Traditionally, in text-aided games, autonomous agents have used either formal language or structured natural language to model state transitions and execute actions using a highly symbolic representation [120, 121]. With the emergence of PLMs, these agents transformed from using symbolic representations to using contextual and neural representations, which capture more complex and high-level semantics of textually stated strategies and communication protocols. As a result, we will elucidate the ways in which language interfaces and PLMs enable autonomous agents in text-aided games to communicate with other agents and make better game plans. Since the impressive release of ChatGPT, some industrial and academic researchers have also been exploring the adaptation of LLMs to enhance the text-aided game experience. *Inworld*[26] claims that LLMs can empower characters in games with distinct personalities and contextual awareness that stay in-world[27] or on-brand,[28] which tremendously improves the games' immersive experience.

---

[26] https://www.inworld.ai/.

[27] The in-world context of a player is the context that the player has with characters controlled by other players and NPCs when she/he is doing role-playing.

[28] The on-brand context of a player is the context that the player has with other players and the game host when she/he is not doing role-playing.

NetEase also announces that they allow Non-Player Characters (NPCs) powered by LLMs in its online game, *Nishuihan*, to communicate with considerable freedom.[29] In addition to communication, language interface and NLP-powered planning have been explored in designing autonomous agents in text-aided games for reward shaping [122], instruction following [123, 124], control policy generalization [125], and representation learning [126]. In addition, language interface plays a major role in training autonomous text-aided game agents. For example, [127, 128] point out that an efficient language-based communication protocol is crucial to a collaboration strategy in multi-agent text-aided games. Jiang et al. [129] suggests that language can naturally compose different sub-skills to enrich the non-compositional abstraction of complex text-aided games' hierarchical strategy. Moreover, [130, 131] verify that language modeling induces representations, which are even useful for offline RL strategy modeling of games. This observation implies a relationship between the high-level semantic coherence of languages and the planning strategy adopted by text-aided games.

11.5 Other Applications

Interactive Natural Language Processing extends the capabilities of language models beyond traditional text processing across various domains. Other than the aforementioned categories, this section explores additional topics where iNLP has shown significant potential, highlighting applications that leverage interaction between language models and users, environments, or other systems.

Reasoning. In fields that demand sophisticated reasoning, such as mathematics and computer programming, NLP techniques have proven particularly helpful and inspiring in conversations with users [132–134]. The recently developed LLMs, for instance, make a great progress on understanding the programming commands from users and respond (or execute simple ones) with appropriate code snippets in real-time [135–138], where the large amount of corpus containing the interactions involving code comments, pull request history, error analysis, and debugging information become the basis of their performances. In the domain of mathematics, requiring interpretation of more abstractive notation and concepts, language models like GPT-4 [139], MAmmoTH [140], InternLM2-Math [141], and LeanDojo [142] have demonstrated the ability to tackle complex arithmetic, algebra, and even some elements of difficult problems through dialogues or question answering. The problems formulated in natural language could be passed to the model for reasoning and answer refining based on user feedback or further clarifications, which underscores the impact of the interactive components in reasoning tasks. By introducing iterative dialogues between the user and the model, the gap between complex computational problems and user-friendly solutions is bridged, allowing users with less mathematical background to learn and solve easier.

---

[29] https://www.youtube.com/watch?v=zGVR5gPgefk.

Social Goods. The social good domain has been boosted and extended by the application of iNLP, where language models contribute across diverse sectors such as psychology, healthcare, and education. These applications leverage the interactive nature of NLP to address specific societal needs through enriched communication and improved access to information. In the field of psychology, interactive language models are used to develop therapeutic tools that offer conversational support, resembling interactions with a human therapist [143–145]. These models are designed to provide psychological assessments or conversational therapy, making mental health support more accessible. One of the notable example applications, Woebot [143], which uses NLP model to deliver cognitive behavioral therapy techniques interactively. NLP applications improve patient interactions with health systems through automated and personalized communication tools [146] when it comes to healthcare area. These applications are utilized for biomedical document analysis [147, 148], symptom assessment, and guide patients through the healthcare system [149, 150], which can improve the overall efficiency of medical consultations and follow-up procedures. Particularly, the chatbots application among the studies use iNLP to provide medical advice based on personal health information and general medical knowledge. Regarding the field of education, interactive language models provide opportunity to create new learning paradigm and experience by introducing, language learning practice tool, personalized tutoring, and enabling adaptable educational content [151]. For instance, tools like Duolingo[30] use iNLP to enhance language learning through conversational practice with AI characters, offering immediate feedback and can adapt to different proficiency levels of the learners.

Personalization. It refers to the process of tailoring a language model's behavior and output, to the unique needs and preferences of each user. This can be achieved through the model's interactions with users, learning from their inputs, demographics, and adapting its behavior accordingly.[31] For example, [152] suggest that ChatGPT has the potential to become more personalized and customized through learning from user interactions and individual preferences. Salemi et al. [153] introduces a personalization benchmark and suggestion to personalize LLMs through retrieval augmentation using user profiles. Wu et al. [154] shows the potential of personalizing PLMs through the use of prompts. Madaan et al. [155] personalizes the PLM via an external memory with human feedback. Personalization can greatly enhance the user experience with language models by providing more preferred responses, improving the model's ability to understand the user's needs and intentions, and ultimately building trust and rapport between the user and the model. However, we should also be aware of the drawbacks brought by personalization. For example, [156] demonstrates that assigning a persona to ChatGPT can potentially magnify its toxicity up to six times.

Model-based Evaluation. Model-based Evaluation enjoys the benefits of PLMs to compute a text quality score for each generated sample and show greater correlation with the human evaluation compared with statistical-based evaluation metrics, such as

---

[30] https://blog.duolingo.com/duolingo-max.

[31] https://www.exponentlabs.io/articles/chatgpt-and-personalization-how-ai-is-changing-the-way-we-interact-with-technology.

BLUERT [157], BERTScore [158], and COMET [159]. Such an evaluation method can be widely implemented and used in various general NLG tasks, and even works for reference-free settings [160–162]. Additionally, existing preliminary research has indicated that LLMs have emerged with the ability to evaluate the AI generated content (AIGC) with human-like judges [163–165]. Some papers also propose the fine-grained analysis of LLMs' ability to evaluate AIGC in specific NLG tasks, including summarization [166, 167], question answering [168], news outlet generation [169], and translation [170]. Moreover, [168, 171] propose to collaborate with LLMs to provide better and cheaper human-like evaluations. Through interactions between models and even humans, we can evaluate LMs in a more effective (accurate) and efficient (automatic) manner. This evaluation process can be akin to a teacher model administering "exams" and "grading" to assess the performance of a student model [172]. Nevertheless, model-based evaluation can exhibit biases, such as preferences for longer or initially presented responses [173, 174]. To mitigate these biases, [175] suggests dividing responses into segments to reduce positional bias, while [176] introduces an evaluation approach using multi-agent debate and discussion.

References

1. Li, X., Thickstun, J., Gulrajani, I., Liang, P.S., Hashimoto, T.B.: Diffusion-lm improves controllable text generation. Advances in Neural Information Processing Systems **35**, 4328–4343 (2022)
2. Zhou, W., Jiang, Y.E., Wilcox, E., Cotterell, R., Sachan, M.: Controlled text generation with natural language instructions (2023)
3. Hokamp, C., Liu, Q.: Lexically constrained decoding for sequence generation using grid beam search. In: Proceedings of the 55th Annual Meeting of the Association for Computational Linguistics (Volume 1: Long Papers), pp. 1535–1546. Association for Computational Linguistics, Vancouver, Canada (2017). https://doi.org/10.18653/v1/P17-1141. https://aclanthology.org/P17-1141
4. Carlsson, F., Öhman, J., Liu, F., Verlinden, S., Nivre, J., Sahlgren, M.: Fine-grained controllable text generation using non-residual prompting. In: Proceedings of the 60th Annual Meeting of the Association for Computational Linguistics (Volume 1: Long Papers), pp. 6837–6857. Association for Computational Linguistics, Dublin, Ireland (2022). https://doi.org/10.18653/v1/2022.acl-long.471. https://aclanthology.org/2022.acl-long.471
5. Dathathri, S., Madotto, A., Lan, J., Hung, J., Frank, E., Molino, P., Yosinski, J., Liu, R.: Plug and play language models: A simple approach to controlled text generation. arXiv preprint arXiv:1912.02164 (2019)
6. Qian, J., Dong, L., Shen, Y., Wei, F., Chen, W.: Controllable natural language generation with contrastive prefixes. In: Findings of the Association for Computational Linguistics: ACL 2022, pp. 2912–2924. Association for Computational Linguistics, Dublin, Ireland (2022). https://doi.org/10.18653/v1/2022.findings-acl.229. https://aclanthology.org/2022.findings-acl.229
7. Hu, Z., Yang, Z., Liang, X., Salakhutdinov, R., Xing, E.P.: Toward controlled generation of text. In: International conference on machine learning, pp. 1587–1596. PMLR (2017)
8. Keskar, N.S., McCann, B., Varshney, L.R., Xiong, C., Socher, R.: Ctrl: A conditional transformer language model for controllable generation. arXiv preprint arXiv:1909.05858 (2019)

9. Lu, X., Welleck, S., Hessel, J., Jiang, L., Qin, L., West, P., Ammanabrolu, P., Choi, Y.: Quark: Controllable text generation with reinforced unlearning. Advances in neural information processing systems **35**, 27591–27609 (2022)

10. Clive, J., Cao, K., Rei, M.: Control prefixes for text generation. arXiv preprint arXiv:2110.08329 (2021)

11. Anderson, P., Fernando, B., Johnson, M., Gould, S.: Guided open vocabulary image captioning with constrained beam search. In: Proceedings of the 2017 Conference on Empirical Methods in Natural Language Processing, pp. 936–945. Association for Computational Linguistics, Copenhagen, Denmark (2017). https://doi.org/10.18653/v1/D17-1098. https://aclanthology.org/D17-1098

12. Post, M., Vilar, D.: Fast lexically constrained decoding with dynamic beam allocation for neural machine translation. In: Proceedings of the 2018 Conference of the North American Chapter of the Association for Computational Linguistics: Human Language Technologies, Volume 1 (Long Papers), pp. 1314–1324. Association for Computational Linguistics, New Orleans, Louisiana (2018). https://doi.org/10.18653/v1/N18-1119. https://aclanthology.org/N18-1119

13. Lu, X., West, P., Zellers, R., Le Bras, R., Bhagavatula, C., Choi, Y.: NeuroLogic decoding: (un)supervised neural text generation with predicate logic constraints. In: Proceedings of the 2021 Conference of the North American Chapter of the Association for Computational Linguistics: Human Language Technologies, pp. 4288–4299. Association for Computational Linguistics, Online (2021). https://doi.org/10.18653/v1/2021.naacl-main.339. https://aclanthology.org/2021.naacl-main.339

14. Lu, X., Welleck, S., West, P., Jiang, L., Kasai, J., Khashabi, D., Le Bras, R., Qin, L., Yu, Y., Zellers, R., Smith, N.A., Choi, Y.: NeuroLogic a*esque decoding: Constrained text generation with lookahead heuristics. In: Proceedings of the 2022 Conference of the North American Chapter of the Association for Computational Linguistics: Human Language Technologies, pp. 780–799. Association for Computational Linguistics, Seattle, United States (2022). https://doi.org/10.18653/v1/2022.naacl-main.57. https://aclanthology.org/2022.naacl-main.57

15. Qin, L., Welleck, S., Khashabi, D., Choi, Y.: COLD decoding: Energy-based constrained text generation with langevin dynamics. In: A.H. Oh, A. Agarwal, D. Belgrave, K. Cho (eds.) Advances in Neural Information Processing Systems (2022). https://openreview.net/forum?id=TiZYrQ-mPup

16. Kumar, S., Paria, B., Tsvetkov, Y.: Gradient-based constrained sampling from language models. In: Proceedings of the 2022 Conference on Empirical Methods in Natural Language Processing, pp. 2251–2277. Association for Computational Linguistics, Abu Dhabi, United Arab Emirates (2022). https://aclanthology.org/2022.emnlp-main.144

17. Zou, X., Yin, D., Zhong, Q., Yang, H., Yang, Z., Tang, J.: Controllable generation from pre-trained language models via inverse prompting. In: Proceedings of the 27th ACM SIGKDD Conference on Knowledge Discovery & Data Mining, pp. 2450–2460 (2021)

18. Zhang, H., Song, H., Li, S., Zhou, M., Song, D.: A survey of controllable text generation using transformer-based pre-trained language models. arXiv preprint arXiv: Arxiv-2201.05337 (2022)

19. Weng, L.: Controllable neural text generation. lilianweng.github.io (2021). https://lilianweng.github.io/posts/2021-01-02-controllable-text-generation/

20. Wu, T.S., Terry, M., Cai, C.J.: Ai chains: Transparent and controllable human-ai interaction by chaining large language model prompts. International Conference On Human Factors In Computing Systems (2021). https://doi.org/10.1145/3491102.3517582

21. Lee, M., Srivastava, M., Hardy, A., Thickstun, J., Durmus, E., Paranjape, A., Gerard-Ursin, I., Li, X.L., Ladhak, F., Rong, F., Wang, R.E., Kwon, M., Park, J.S., Cao, H., Lee, T., Bommasani, R., Bernstein, M.S., Liang, P.: Evaluating human-language model interaction. CoRR

abs/2212.09746 (2022). https://doi.org/10.48550/arXiv.2212.09746. https://doi.org/10.48550/arXiv.2212.09746

22. Christiano, P., Leike, J., Brown, T.B., Martic, M., Legg, S., Amodei, D.: Deep reinforcement learning from human preferences. arXiv preprint arXiv: Arxiv-1706.03741 (2017)

23. Ouyang, L., Wu, J., Jiang, X., Almeida, D., Wainwright, C.L., Mishkin, P., Zhang, C., Agarwal, S., Slama, K., Ray, A., et al.: Training language models to follow instructions with human feedback. arXiv preprint arXiv:2203.02155 (2022)

24. Fu Yao; Peng, H., Khot, T.: How does GPT obtain its ability? tracing emergent abilities of language models to their sources. Yao Fu's Notion (2022). https://yaofu.notion.site/How-does-GPT-Obtain-its-Ability-Tracing-Emergent-Abilities-of-Language-Models-to-their-Sources-b9a57ac0fcf74f30a1ab9e3e36fa1dc1

25. Li, D., Rawat, A.S., Zaheer, M., Wang, X., Lukasik, M., Veit, A., Yu, F., Kumar, S.: Large language models with controllable working memory. arXiv preprint arXiv: Arxiv-2211.05110 (2022)

26. Madaan, A., Tandon, N., Gupta, P., Hallinan, S., Gao, L., Wiegreffe, S., Alon, U., Dziri, N., Prabhumoye, S., Yang, Y., Welleck, S., Majumder, B.P., Gupta, S., Yazdanbakhsh, A., Clark, P.: Self-refine: Iterative refinement with self-feedback. arXiv preprint arXiv: Arxiv-2303.17651 (2023)

27. Ahn, M., Brohan, A., Brown, N., Chebotar, Y., Cortes, O., David, B., Finn, C., Gopalakrishnan, K., Hausman, K., Herzog, A., et al.: Do as i can, not as i say: Grounding language in robotic affordances. arXiv preprint arXiv:2204.01691 (2022)

28. Dang, H., Benharrak, K., Lehmann, F., Buschek, D.: Beyond text generation: Supporting writers with continuous automatic text summaries. In: Proceedings of the 35th Annual ACM Symposium on User Interface Software and Technology, pp. 1–13 (2022)

29. Singh, N., Bernal, G., Savchenko, D., Glassman, E.L.: Where to hide a stolen elephant: Leaps in creative writing with multimodal machine intelligence. ACM Transactions on Computer-Human Interaction (2022)

30. Arnold, K.C., Volzer, A.M., Madrid, N.G.: Generative models can help writers without writing for them. In: IUI Workshops (2021)

31. Jakesch, M., Bhat, A., Buschek, D., Zalmanson, L., Naaman, M.: Co-writing with opinionated language models affects users' views. In: Proceedings of the 2023 CHI Conference on Human Factors in Computing Systems, pp. 1–15 (2023)

32. Dang, H., Goller, S., Lehmann, F., Buschek, D.: Choice over control: How users write with large language models using diegetic and non-diegetic prompting. In: Proceedings of the 2023 CHI Conference on Human Factors in Computing Systems, pp. 1–17 (2023)

33. Shi, S., Zhao, E., Tang, D., Wang, Y., Li, P., Bi, W., Jiang, H., Huang, G., Cui, L., Huang, X., et al.: Effidit: Your ai writing assistant. arXiv preprint arXiv:2208.01815 (2022)

34. Kim, Z.M., Du, W., Raheja, V., Kumar, D., Kang, D.: Improving iterative text revision by learning where to edit from other revision tasks. In: Proceedings of the 2022 Conference on Empirical Methods in Natural Language Processing, pp. 9986–9999. Association for Computational Linguistics, Abu Dhabi, United Arab Emirates (2022). https://aclanthology.org/2022.emnlp-main.678

35. Du, W., Kim, Z.M., Raheja, V., Kumar, D., Kang, D.: Read, revise, repeat: A system demonstration for human-in-the-loop iterative text revision. In: Proceedings of the First Workshop on Intelligent and Interactive Writing Assistants (In2Writing 2022), pp. 96–108. Association for Computational Linguistics, Dublin, Ireland (2022). https://doi.org/10.18653/v1/2022.in2writing-1.14. https://aclanthology.org/2022.in2writing-1.14

36. Du, W., Raheja, V., Kumar, D., Kim, Z.M., Lopez, M., Kang, D.: Understanding iterative revision from human-written text. In: Proceedings of the 60th Annual Meeting of the Association

for Computational Linguistics (Volume 1: Long Papers), pp. 3573–3590. Association for Computational Linguistics, Dublin, Ireland (2022). https://doi.org/10.18653/v1/2022.acl-long.250. https://aclanthology.org/2022.acl-long.250

37. Zhang, J., Zhao, Y., Saleh, M., Liu, P.: Pegasus: Pre-training with extracted gap-sentences for abstractive summarization. In: International Conference on Machine Learning, pp. 11328–11339. PMLR (2020)

38. Van, H., Kauchak, D., Leroy, G.: AutoMeTS: The autocomplete for medical text simplification. In: Proceedings of the 28th International Conference on Computational Linguistics, pp. 1424–1434. International Committee on Computational Linguistics, Barcelona, Spain (Online) (2020). https://doi.org/10.18653/v1/2020.coling-main.122. https://aclanthology.org/2020.coling-main.122

39. Sun, S., Zhao, W., Manjunatha, V., Jain, R., Morariu, V., Dernoncourt, F., Srinivasan, B.V., Iyyer, M.: IGA: An intent-guided authoring assistant. In: Proceedings of the 2021 Conference on Empirical Methods in Natural Language Processing, pp. 5972–5985. Association for Computational Linguistics, Online and Punta Cana, Dominican Republic (2021). https://doi.org/10.18653/v1/2021.emnlp-main.483. https://aclanthology.org/2021.emnlp-main.483

40. Li, H., Liu, L., Huang, G., Shi, S.: GWLAN: General word-level AutocompletioN for computer-aided translation. In: Proceedings of the 59th Annual Meeting of the Association for Computational Linguistics and the 11th International Joint Conference on Natural Language Processing (Volume 1: Long Papers), pp. 4792–4802. Association for Computational Linguistics, Online (2021). https://doi.org/10.18653/v1/2021.acl-long.370. https://aclanthology.org/2021.acl-long.370

41. Casacuberta, F., Foster, G., Huang, G., Koehn, P., Kovacs, G., Liu, L., Shi, S., Watanabe, T., Zong, C.: Findings of the word-level AutoCompletion shared task in WMT 2022. In: Proceedings of the Seventh Conference on Machine Translation (WMT), pp. 812–820. Association for Computational Linguistics, Abu Dhabi, United Arab Emirates (Hybrid) (2022). https://aclanthology.org/2022.wmt-1.75

42. Miao, N., Zhou, H., Mou, L., Yan, R., Li, L.: Cgmh: Constrained sentence generation by metropolis-hastings sampling. In: Proceedings of the AAAI Conference on Artificial Intelligence, vol. 33, pp. 6834–6842 (2019)

43. Sha, L.: Gradient-guided unsupervised lexically constrained text generation. In: Proceedings of the 2020 Conference on Empirical Methods in Natural Language Processing (EMNLP), pp. 8692–8703. Association for Computational Linguistics, Online (2020). https://doi.org/10.18653/v1/2020.emnlp-main.701. https://aclanthology.org/2020.emnlp-main.701

44. Nie, J., Yang, L., Chen, Y., Kong, C., Zhu, J., Yang, E.: Lexical complexity controlled sentence generation. arXiv preprint arXiv:2211.14540 (2022)

45. Zheng, K., Sun, Q., Yang, Y., Xu, F.: Knowledge stimulated contrastive prompting for low-resource stance detection. In: Findings of the Association for Computational Linguistics: EMNLP 2022, pp. 1168–1178. Association for Computational Linguistics, Abu Dhabi, United Arab Emirates (2022). https://aclanthology.org/2022.findings-emnlp.83

46. Lane, W., Bird, S.: Interactive word completion for morphologically complex languages. In: Proceedings of the 28th International Conference on Computational Linguistics, pp. 4600–4611 (2020)

47. H. Kumar, S., Su, H., Manuvinakurike, R., Pinaroc, M., Prasad, S., Sahay, S., Nachman, L.: CueBot: Cue-controlled response generation for assistive interaction usages. In: Ninth Workshop on Speech and Language Processing for Assistive Technologies (SLPAT-2022), pp. 66–79. Association for Computational Linguistics, Dublin, Ireland (2022). https://doi.org/10.18653/v1/2022.slpat-1.9. https://aclanthology.org/2022.slpat-1.9

48. Manjavacas, E., Karsdorp, F., Burtenshaw, B., Kestemont, M.: Synthetic literature: Writing science fiction in a co-creative process. In: Proceedings of the Workshop on Computational Creativity in Natural Language Generation (CC-NLG 2017), pp. 29–37. Association for Computational Linguistics, Santiago de Compostela, Spain (2017). https://doi.org/10.18653/v1/W17-3904. https://aclanthology.org/W17-3904

49. Tapscott, A., León, C., Gervás, P.: Generating stories using role-playing games and simulated human-like conversations. In: Proceedings of the 3rd Workshop on Computational Creativity in Natural Language Generation (CC-NLG 2018), pp. 34–42. Association for Computational Linguistics, Tilburg, the Netherlands (2018). https://doi.org/10.18653/v1/W18-6606. https://aclanthology.org/W18-6606

50. Mirowski, P., Mathewson, K.W., Pittman, J., Evans, R.: Co-writing screenplays and theatre scripts with language models: An evaluation by industry professionals. arXiv preprint arXiv:2209.14958 (2022)

51. Fok, R., Weld, D.S.: What can't large language models do? the future of ai-assisted academic writing

52. Astigarraga, A., María Martínez-Otzeta, J., Rodriguez, I., Sierra, B., Lazkano, E.: Poet's little helper: A methodology for computer-based poetry generation. a case study for the Basque language. In: Proceedings of the Workshop on Computational Creativity in Natural Language Generation (CC-NLG 2017), pp. 2–10. Association for Computational Linguistics, Santiago de Compostela, Spain (2017). https://doi.org/10.18653/v1/W17-3901. https://aclanthology.org/W17-3901

53. Oliveira, H.G., Mendes, T., Boavida, A.: Co-poetryme: a co-creative interface for the composition of poetry. In: Proceedings of the 10th International Conference on Natural Language Generation, pp. 70–71 (2017)

54. Hämäläinen, M.: Poem machine - a co-creative NLG web application for poem writing. In: Proceedings of the 11th International Conference on Natural Language Generation, pp. 195–196. Association for Computational Linguistics, Tilburg University, The Netherlands (2018). https://doi.org/10.18653/v1/W18-6525. https://aclanthology.org/W18-6525

55. Yang, K., Klein, D., Peng, N., Tian, Y.: Doc: Improving long story coherence with detailed outline control. arXiv preprint arXiv:2212.10077 (2022)

56. Chakrabarty, T., Padmakumar, V., He, H.: Help me write a poem: Instruction tuning as a vehicle for collaborative poetry writing. arXiv preprint arXiv:2210.13669 (2022)

57. Lee, Y., Ka, S., Son, B., Kang, P., Kang, J.: Navigating the path of writing: Outline-guided text generation with large language models. arXiv preprint arXiv:2404.13919 (2024)

58. Wang, T., Chen, J., Jia, Q., Wang, S., Fang, R., Wang, H., Gao, Z., Xie, C., Xu, C., Dai, J., et al.: Weaver: Foundation models for creative writing. arXiv preprint arXiv:2401.17268 (2024)

59. Ippolito, D., Yuan, A., Coenen, A., Burnam, S.: Creative writing with an ai-powered writing assistant: Perspectives from professional writers. arXiv preprint arXiv:2211.05030 (2022)

60. Yuan, A., Coenen, A., Reif, E., Ippolito, D.: Wordcraft: Story writing with large language models. In: 27th International Conference on Intelligent User Interfaces, IUI '22, p. 841–852. Association for Computing Machinery, New York, NY, USA (2022). https://doi.org/10.1145/3490099.3511105. https://doi.org/10.1145/3490099.3511105

61. Chen, H., Ji, Y., Evans, D.: Balanced adversarial training: Balancing tradeoffs between fickleness and obstinacy in NLP models. In: Proceedings of the 2022 Conference on Empirical Methods in Natural Language Processing, pp. 632–647. Association for Computational Linguistics, Abu Dhabi, United Arab Emirates (2022). https://aclanthology.org/2022.emnlp-main.40

62. Lee, M., Liang, P., Yang, Q.: Coauthor: Designing a human-ai collaborative writing dataset for exploring language model capabilities. In: Proceedings of the 2022 CHI Conference on Human Factors in Computing Systems, pp. 1–19 (2022)

63. Zhou, P., Zhu, A., Hu, J., Pujara, J., Ren, X., Callison-Burch, C., Choi, Y., Ammanabrolu, P.: An ai dungeon master's guide: Learning to converse and guide with intents and theory-of-mind in dungeons and dragons. ARXIV.ORG (2022). https://doi.org/10.48550/arXiv.2212.10060

64. Shen, H., Wu, T.: Parachute: Evaluating interactive human-lm co-writing systems. arXiv preprint arXiv:2303.06333 (2023)

65. Gómez-Rodríguez, C., Williams, P.: A confederacy of models: A comprehensive evaluation of LLMs on creative writing. arXiv preprint arXiv:2310.08433 (2023)

66. Zhao, X., Li, M., Weber, C., Hafez, M.B., Wermter, S.: Chat with the environment: Interactive multimodal perception using large language models. arXiv preprint arXiv:2303.08268 (2023)

67. Lynch, C., Wahid, A., Tompson, J., Ding, T., Betker, J., Baruch, R., Armstrong, T., Florence, P.: Interactive language: Talking to robots in real time (2022)

68. Huang, W., Xia, F., Xiao, T., Chan, H., Liang, J., Florence, P., Zeng, A., Tompson, J., Mordatch, I., Chebotar, Y., et al.: Inner monologue: Embodied reasoning through planning with language models. arXiv preprint arXiv:2207.05608 (2022)

69. Paranjape, B., Lundberg, S., Singh, S., Hajishirzi, H., Zettlemoyer, L., Ribeiro, M.T.: Art: Automatic multi-step reasoning and tool-use for large language models. arXiv preprint arXiv: Arxiv-2303.09014 (2023)

70. Jiang, Y., Huang, Z., Pan, X., Loy, C.C., Liu, Z.: Talk-to-edit: Fine-grained facial editing via dialog. In: Proceedings of the IEEE/CVF International Conference on Computer Vision, pp. 13799–13808 (2021)

71. Elgohary, A., Meek, C., Richardson, M., Fourney, A., Ramos, G., Awadallah, A.H.: NL-EDIT: Correcting semantic parse errors through natural language interaction. In: Proceedings of the 2021 Conference of the North American Chapter of the Association for Computational Linguistics: Human Language Technologies, pp. 5599–5610. Association for Computational Linguistics, Online (2021). https://doi.org/10.18653/v1/2021.naacl-main.444. https://aclanthology.org/2021.naacl-main.444

72. Huang, C., Mees, O., Zeng, A., Burgard, W.: Audio visual language maps for robot navigation. arXiv preprint arXiv:2303.07522 (2023)

73. Gan, C., Zhang, Y., Wu, J., Gong, B., Tenenbaum, J.B.: Look, listen, and act: Towards audio-visual embodied navigation. In: 2020 IEEE International Conference on Robotics and Automation (ICRA), pp. 9701–9707. IEEE (2020)

74. Sharma, P., Sundaralingam, B., Blukis, V., Paxton, C., Hermans, T., Torralba, A., Andreas, J., Fox, D.: Correcting robot plans with natural language feedback. arXiv preprint arXiv:2204.05186 (2022)

75. Gao, X., Gao, Q., Gong, R., Lin, K., Thattai, G., Sukhatme, G.S.: Dialfred: Dialogue-enabled agents for embodied instruction following. IEEE Robotics and Automation Letters 7(4), 10049–10056 (2022)

76. Ma, Z., VanDerPloeg, B., Bara, C.P., Huang, Y., Kim, E.I., Gervits, F., Marge, M., Chai, J.: DOROTHIE: Spoken dialogue for handling unexpected situations in interactive autonomous driving agents. In: Findings of the Association for Computational Linguistics: EMNLP 2022, pp. 4800–4822. Association for Computational Linguistics, Abu Dhabi, United Arab Emirates (2022). https://aclanthology.org/2022.findings-emnlp.354

77. Suh, J., Bennett, C.C., Weiss, B., Yoon, E., Jeong, J., Chae, Y.: Development of speech dialogue systems for social ai in cooperative game environments. In: 2021 IEEE Region 10 Symposium (TENSYMP), pp. 1–4. IEEE (2021)

78. Lai, B., Zhang, H., Liu, M., Pariani, A., Ryan, F., Jia, W., Hayati, S.A., Rehg, J.M., Yang, D.: Werewolf among us: A multimodal dataset for modeling persuasion behaviors in social deduction games. arXiv preprint arXiv:2212.08279 (2022)

79. Krishnaswamy, N., Alalyani, N.: Embodied multimodal agents to bridge the understanding gap. In: Proceedings of the First Workshop on Bridging Human–Computer Interaction and Natural Language Processing, pp. 41–46. Association for Computational Linguistics, Online (2021). https://aclanthology.org/2021.hcinlp-1.7

80. Puig, X., Shu, T., Li, S., Wang, Z., Liao, Y.H., Tenenbaum, J.B., Fidler, S., Torralba, A.: Watch-and-help: A challenge for social perception and human-ai collaboration. arXiv preprint arXiv:2010.09890 (2020)

81. Patel, S., Wani, S., Jain, U., Schwing, A.G., Lazebnik, S., Savva, M., Chang, A.X.: Interpretation of emergent communication in heterogeneous collaborative embodied agents. In: Proceedings of the IEEE/CVF International Conference on Computer Vision, pp. 15953–15963 (2021)

82. Xiao, Z., Zhou, M.X., Liao, Q.V., Mark, G., Chi, C., Chen, W., Yang, H.: Tell me about yourself: Using an ai-powered chatbot to conduct conversational surveys with open-ended questions. ACM Transactions on Computer-Human Interaction (TOCHI) **27**(3), 1–37 (2020)

83. Verma, S., Fu, J., Yang, S., Levine, S.: CHAI: A CHatbot AI for task-oriented dialogue with offline reinforcement learning. In: Proceedings of the 2022 Conference of the North American Chapter of the Association for Computational Linguistics: Human Language Technologies, pp. 4471–4491. Association for Computational Linguistics, Seattle, United States (2022). https://doi.org/10.18653/v1/2022.naacl-main.332. https://aclanthology.org/2022.naacl-main.332

84. Nawaz, N., Gomes, A.M.: Artificial intelligence chatbots are new recruiters. IJACSA) International Journal of Advanced Computer Science and Applications **10**(9) (2019)

85. Bittner, E.A., Oeste-Reiß, S., Leimeister, J.M.: Where is the bot in our team? toward a taxonomy of design option combinations for conversational agents in collaborative work. In: Hawaii International Conference on System Sciences (HICSS) (2019)

86. Xu, Y., Fang, M., Chen, L., Du, Y., Zhou, J., Zhang, C.: Perceiving the world: Question-guided reinforcement learning for text-based games. In: Proceedings of the 60th Annual Meeting of the Association for Computational Linguistics (Volume 1: Long Papers), pp. 538–560. Association for Computational Linguistics, Dublin, Ireland (2022). https://doi.org/10.18653/v1/2022.acl-long.41. https://aclanthology.org/2022.acl-long.41

87. Bara, C.P., CH-Wang, S., Chai, J.: MindCraft: Theory of mind modeling for situated dialogue in collaborative tasks. In: Proceedings of the 2021 Conference on Empirical Methods in Natural Language Processing, pp. 1112–1125. Association for Computational Linguistics, Online and Punta Cana, Dominican Republic (2021). https://doi.org/10.18653/v1/2021.emnlp-main.85. https://aclanthology.org/2021.emnlp-main.85

88. O'Leary, D.E.: Google's duplex: Pretending to be human. Intelligent Systems in Accounting, Finance and Management **26**(1), 46–53 (2019)

89. Bakhtin, A., Brown, N., Dinan, E., Farina, G., Flaherty, C., Fried, D., Goff, A., Gray, J., Hu, H., et al.: Human-level play in the game of diplomacy by combining language models with strategic reasoning. Science **378**(6624), 1067–1074 (2022)

90. Osborne, P., Nõmm, H., Freitas, A.: A survey of text games for reinforcement learning informed by natural language. Transactions of the Association for Computational Linguistics **10**, 873–887 (2022). https://doi.org/10.1162/tacl_a_00495. https://aclanthology.org/2022.tacl-1.51

91. Kramár, J., Eccles, T., Gemp, I., Tacchetti, A., McKee, K.R., Malinowski, M., Graepel, T., Bachrach, Y.: Negotiation and honesty in artificial intelligence methods for the board game of diplomacy. Nature Communications **13**(1), 7214 (2022)

92. Fan, L., Wang, G., Jiang, Y., Mandlekar, A., Yang, Y., Zhu, H., Tang, A., Huang, D.A., Zhu, Y., Anandkumar, A.: Minedojo: Building open-ended embodied agents with internet-scale knowledge. arXiv preprint arXiv:2206.08853 (2022)

93. Yuan, H., Zhang, C., Wang, H., Xie, F., Cai, P., Dong, H., Lu, Z.: Plan4mc: Skill reinforcement learning and planning for open-world minecraft tasks. arXiv preprint arXiv:2303.16563 (2023)

94. Sironi, C.F., Winands, M.H.: Adaptive general search framework for games and beyond. In: 2021 IEEE Conference on Games (CoG), pp. 1–8. IEEE (2021)

95. Tennenholtz, G., Mannor, S.: The natural language of actions. In: International Conference on Machine Learning, pp. 6196–6205. PMLR (2019)

96. Ammanabrolu, P., Hausknecht, M.: Graph constrained reinforcement learning for natural language action spaces. arXiv preprint arXiv:2001.08837 (2020)

97. Zhang, Y., Yang, J., Pan, J., Storks, S., Devraj, N., Ma, Z., Yu, K.P., Bao, Y., Chai, J.: Danli: Deliberative agent for following natural language instructions. arXiv preprint arXiv:2210.12485 (2022)

98. Park, J.S., O'Brien, J.C., Cai, C.J., Morris, M.R., Liang, P., Bernstein, M.S.: Generative agents: Interactive simulacra of human behavior. arXiv preprint arXiv:2304.03442 (2023)

99. Ammanabrolu, P., Broniec, W., Mueller, A., Paul, J., Riedl, M.O.: Toward automated quest generation in text-adventure games. arXiv preprint arXiv:1909.06283 (2019)

100. Côté, M.A., Kádár, A., Yuan, X., Kybartas, B., Barnes, T., Fine, E., Moore, J., Hausknecht, M., El Asri, L., Adada, M., et al.: Textworld: A learning environment for text-based games. In: Computer Games: 7th Workshop, CGW 2018, Held in Conjunction with the 27th International Conference on Artificial Intelligence, IJCAI 2018, Stockholm, Sweden, July 13, 2018, Revised Selected Papers 7, pp. 41–75. Springer (2019)

101. Hausknecht, M., Ammanabrolu, P., Côté, M.A., Yuan, X.: Interactive fiction games: A colossal adventure. In: Proceedings of the AAAI Conference on Artificial Intelligence, vol. 34, pp. 7903–7910 (2020)

102. Lin, Y., Kasamatsu, M., Chen, T., Fujita, T., Deng, H., Utsuro, T.: Automatic annotation of werewolf game corpus with players revealing oneselves as seer/medium and divination/medium results. In: Workshop on Games and Natural Language Processing, pp. 85–93 (2020)

103. Tuin, H., Rooijackers, M.: Automatically detecting player roles in among us. In: 2021 IEEE Conference on Games (CoG), pp. 1–5. IEEE (2021)

104. Chen, N., Wang, Y., Jiang, H., Cai, D., Chen, Z., Li, J.: What would harry say? building dialogue agents for characters in a story. arXiv preprint arXiv:2211.06869 (2022)

105. Callison-Burch, C., Tomar, G.S., Martin, L.J., Ippolito, D., Bailis, S., Reiter, D.: Dungeons and dragons as a dialog challenge for artificial intelligence. arXiv preprint arXiv:2210.07109 (2022)

106. Rameshkumar, R., Bailey, P.: Storytelling with dialogue: A critical role dungeons and dragons dataset. In: Proceedings of the 58th Annual Meeting of the Association for Computational Linguistics, pp. 5121–5134 (2020)

107. Peiris, A., de Silva, N.: Synthesis and evaluation of a domain-specific large data set for dungeons & dragons. arXiv preprint arXiv:2212.09080 (2022)

108. Zhu, A., Aggarwal, K., Feng, A., Martin, L.J., Callison-Burch, C.: Fireball: A dataset of dungeons and dragons actual-play with structured game state information. arXiv preprint arXiv:2305.01528 (2023)

109. Mehta, N., Teruel, M., Sanz, P.F., Deng, X., Awadallah, A.H., Kiseleva, J.: Improving grounded language understanding in a collaborative environment by interacting with agents through help feedback. arXiv preprint arXiv:2304.10750 (2023)

110. Li, G., Hammoud, H.A.A.K., Itani, H., Khizbullin, D., Ghanem, B.: Camel: Communicative agents for" mind" exploration of large scale language model society. arXiv preprint arXiv:2303.17760 (2023)

111. Wei, J., Shuster, K., Szlam, A., Weston, J., Urbanek, J., Komeili, M.: Multi-party chat: Conversational agents in group settings with humans and models. arXiv preprint arXiv:2304.13835 (2023)

112. Yuan, X., Côté, M.A., Sordoni, A., Laroche, R., Combes, R.T.d., Hausknecht, M., Trischler, A.: Counting to explore and generalize in text-based games. arXiv preprint arXiv:1806.11525 (2018)
113. Yao, S., Narasimhan, K., Hausknecht, M.: Reading and acting while blindfolded: The need for semantics in text game agents. arXiv preprint arXiv:2103.13552 (2021)
114. Li, W., Bai, Y., Lu, J., Yi, K.: Immersive text game and personality classification. arXiv preprint arXiv:2203.10621 (2022)
115. Yin, X., May, J.: Zero-shot learning of text adventure games with sentence-level semantics. arXiv preprint arXiv:2004.02986 (2020)
116. Mnih, V., Kavukcuoglu, K., Silver, D., Graves, A., Antonoglou, I., Wierstra, D., Riedmiller, M.: Playing atari with deep reinforcement learning. arXiv preprint arXiv:1312.5602 (2013)
117. Mnih, V., Kavukcuoglu, K., Silver, D., Rusu, A.A., Veness, J., Bellemare, M.G., Graves, A., Riedmiller, M., Fidjeland, A.K., Ostrovski, G., et al.: Human-level control through deep reinforcement learning. nature **518**(7540), 529–533 (2015)
118. Xu, Y., Chen, L., Fang, M., Wang, Y., Zhang, C.: Deep reinforcement learning with transformers for text adventure games. In: 2020 IEEE Conference on Games (CoG), pp. 65–72. IEEE (2020)
119. Todd, G., Cheng, Z., Liu, Y., Togelius, J.: Towards knowledge-graph constrained generation for text adventure games. In: The Third Wordplay: When Language Meets Games Workshop (2022)
120. Branavan, S.R., Chen, H., Zettlemoyer, L., Barzilay, R.: Reinforcement learning for mapping instructions to actions. In: Proceedings of the Joint Conference of the 47th Annual Meeting of the ACL and the 4th International Joint Conference on Natural Language Processing of the AFNLP, pp. 82–90 (2009)
121. Vinyals, O., Ewalds, T., Bartunov, S., Georgiev, P., Vezhnevets, A.S., Yeo, M., Makhzani, A., Küttler, H., Agapiou, J., Schrittwieser, J., et al.: Starcraft ii: A new challenge for reinforcement learning. arXiv preprint arXiv:1708.04782 (2017)
122. Goyal, P., Niekum, S., Mooney, R.J.: Using natural language for reward shaping in reinforcement learning. arXiv preprint arXiv:1903.02020 (2019)
123. Tuli, M., Li, A., Vaezipoor, P., Klassen, T., Sanner, S., McIlraith, S.: Learning to follow instructions in text-based games. Advances in Neural Information Processing Systems **35**, 19441–19455 (2022)
124. Chen, V., Gupta, A., Marino, K.: Ask your humans: Using human instructions to improve generalization in reinforcement learning. arXiv preprint arXiv:2011.00517 (2020)
125. Hanjie, A.W., Zhong, V.Y., Narasimhan, K.: Grounding language to entities and dynamics for generalization in reinforcement learning. In: M. Meila, T. Zhang (eds.) Proceedings of the 38th International Conference on Machine Learning, *Proceedings of Machine Learning Research*, vol. 139, pp. 4051–4062. PMLR (2021). https://proceedings.mlr.press/v139/hanjie21a.html
126. Karamcheti, S., Nair, S., Chen, A.S., Kollar, T., Finn, C., Sadigh, D., Liang, P.: Language-driven representation learning for robotics. arXiv preprint arXiv:2302.12766 (2023)
127. Havrylov, S., Titov, I.: Emergence of language with multi-agent games: Learning to communicate with sequences of symbols. Advances in neural information processing systems **30** (2017)
128. Wong, A., Bäck, T., Kononova, A.V., Plaat, A.: Deep multiagent reinforcement learning: Challenges and directions. Artificial Intelligence Review pp. 1–34 (2022)
129. Jiang, Y., Gu, S.S., Murphy, K.P., Finn, C.: Language as an abstraction for hierarchical deep reinforcement learning. In: H. Wallach, H. Larochelle, A. Beygelzimer, F. d' Alché-Buc, E. Fox, R. Garnett (eds.) Advances in Neural Information Processing Systems, vol. 32. Curran Associates, Inc. (2019)
130. Reid, M., Yamada, Y., Gu, S.S.: Can wikipedia help offline reinforcement learning? arXiv preprint arXiv:2201.12122 (2022)

131. Li, S., Puig, X., Paxton, C., Du, Y., Wang, C., Fan, L., Chen, T., Huang, D.A., Akyürek, E., Anandkumar, A., et al.: Pre-trained language models for interactive decision-making. Advances in Neural Information Processing Systems **35**, 31199–31212 (2022)

132. Zhang, Z., Chen, C., Liu, B., Liao, C., Gong, Z., Yu, H., Li, J., Wang, R.: Unifying the perspectives of nlp and software engineering: A survey on language models for code. arXiv preprint arXiv:2311.07989 (2023)

133. Qiao, S., Ou, Y., Zhang, N., Chen, X., Yao, Y., Deng, S., Tan, C., Huang, F., Chen, H.: Reasoning with language model prompting: A survey. arXiv preprint arXiv: Arxiv-2212.09597 (2022)

134. Huang, J., Chang, K.: Towards reasoning in large language models: A survey. ARXIV.ORG (2022). 10.48550/arXiv.2212.10403

135. Chen, M., Tworek, J., Jun, H., Yuan, Q., de Oliveira Pinto, H.P., et al.: Evaluating large language models trained on code. arXiv preprint arXiv: Arxiv-2107.03374 (2021)

136. Roziere, B., Gehring, J., Gloeckle, F., Sootla, S., Gat, I., Tan, X.E., Adi, Y., Liu, J., Remez, T., Rapin, J., et al.: Code llama: Open foundation models for code. arXiv preprint arXiv:2308.12950 (2023)

137. Li, R., Allal, L.B., Zi, Y., Muennighoff, N., Kocetkov, D., Mou, C., Marone, M., Akiki, C., Li, J., Chim, J., et al.: Starcoder: may the source be with you! arXiv preprint arXiv:2305.06161 (2023)

138. Guo, D., Zhu, Q., Yang, D., Xie, Z., Dong, K., Zhang, W., Chen, G., Bi, X., Wu, Y., Li, Y., et al.: Deepseek-coder: When the large language model meets programming–the rise of code intelligence. arXiv preprint arXiv:2401.14196 (2024)

139. OpenAI: GPT-4 technical report. PREPRINT (2023)

140. Yue, X., Qu, X., Zhang, G., Fu, Y., Huang, W., Sun, H., Su, Y., Chen, W.: Mammoth: Building math generalist models through hybrid instruction tuning. arXiv preprint arXiv: 2309.05653 (2023). https://arxiv.org/abs/2309.05653v1

141. Ying, H., Zhang, S., Li, L., Zhou, Z., Shao, Y., Fei, Z., Ma, Y., Hong, J., Liu, K., Wang, Z., et al.: Internlm-math: Open math large language models toward verifiable reasoning. arXiv preprint arXiv:2402.06332 (2024)

142. Yang, K., Swope, A., Gu, A., Chalamala, R., Song, P., Yu, S., Godil, S., Prenger, R.J., Anandkumar, A.: Leandojo: Theorem proving with retrieval-augmented language models. Advances in Neural Information Processing Systems **36** (2024)

143. Fitzpatrick, K.K., Darcy, A., Vierhile, M.: Delivering cognitive behavior therapy to young adults with symptoms of depression and anxiety using a fully automated conversational agent (woebot): a randomized controlled trial. JMIR mental health **4**(2), e7785 (2017)

144. Das, A., Selek, S., Warner, A.R., Zuo, X., Hu, Y., Keloth, V.K., Li, J., Zheng, W.J., Xu, H.: Conversational bots for psychotherapy: a study of generative transformer models using domain-specific dialogues. In: Proceedings of the 21st Workshop on Biomedical Language Processing, pp. 285–297 (2022)

145. Nie, J., Shao, H., Fan, Y., Shao, Q., You, H., Preindl, M., Jiang, X.: LLM-based conversational ai therapist for daily functioning screening and psychotherapeutic intervention via everyday smart devices. arXiv preprint arXiv:2403.10779 (2024)

146. Locke, S., Bashall, A., Al-Adely, S., Moore, J., Wilson, A., Kitchen, G.B.: Natural language processing in medicine: a review. Trends in Anaesthesia and Critical Care **38**, 4–9 (2021)

147. Lee, J., Yoon, W., Kim, S., Kim, D., Kim, S., So, C.H., Kang, J.: Biobert: a pre-trained biomedical language representation model for biomedical text mining. Bioinformatics **36**(4), 1234–1240 (2020)

148. Li, Y., Wehbe, R.M., Ahmad, F.S., Wang, H., Luo, Y.: Clinical-longformer and clinical-bigbird: Transformers for long clinical sequences. arXiv preprint arXiv:2201.11838 (2022)

149. Singhal, K., Azizi, S., Tu, T., Mahdavi, S.S., Wei, J., Chung, H.W., Scales, N., Tanwani, A., Cole-Lewis, H., Pfohl, S., et al.: Large language models encode clinical knowledge. arXiv preprint arXiv:2212.13138 (2022)

150. Li, Y., Li, Z., Zhang, K., Dan, R., Jiang, S., Zhang, Y.: Chatdoctor: A medical chat model fine-tuned on a large language model meta-ai (llama) using medical domain knowledge. Cureus **15**(6) (2023)

151. Litman, D.: Natural language processing for enhancing teaching and learning. In: Proceedings of the AAAI conference on artificial intelligence, vol. 30 (2016)

152. Rao, H., Leung, C., Miao, C.: Can ChatGPT assess human personalities? a general evaluation framework. arXiv preprint arXiv:2303.01248 (2023)

153. Salemi, A., Mysore, S., Bendersky, M., Zamani, H.: Lamp: When large language models meet personalization. arXiv preprint arXiv:2304.11406 (2023)

154. Wu, Y., Xie, R., Zhu, Y., Zhuang, F., Zhang, X., Lin, L., He, Q.: Personalized prompt for sequential recommendation. arXiv preprint arXiv: Arxiv-2205.09666 (2022)

155. Madaan, A., Tandon, N., Clark, P., Yang, Y.: Memory-assisted prompt editing to improve GPT-3 after deployment. arXiv preprint arXiv:2201.06009 (2022)

156. Deshpande, A., Murahari, V., Rajpurohit, T., Kalyan, A., Narasimhan, K.: Toxicity in ChatGPT: Analyzing persona-assigned language models (2023)

157. Sellam, T., Das, D., Parikh, A.P.: BLEURT: learning robust metrics for text generation. In: D. Jurafsky, J. Chai, N. Schluter, J.R. Tetreault (eds.) Proceedings of the 58th Annual Meeting of the Association for Computational Linguistics, ACL 2020, Online, July 5-10, 2020, pp. 7881–7892. Association for Computational Linguistics (2020). https://doi.org/10.18653/v1/2020.acl-main.704. https://doi.org/10.18653/v1/2020.acl-main.704

158. Zhang, T., Kishore, V., Wu, F., Weinberger, K.Q., Artzi, Y.: Bertscore: Evaluating text generation with BERT. In: 8th International Conference on Learning Representations, ICLR 2020, Addis Ababa, Ethiopia, April 26-30, 2020. OpenReview.net (2020). https://openreview.net/forum?id=SkeHuCVFDr

159. Rei, R., Stewart, C., Farinha, A.C., Lavie, A.: COMET: A neural framework for MT evaluation. In: B. Webber, T. Cohn, Y. He, Y. Liu (eds.) Proceedings of the 2020 Conference on Empirical Methods in Natural Language Processing, EMNLP 2020, Online, November 16-20, 2020, pp. 2685–2702. Association for Computational Linguistics (2020). https://doi.org/10.18653/v1/2020.emnlp-main.213. https://doi.org/10.18653/v1/2020.emnlp-main.213

160. Zhou, W., Xu, K.: Learning to compare for better training and evaluation of open domain natural language generation models. In: AAAI, pp. 9717–9724. AAAI Press (2020)

161. Wan, Y., Liu, D., Yang, B., Zhang, H., Chen, B., Wong, D.F., Chao, L.S.: Unite: Unified translation evaluation. CoRR **abs/2204.13346** (2022). https://doi.org/10.48550/arXiv.2204.13346. https://doi.org/10.48550/arXiv.2204.13346

162. Zouhar, V., Dhuliawala, S., Zhou, W., Daheim, N., Kocmi, T., Jiang, Y.E., Sachan, M.: Poor man's quality estimation: Predicting reference-based MT metrics without the reference. CoRR **abs/2301.09008** (2023)

163. Gilardi, F., Alizadeh, M., Kubli, M.: ChatGPT outperforms crowd-workers for text-annotation tasks. arXiv preprint arXiv:2303.15056 (2023)

164. Wang, J., Liang, Y., Meng, F., Shi, H., Li, Z., Xu, J., Qu, J., Zhou, J.: Is ChatGPT a good nlg evaluator? a preliminary study. arXiv preprint arXiv:2303.04048 (2023)

165. Chen, Y., Wang, R., Jiang, H., Shi, S., Xu, R.: Exploring the use of large language models for reference-free text quality evaluation: A preliminary empirical study. arXiv preprint arXiv:2304.00723 (2023)

166. Luo, Z., Xie, Q., Ananiadou, S.: ChatGPT as a factual inconsistency evaluator for abstractive text summarization. arXiv preprint arXiv:2303.15621 (2023)

167. Gao, M., Ruan, J., Sun, R., Yin, X., Yang, S., Wan, X.: Human-like summarization evaluation with ChatGPT. arXiv preprint arXiv:2304.02554 (2023)

168. He, X., Lin, Z., Gong, Y., Jin, A., Zhang, H., Lin, C., Jiao, J., Yiu, S.M., Duan, N., Chen, W., et al.: Annollm: Making large language models to be better crowdsourced annotators. arXiv preprint arXiv:2303.16854 (2023)

169. Yang, K.C., Menczer, F.: Large language models can rate news outlet credibility. arXiv preprint arXiv:2304.00228 (2023)

170. Lu, Q., Qiu, B., Ding, L., Xie, L., Tao, D.: Error analysis prompting enables human-like translation evaluation in large language models: A case study on ChatGPT. arXiv preprint arXiv:2303.13809 (2023)

171. Liu, Y., Iter, D., Xu, Y., Wang, S., Xu, R., Zhu, C.: Gpteval: Nlg evaluation using GPT-4 with better human alignment. arXiv preprint arXiv:2303.16634 (2023)

172. ter Hoeve, M., Kharitonov, E., Hupkes, D., Dupoux, E.: Towards interactive language modeling. arXiv preprint arXiv: Arxiv-2112.11911 (2021)

173. Wang, P., Li, L., Chen, L., Cai, Z., Zhu, D., Lin, B., Cao, Y., Liu, Q., Liu, T., Sui, Z.: Large language models are not fair evaluators. arXiv preprint arXiv: 2305.17926 (2023)

174. Zheng, C., Zhou, H., Meng, F., Zhou, J., Huang, M.: Large language models are not robust multiple choice selectors. arXiv preprint arXiv: 2309.03882 (2023)

175. Li, Z., Wang, C., Ma, P., Wu, D., Wang, S., Gao, C., Liu, Y.: Split and merge: Aligning position biases in large language model based evaluators. arXiv preprint arXiv: 2310.01432 (2023)

176. Chan, C.M., Chen, W., Su, Y., Yu, J., Xue, W., Zhang, S., Fu, J., Liu, Z.: Chateval: Towards better LLM-based evaluators through multi-agent debate. arXiv preprint arXiv: 2308.07201 (2023). https://arxiv.org/abs/2308.07201v1

In this part, we provide further discussions on the topic of Interactive NLP. As Interactive NLP systems become increasingly integrated into society, their influence extends across various domains, including education, legal services, and public administration. This widespread impact necessitates a thorough examination of the ethical considerations involved in deploying these systems. Therefore, in the "Ethics and Safety" chapter, we discuss the ethical challenges associated with these systems, focusing on issues such as bias, privacy, transparency, and the potential for misuse, which are critical for ensuring that Interactive NLP is developed and applied responsibly.

Another important aspect worth discussing is the future directions of Interactive NLP. Interactive NLP systems represent the next stage in the evolution of large language models. Given this progression, what future directions should we focus on in the era of Interactive NLP? In the "Future Directions" chapter, we explore various topics that not only address the current limitations of LLMs, which significantly affect the efficiency and quality of interaction, but also identify the challenges that must be overcome to advance LLMs to the next era. These topics include improving factual and value alignment, enhancing the realism of social embodiment, developing more efficient and accessible models, among others. Addressing these issues is essential for the continued progress and effectiveness of Interactive NLP systems in the future.

Chenghua Lin and Mong Yuan Sim

LLMs have demonstrated remarkable capabilities to understand, interpret, and generate human-like text. A plethora of LLM-based applications has emerged and been adopted prevalently in our daily lives. As a result, the utilization of these models also presents profound challenges across many societal domains. Therefore, it is crucial to consider the ethical implications of using LLMs, especially around the impact on education, bias and fairness, privacy, harmful content and misinformation.

12.1 Impact on Education

The advent of LLMs, exemplified by ChatGPT, has introduced substantial challenges to the existing education systems. One primary concern is the misuse of ChatGPT for academic assignments such as writing essays and solving scientific problems, which has raised deep concerns among K-12 educators, who perceive it as a potential threat to the education system [1]. To address this issue, plagiarism detection tools such as GPTZero,[1] AI Classifier,[2] and

---

[1] https://gptzero.me.
[2] https://platform.openai.com/ai-text-classifier.

C. Lin (✉)
Department of Computer Science, University of Manchester, Manchester, UK
e-mail: chenghua.lin@manchester.ac.uk

M. Y. Sim
University of Adelaide, Adelaide, SA, Australia
e-mail: mongyuan.sim@student.adelaide.edu.au

© The Author(s), under exclusive license to Springer Nature Switzerland AG 2026
Z. Wang et al. (eds.), *Interactive Natural Language Processing*, Synthesis Lectures
on Human Language Technologies, https://doi.org/10.1007/978-3-032-06264-2_12

DetectGPT[3] have been developed for detecting AI-generated content. Most of these AI detection tools focus on perplexity (text randomness) and burstiness (use of non-common terms). Nevertheless, these tools have yet to demonstrate their effectiveness in capturing AI-generated content in a real-world setting. Last but not least, computer-assisted writing tools, including ChatGPT, have limited capacities to assist users in learning and acquiring writing skills and principles. Their primary focus is on enhancing productivity rather than facilitating skill development, which is crucial for educational purposes.

12.2   Social Bias

As language models are typically trained with large-scale web corpus, it becomes highly susceptible to societal biases. It is known to further amplify the discrimination [2], including the potential downgrading of resumes [3] and the generation of texts that contain stereotypes, toxicity, and racism [4]. Resume "whitening" has always been an issue where job applicants are forced to hide their identity as a minority gender, racial, religion, or region group to land a job interview. This problem still exists even though many companies started to use AI-supported tools to rank and filter resumes. Social bias in word embeddings is reflected when the word *man* is closer to *programmer* compared to *woman* and *programmer* [5]. Applications that utilize pretrained word embeddings for downstream tasks such as classification and analysis will then obtain results with social bias, causing fairness issues of the output. Hutchinson et al. (2020) use toxicity prediction and sentiment analysis to assess language models' bias towards people with disabilities. Results showed that the sentence *I am a person with mental illness* and *I will fight for people with mental illnesses* are more toxic than *I am a tall person*. HERB [6], a bias evaluation metric which utilizes bias in a sub-region to evaluate language model's bias in a region on contextualized level, brings researchers' attention to not only societal bias in the whole of a region but also sub-regions. You might need to add how this is related to this section. The aforementioned findings suggest that societal biases observed in language models could serve as an indication that stereotypes should be addressed and mitigated, rather than leaving them to harm minorities. Considering that everyone can now easily access LLMs, biases should be filtered out or mitigated to ensure that they are not amplified and further affect people's thoughts in making decisions such as recruiting and assessing individuals.

12.3   Privacy Concern

Large Language Models (LLMs) also raise concerns regarding user privacy. Training these models necessitates access to large amounts of data, often entailing the personal details of individuals. This information is typically derived from licensed or publicly accessible

---

[3] https://detectgpt.com.

datasets and can be utilized for a range of purposes, such as deducing geographical locations from phone codes in the data. There are already studies showing the possibilities of distilling sensitive information from large language models through prompting [7, 8]. In the era of interactive NLP (iNLP), humans are more actively interacting with these foundation models, which could potentially lead to more frequent user information leakage. Therefore, it is pressing to establish relevant policies for collecting and storing personal data. Furthermore, the practice of data anonymization is crucial to maintaining ethical standards in dealing with privacy matters. There have been some pioneering studies that investigate privacy-preserving issues [9, 10]. We believe that more research efforts should be dedicated to privacy preservation in large language models, which will play a central role in the era of interactive Natural Language Processing (iNLP).

12.4 Transparency and Explainability

LLMs function as black-box mechanisms, meaning their internal decision-making processes are not easily observable or understandable. This inherent opacity leads to transparency and explainability concerns. The behavior of these large models is often unpredictable, making their results difficult to trust and verify. As these systems become more integrated into various applications, ensuring that their processes and outputs are transparent and understandable is essential for gaining user trust, and complying with regulatory requirements. Firstly, the unpredictable nature of LLMs comes from their complexity and the vast amounts of data they process. It is difficult to trace how specific inputs lead to specific outputs. This lack of predictability can result in outputs that are not only surprising but potentially harmful or biased, thereby eroding user trust and making it challenging to rely on the system's decisions, especially in high-stakes domains like healthcare [11], education [12], legal services [13], etc. Explainability is equally crucial, as users and stakeholders need to understand not just what decisions the LLM system makes, but why it makes them. Recognizing these challenges, there is a growing trend towards addressing these concerns. One possible approach is to integrate user involvement and feedback [14]. Engaging users in the development and deployment phases of iNLP systems can provide valuable insights into what kind of transparency and explanations they find useful. Alternatively, [15] presents a framework that extracts detailed information about API-protected LLMs by exploiting their low-dimensional output subspace. This framework efficiently discovers the LLM's embedding size, captures full vocabulary outputs, and detects and distinguishes between model updates. By enabling LLM API users to detect these changes, the framework could enhance transparency and accountability for the providers. Future research and development efforts should continue to explore how to enhance the transparency and explainability of LLMs, as well as how to responsibly integrate them into critical domains of society.

12.5   Legal and Regulatory Issues

iNLP presents a range of legal and regulatory challenges that must be carefully navigated to ensure compliance and maintain public trust. One of the primary concerns is intellectual property rights. iNLP systems frequently utilize large datasets and pre-trained models, which may contain copyrighted material. Ensuring proper licensing, understanding and applying fair use principles, and attributing sources of data and models are necessary steps to respect the intellectual property rights of creators. [16] promotes a framework of ethical responsibility and information integrity by incorporating a citation mechanism to ensure proper attribution. Establishing clear lines of responsibility for the design, deployment, and operation of iNLP systems is crucial. This involves understanding how existing liability frameworks apply, and addressing any gaps through updated policies and regulations. It is also important to provide users with mechanisms to contest and seek redress for adverse decisions made by iNLP systems, ensuring that there is a transparent process for handling grievances and correcting errors. Furthermore, different industries have unique regulatory requirements that iNLP systems must comply with. For example, [11] summarize additional regulatory adaptations for the practice of medicine and healthcare. iNLP systems are often deployed across multiple jurisdictions, each with its own set of laws and regulatory frameworks. Navigating regulations on cross-border data transfers, such as the EU-U.S. Privacy Shield framework, adapting systems to comply with local laws and cultural norms, and following international standards and best practices are essential to ensure broad compliance and interoperability.

12.6   Manipulation and Abuse

LLMs have the ability to generate text that is highly convincing and human-like, which can be exploited for malicious purposes, such as creating and spreading harmful content and misinformation. One of the primary concerns is the use of LLMs to fabricate rumors and disseminate false information. The sophisticated language capabilities of these models allow them to create deceptive content that is indistinguishable from human-generated text, leading to the rapid spread of misinformation on social media and other online platforms. This "deepfake" technology can cause public panic, harm reputations, and undermine trust in legitimate information sources. Digital fraud is another significant threat posed by LLMs. Cybercriminals can leverage these models to craft highly personalized phishing emails, impersonate individuals in online communications, and generate fake reviews or testimonials. The realistic and persuasive nature of the text produced by LLMs makes it easier for scammers to deceive individuals and organizations, leading to financial losses and breaches of sensitive information. Moreover, LLMs can be used to amplify the dissemination of false information. Automated bots powered by these models can flood social media with misleading narratives, manipulate public opinion, and sway political outcomes.

The ability of LLMs to generate content at scale exacerbates the challenge of combating misinformation and maintaining the integrity of information ecosystems. Recent studies, such as those by [17] have demonstrated the potential of using images and sounds to inject prompts and instructions into multi-modal LLMs, manipulating models like LLaVA [18] and PandaGPT [19] to produce attacker-chosen text or follow specific instructions through adversarial perturbations. Reference [20] further built upon these concerns with prompt injection attacks, proposing benchmarks for evaluating vulnerabilities and defenses against such adversarial manipulations of LLMs. Addressing these challenges requires concerted efforts in developing robust defense mechanisms to safeguard against the misuse of LLM capabilities [21]. Future research should focus on enhancing the security and reliability of LLMs in interactive contexts.

12.7 Environmental Impact

The deployment of LLMs has significant environmental implications due to their high energy consumption for preparation, training, and inference during each user interaction. Specifically, data preparation involves energy-intensive storage and transmission. Training LLMs requires running complex algorithms on large datasets, demanding substantial computational power. High-performance hardware like graphics processing units (GPUs) and tensor processing units (TPUs) are often used for these tasks. Inference also consumes considerable energy each time a user interacts with the model, particularly in large-scale deployments with numerous users. The high energy consumption associated with LLM deployment can lead to increased greenhouse gas emissions, especially if the energy source is fossil-fuel-based. Additionally, the manufacture and disposal of specialized hardware contribute to the overall carbon footprint and resource depletion. The environmental impact is further amplified in iNLP systems due to the interactive nature of these applications. Each feedback in a loop with external objects triggers real-time processing, which demands continuous computational resources. This constant activity results in a higher frequency of energy consumption compared to non-iNLP models. Moreover, the necessity for high uptime and low latency in iNLP systems means that energy-efficient practices must be balanced with performance requirements. Several studies have estimated the compute and energy costs of training and inference for LMs [22–24], highlighting the need for more sustainable AI practices. Strategies based on adaptive resource allocation, configuration changes, and hybrid data center models have already been proposed to mitigate the issue [25, 26]. To address these environmental concerns, more efforts should be directed toward ensuring the responsible and sustainable development of iNLP. This includes optimizing algorithms to reduce energy consumption, utilizing renewable energy sources, and implementing energy-efficient hardware designs, etc.

References

1. Rudolph, J., Tan, S., Tan, S.: ChatGPT: Bullshit spewer or the end of traditional assessments in higher education **6**(1) (2023)
2. Leino, K., Fredrikson, M., Black, E., Sen, S., Datta, A.: Feature-wise bias amplification. CoRR **abs/1812.08999** (2018). URL http://arxiv.org/abs/1812.08999
3. Dastin, J.: Amazon scraps secret ai recruiting tool that showed bias against women (2018). URL https://www.reuters.com/article/us-amazon-com-jobs-automation-insight/amazon-scraps-secret-ai-recruiting-tool-that-showed-bias-against-women-idUSKCN1MK08G
4. Hutchinson, B., Prabhakaran, V., Denton, E., Webster, K., Zhong, Y., Denuyl, S.: Social biases in NLP models as barriers for persons with disabilities. In: Proceedings of the 58th Annual Meeting of the Association for Computational Linguistics, pp. 5491–5501. Association for Computational Linguistics, Online (2020). https://doi.org/10.18653/v1/2020.acl-main.487. URL https://aclanthology.org/2020.acl-main.487
5. Bolukbasi, T., Chang, K.W., Zou, J., Saligrama, V., Kalai, A.: Man is to computer programmer as woman is to homemaker? debiasing word embeddings. In: Proceedings of the 30th International Conference on Neural Information Processing Systems, NIPS'16, p. 4356–4364. Curran Associates Inc., Red Hook, NY, USA (2016)
6. Li, Y., Zhang, G., Yang, B., Lin, C., Ragni, A., Wang, S., Fu, J.: HERB: Measuring hierarchical regional bias in pre-trained language models. In: Findings of the Association for Computational Linguistics: AACL-IJCNLP 2022, pp. 334–346. Association for Computational Linguistics, Online only (2022). URL https://aclanthology.org/2022.findings-aacl.32
7. Carlini, N., Tramer, F., Wallace, E., Jagielski, M., Herbert-Voss, A., Lee, K., Roberts, A., Brown, T.B., Song, D., Erlingsson, U., et al.: Extracting training data from large language models. In: USENIX Security Symposium, vol. 6 (2021)
8. Carlini, N., Hayes, J., Nasr, M., Jagielski, M., Sehwag, V., Tramer, F., Balle, B., Ippolito, D., Wallace, E.: Extracting training data from diffusion models. arXiv preprint arXiv:2301.13188 (2023)
9. Li, Y., Tan, Z., Liu, Y.: Privacy-preserving prompt tuning for large language model services. arXiv preprint arXiv:2305.06212 (2023)
10. Shi, W., Chen, S., Zhang, C., Jia, R., Yu, Z.: Just fine-tune twice: Selective differential privacy for large language models. arXiv preprint arXiv:2204.07667 (2022)
11. Meskó, B., Topol, E.J.: The imperative for regulatory oversight of large language models (or generative ai) in healthcare. NPJ digital medicine **6**(1), 120 (2023)
12. Wang, S., Xu, T., Li, H., Zhang, C., Liang, J., Tang, J., Yu, P.S., Wen, Q.: Large language models for education: A survey and outlook. arXiv preprint arXiv:2403.18105 (2024)
13. Lai, J., Gan, W., Wu, J., Qi, Z., Yu, P.S.: Large language models in law: A survey. arXiv preprint arXiv:2312.03718 (2023)
14. Wu, T.S., Terry, M., Cai, C.J.: Ai chains: Transparent and controllable human-ai interaction by chaining large language model prompts. International Conference On Human Factors In Computing Systems (2021). https://doi.org/10.1145/3491102.3517582
15. Finlayson, M., Swayamdipta, S., Ren, X.: Logits of api-protected LLMs leak proprietary information. arXiv preprint arXiv:2403.09539 (2024)
16. Huang, J., Chang, K.C.C.: Citation: A key to building responsible and accountable large language models. arXiv preprint arXiv:2307.02185 (2023)
17. Bagdasaryan, E., Hsieh, T.Y., Nassi, B., Shmatikov, V.: (ab) using images and sounds for indirect instruction injection in multi-modal LLMs. arXiv preprint arXiv:2307.10490 (2023)
18. Liu, H., Li, C., Wu, Q., Lee, Y.J.: Visual instruction tuning. In: NeurIPS (2023)

19. Su, Y., Lan, T., Li, H., Xu, J., Wang, Y., Cai, D.: Pandagpt: One model to instruction-follow them all. arXiv preprint arXiv:2305.16355 (2023)

20. Liu, Y., Jia, Y., Geng, R., Jia, J., Gong, N.Z.: Formalizing and benchmarking prompt injection attacks and defenses. In: USENIX Security Symposium (2024)

21. Sun, L., Huang, Y., Wang, H., Wu, S., Zhang, Q., Gao, C., Huang, Y., Lyu, W., Zhang, Y., Li, X., et al.: Trustllm: Trustworthiness in large language models. arXiv preprint arXiv:2401.05561 (2024)

22. Samsi, S., Zhao, D., McDonald, J., Li, B., Michaleas, A., Jones, M., Bergeron, W., Kepner, J., Tiwari, D., Gadepally, V.: From words to watts: Benchmarking the energy costs of large language model inference. In: 2023 IEEE High Performance Extreme Computing Conference (HPEC), pp. 1–9 (2023). https://doi.org/10.1109/HPEC58863.2023.10363447

23. Gultekin, S., Globo, A., Zugarini, A., Ernandes, M., Rigutini, L.: An energy-based comparative analysis of common approaches to text classification in the legal domain. arXiv preprint arXiv:2311.01256 (2023)

24. Strubell, E., Ganesh, A., McCallum, A.: Energy and policy considerations for modern deep learning research. In: Proceedings of the AAAI conference on artificial intelligence, vol. 34, pp. 13693–13696 (2020)

25. Stojkovic, J., Choukse, E., Zhang, C., Goiri, I., Torrellas, J.: Towards greener LLMs: Bringing energy-efficiency to the forefront of LLM inference. arXiv preprint arXiv:2403.20306 (2024)

26. Wilkins, G., Keshav, S., Mortier, R.: Hybrid heterogeneous clusters can lower the energy consumption of LLM inference workloads. In: Proceedings of the 15th ACM International Conference on Future and Sustainable Energy Systems, pp. 506–513 (2024)

Future Directions

Zekun Wang, Wenhu Chen, Ruibo Liu, Kexin Yang, Wangchunshu Zhou, Chenghua Lin, Qi Liu, Mong Yuan Sim, Ge Zhang, Xiuying Chen, Ke Xu and Jie Fu

Alignment. Alignment for language models can be categorized into factual alignment and value alignment. Factual alignment requires the model to tell it is not capable of answering the question when it does not perpetuate the needed knowledge [1]. However, factual alignment is challenging in practice since (1) it is hard to verify what knowledge has been contained in the pre-trained model [2], and (2) we still lack a convenient knowledge editing method to update certain knowledge while do not impair others [3, 4]. Future work can

Z. Wang
Beihang University, Haidian District, Beijing, China
e-mail: zenmoore@buaa.edu.cn

W. Chen
University of Waterloo, Cheriton School of Computer Science, Waterloo, Canada
e-mail: wenhuchen@uwaterloo.ca

R. Liu
Dartmouth College, Department of Computer Science, Class of 1982 Engineering & Computer Science Center, Hanover, USA
e-mail: ruibo.liu.gr@dartmouth.edu

K. Yang
No. 24 South Sect. 1, Chengdu, China

W. Zhou
AI Waves, Haichuanglvgu 6-213, Yuhang District, Hangzhou, Zhejiang, China
e-mail: chunshu@aiwaves.cn

C. Lin
Department of Computer Science, University of Manchester, Manchester, UK
e-mail: chenghua.lin@manchester.ac.uk

Q. Liu
City University of Hong Kong, G2328, Yeung Kin Man Academic Building, Kowloon Tong, Hong Kong, China

© The Author(s), under exclusive license to Springer Nature Switzerland AG 2026
Z. Wang et al. (eds.), *Interactive Natural Language Processing*, Synthesis Lectures on Human Language Technologies, https://doi.org/10.1007/978-3-032-06264-2_13

consider developing tools to detect the "blind spot of knowledge" by analyzing the probability confidence in the model predictions, and efficient approaches to edit the knowledge in pre-trained models at scale. For value alignment, existing work mainly focuses on using human [5] or AI feedback [6] to train a reward model as the proxy of human judgment. During training, this reward model will continuously interact with the generative LM to enhance desired behaviors and inhibits undesired ones [7]. RLHF is the representative approach in this manner, which has been widely used in products such as OpenAI ChatGPT. However, recent works have shown that inaccurate reward modeling can be exploited by the RL optimization [8], which is also called "reward hacking" problem in the RL formalization [9, 10]. Future work can seek more diverse and fine-grained signals to replace scalar form rewards to aid a more stable and efficient alignment training.

Social Embodiment. NLP models should incorporate a more comprehensive view of the world, including an embodied and social context, to simulate realistic human behavior [11, 12]. This is because social and cultural factors heavily influence human behavior. Recently, generative agents have been introduced as a way to simulate believable human behavior by incorporating a Large Language Model (LLM) with a complete record of the agent's experiences [13]. However, there are still challenges to be addressed to improve the accuracy and complexity of such social agents. Scaling the iNLP to handle larger and more complex environments is a potential future direction. This would enable the agent to handle more ambitious simulations of human behaviors and generate more realistic responses to user interactions. We refer readers to [14, 15] for further information on Embodied AI with LLMs, which serves as an important instantiation of iNLP.

Plasticity. A significant challenge encountered with iNLP is the constant need for updates to adapt to changes in the real world. The prevalent approach in the academic literature typically utilizes gradient-based fine-tuning methods. These methods adjust an extensive number of parameters in the pre-trained models simultaneously, which can be overkill.

M. Y. Sim
University of Adelaide, Adelaide, SA, Australia
e-mail: mongyuan.sim@student.adelaide.edu.au

G. Zhang
University of Michigan, Haidian District, Beijing, China
e-mail: gezhang@umich.edu

X. Chen
King Abdullah University of Science and Technology, Thuwal, Saudi Arabia
e-mail: xiuying.chen@kaust.edu.sa

K. Xu
Beihang University, Beijing, China
e-mail: kexu@nlsde.buaa.edu.cn

J. Fu (✉)
Hong Kong University of Science and Technology, Hong Kong, China
e-mail: jiefu@ust.hk

Nonetheless, if an insufficient number of parameters are adjusted, the models may not effectively adapt to changes in real-world scenarios. Consequently, identifying methods for effective updates to iNLP models is essential for practical applicability [16–18]. In recent years, burgeoning interest among researchers in the field of continual learning has emerged, aiming to enhance a model's capacity to learn persistently over time while minimizing the loss of previously acquired information. Continual learning enables the iNLP model to adapt to new data and dynamic situations without necessitating the retraining of the model from scratch. It is worth noting that biological neural networks acquire new skills continually within their lifetime based on neuronal plasticity [19]. Future research focusing on more human-like models [20, 21] is anticipated to expedite advancements in continual learning for iNLP.

Speed and Efficiency. iNLP usually requires large language models as the backbone, and thus suffers from their high latency and huge computational cost [22, 23]. The high latency issue is even more crucial for iNLP compared to conventional NLP due to the need for frequent iterative calls. A large amount of work has been done on improving the speed and efficiency of large language models, including both *static* methods such as knowledge distillation [24, 25], pruning [26, 27], quantization [28, 29] and module replacing [30]; and *dynamic* methods such as adaptive computation [31], early-exiting [32, 33], speculative decoding [34], and model cascade [35, 36]. However, most of the aforementioned methods require access to the model parameters, which may not be possible in the future since most state-of-the-art generalist models such as ChatGPT and PaLM [37, 38] are closed-sourced. Therefore, developing techniques that can accelerate inference for LLMs without access to their parameters is a promising future for efficient iNLP. Moreover, it is important to consider not only the acceleration ratio or preserved performance of accelerated models but also their robustness, biases, and alignment [39].

Context Length. Context length refers to the maximum numbers of input tokens permitted by a language model. For example, ChatGPT has a context window of 8K tokens, while GPT-4 [40] extends it to 32K tokens. iNLP can greatly benefit from a long context window. The reason is three-fold: (1) It allows for maintaining and understanding a more extensive conversational history. (2) The ability to process a longer context is crucial for tasks that involve large pieces of text, such as long document-based QA and a detailed observation in the environment. (3) It can also facilitate the generation of long-form content. Recent studies on memorizing Transformers [41–44] have illustrated the potential to scale the context window to tens of thousands of tokens using memory augmentation techniques. Additionally, Anthropic has introduced a chatbot with a 100K context window.[1] Recently, common approaches for extending context length include positional interpolation [45] and efficient attention mechanisms such as Lightning Attention [46] and Attention Sink [47]. However, despite these advancements, more research is needed to investigate the challenges associated with significantly increasing the context length. We refer readers to [48] for a comprehensive overview of existing methodologies for extending the context window of LLMs.

---

[1] https://www.anthropic.com/index/100k-context-windows.

Long Text Generation. The capability to generate long text is crucial in iNLP contexts. For example, in real-life conversations, humans frequently convey intricate ideas and participate in extremely long discussions that necessitate numerous rounds of information exchanges. Moreover, for long-horizon robotic manipulation tasks, LMs need to generate a long action plan for execution. However, as the generated text lengthens, current language models have the propensity to produce content that may lack structure, coherence, quality, and even the relevance to the input prompts. Consequently, more sophisticated natural language processing techniques are needed to accurately capture the subtleties of language and produce text that is both coherent and useful. This challenge is closely related to the context length extension problem [48], as a generative LLM capable of long-context modeling can, in principle, support long-form text generation. Some studies focus on training LLMs to generate ultra-long outputs using dedicated data pipelines [49], while others leverage agentic workflows to facilitate long-text generation [50]. HelloBench [51] serves as an important benchmark for evaluating the long-text generation capabilities of LLMs.

Accessibility. In the realm of large language model deployment, accessibility emerges as a critical concern. The most prominent LLMs, such as the GPT-family models [5, 40, 52] and Bard,[2] are predominantly closed-source, creating a significant barrier for those seeking to utilize them for specific purposes. Recently, researchers have shifted their focus to developing open-source LLMs, including LLaMA [53], Pythia [54], and GLM [55]. The movement towards open-sourcing large language models is expected to gain momentum in the future. Another emerging trend that has received limited research attention so far is the accessibility of deploying LLMs on edge devices such as smartphones, laptops, and automobiles, despite the existence of several previous works on the topic [56, 57].[3] Research towards more accessible language models can expand the possibilities for iNLP. For instance, it can be particularly beneficial in scenarios that involve offline interaction.[4]

Interpretability. Although the interactive language models have shown a powerful ability to understand and generate complex language across a wide range of topics and contexts, their "inner workings" are still a black box for both the users and the researchers. We assume that gaining a deeper understanding of LMs and their interpretability can lead to improved interaction behaviors exhibited by LM agents. For example, [58] utilizes GPT-4 to provide explanations for all the neurons in GPT-2 [59] by analyzing their activations in response to input text. Intuitively, such explainability can facilitate knowledge updates in language models within the context of iNLP [4, 60]. Additionally, the analysis of scaling laws [61–63], emergent abilities [64, 65], grokking [65, 66], scaling-up performance prediction [40, 63], double descent phenomenon [65, 67], attention sink [47, 68], trade-offs between alignment and general performance [8], and interpretability for the interaction behavior of LMs

---

[2] https://bard.google.com/.

[3] https://github.com/mlc-ai/mlc-llm.

[4] In situations where network connectivity is unavailable, relying on closed-source language models accessed through the Internet becomes infeasible. Therefore, the use of a local language model becomes necessary.

[13, 69] are also promising avenues for future research. [70] serves as an excellent primer on LLM interpretability.

Creativity. Contrary to the prevailing language modeling approach, which relies on learning statistical relationships, creativity involves the generation of original ideas, concepts, or perspectives that deviate from conventional patterns. The pursuit of creativity has long been a significant challenge in the AI community, driven by the desire to develop human-level agents [71] capable of generating novel knowledge and contributing original ideas across various domains. To effectively generate more creative content, it is crucial to establish a detailed definition or judging criteria of creativity. For instance, generating novel metaphors requires establishing conceptual mappings between the source and target domains [72, 73], whereas story generation involves the creation of original, coherent, and engaging narratives and plotlines [74, 75]. Additionally, the ability to generate new knowledge in the generated text, rather than solely extracting existing knowledge, can contribute to enhancing creativity. Furthermore, it is essential to explore approaches for enhancing creativity in generated content to ensure practical utility. For instance, enabling language models to discover theories or laws based on observed phenomena requires dedicated efforts and potentially entails exploring new paradigms for language models to engage in conscious thinking [76]. Research towards more creative language models may unlock a range of complex interactive properties or behaviors of LMs, such as the development of a more creative writing assistant or even the emergence of sophisticated debates between language model agents.

Evaluation. As shown in Sect. 10, evaluation for iNLP is still barren and lacks diversity. How to design a better evaluation method will be one of the most important research topics in the future, which will profoundly affect the design and optimization direction of the iNLP frameworks. Specifically, evaluation methods under interactive settings may develop in the following aspects: (1) Pay more attention to the evaluation of the interaction process, rather than just the result [77]. (2) Design a more standard evaluation benchmark to support the comparison of different interactive models. (3) Evaluate the interactivity of large language models through tasks of growing complexity.

Real-Time AI. This refers to the capability of AI systems to process and respond to inputs with minimal latency, enabling seamless, human-like interactions. This is critical for iNLP, as it supports applications like streaming video understanding [78], streaming speech generation (e.g., GLM-4-Voice [79]), and real-time multimodal conversational systems (e.g., GPT-4o [80]). The significance of real-time AI lies in its ability to (1) provide instant feedback for natural conversational flow, (2) handle dynamic, multimodal inputs (e.g., text, audio, video) for tasks like real-time translation or video analysis, and (3) enhance user experience in time-sensitive applications such as virtual assistants or robotic control. However, achieving real-time performance faces challenges, including high computational complexity, latency in sequential token generation, efficient multimodal data feature modeling, among others. Current solutions include model optimization techniques like quantization and pruning, hardware acceleration, and efficient algorithms such as continuous batching [81]. Despite these advancements, trade-offs between latency and accuracy remain, and significant gaps persist between current capabilities and users' requirements.

References

1. Kadavath, S., Conerly, T., Askell, A., Henighan, T., Drain, D., Perez, E., Schiefer, N., Dodds, Z.H., DasSarma, N., Tran-Johnson, E., et al.: Language models (mostly) know what they know. arXiv preprint arXiv:2207.05221 (2022)
2. Lin, S., Hilton, J., Evans, O.: Teaching models to express their uncertainty in words. arXiv preprint arXiv:2205.14334 (2022)
3. De Cao, N., Aziz, W., Titov, I.: Editing factual knowledge in language models. arXiv preprint arXiv:2104.08164 (2021)
4. Meng, K., Bau, D., Andonian, A., Belinkov, Y.: Locating and editing factual associations in GPT. Advances in Neural Information Processing Systems **36** (2022)
5. Ouyang, L., Wu, J., Jiang, X., Almeida, D., Wainwright, C.L., Mishkin, P., Zhang, C., Agarwal, S., Slama, K., Ray, A., et al.: Training language models to follow instructions with human feedback. arXiv preprint arXiv:2203.02155 (2022)
6. Bai, Y., Kadavath, S., Kundu, S., Askell, A., Kernion, J., Jones, A., Chen, A., Goldie, A., Mirhoseini, A., McKinnon, C., et al.: Constitutional ai: Harmlessness from ai feedback. arXiv preprint arXiv:2212.08073 (2022)
7. Liu, R., Zhang, G., Feng, X., Vosoughi, S.: Aligning generative language models with human values. In: Findings of the Association for Computational Linguistics: NAACL 2022, pp. 241–252. Association for Computational Linguistics, Seattle, United States (2022). https://doi.org/10.18653/v1/2022.findings-naacl.18. URL https://aclanthology.org/2022.findings-naacl.18
8. Wolf, Y., Wies, N., Levine, Y., Shashua, A.: Fundamental limitations of alignment in large language models. arXiv preprint arXiv:2304.11082 (2023)
9. Ibarz, B., Leike, J., Pohlen, T., Irving, G., Legg, S., Amodei, D.: Reward learning from human preferences and demonstrations in atari. Advances in neural information processing systems **31** (2018)
10. Hadfield-Menell, D., Milli, S., Abbeel, P., Russell, S.J., Dragan, A.: Inverse reward design. Advances in neural information processing systems **30** (2017)
11. Bisk, Y., Holtzman, A., Thomason, J., Andreas, J., Bengio, Y., Chai, J., Lapata, M., Lazaridou, A., May, J., Nisnevich, A., Pinto, N., Turian, J.: Experience grounds language. arXiv preprint arXiv: Arxiv-2004.10151 (2020)
12. Bolotta, S., Dumas, G.: Social neuro ai: Social interaction as the "dark matter" of ai. Frontiers in Computer Science **4** (2022). https://doi.org/10.3389/fcomp.2022.846440. URL https://www.frontiersin.org/articles/10.3389/fcomp.2022.846440
13. Park, J.S., O'Brien, J.C., Cai, C.J., Morris, M.R., Liang, P., Bernstein, M.S.: Generative agents: Interactive simulacra of human behavior. arXiv preprint arXiv:2304.03442 (2023)
14. Durante, Z., Huang, Q., Wake, N., Gong, R., Park, J.S., Sarkar, B., Taori, R., Noda, Y., Terzopoulos, D., Choi, Y., Ikeuchi, K., Vo, H., Fei-Fei, L., Gao, J.: Agent ai: Surveying the horizons of multimodal interaction. arXiv preprint arXiv: 2401.03568 (2024)
15. Liu, Y., Chen, W., Bai, Y., Liang, X., Li, G., Gao, W., Lin, L.: Aligning cyber space with physical world: A comprehensive survey on embodied ai. arXiv preprint arXiv: 2407.06886 (2024)
16. Mitchell, E., Lin, C., Bosselut, A., Finn, C., Manning, C.D.: Fast model editing at scale. arXiv preprint arXiv:2110.11309 (2021)

17. Mitchell, E., Lin, C., Bosselut, A., Manning, C.D., Finn, C.: Memory-based model editing at scale. In: International Conference on Machine Learning, pp. 15817–15831. PMLR (2022)
18. Mitchell, E., Noh, J.J., Li, S., Armstrong, W.S., Agarwal, A., Liu, P., Finn, C., Manning, C.D.: Enhancing self-consistency and performance of pre-trained language models through natural language inference. arXiv preprint arXiv:2211.11875 (2022)
19. Hebb, D.O.: The organization of behavior: A neuropsychological theory. Psychology press (2005)
20. Zador, A., Escola, S., Richards, B., Ölveczky, B., Bengio, Y., Boahen, K., Botvinick, M., Chklovskii, D., Churchland, A., Clopath, C., et al.: Catalyzing next-generation artificial intelligence through neuroai. Nature Communications **14**(1), 1597 (2023)
21. Wang, F., Tian, H., Xiong, H., Wu, H., Fu, J., Cao, Y., Kang, Y., Wang, H.: Evolving decomposed plasticity rules for information-bottlenecked meta-learning. arXiv preprint arXiv:2109.03554 (2021)
22. Schwartz, R., Dodge, J., Smith, N.A., Etzioni, O.: Green AI. Commun. ACM **63**(12), 54–63 (2020)
23. Xu, J., Zhou, W., Fu, Z., Zhou, H., Li, L.: A survey on green deep learning. CoRR **abs/2111.05193** (2021)
24. Sanh, V., Debut, L., Chaumond, J., Wolf, T.: Distilbert, a distilled version of bert: smaller, faster, cheaper and lighter. arXiv preprint arXiv:1910.01108 (2019)
25. Zhou, W., Xu, C., McAuley, J.: BERT learns to teach: Knowledge distillation with meta learning. In: Proceedings of the 60th Annual Meeting of the Association for Computational Linguistics (Volume 1: Long Papers), pp. 7037–7049. Association for Computational Linguistics, Dublin, Ireland (2022). https://doi.org/10.18653/v1/2022.acl-long.485. URL https://aclanthology.org/2022.acl-long.485
26. Michel, P., Levy, O., Neubig, G.: Are sixteen heads really better than one? In: H. Wallach, H. Larochelle, A. Beygelzimer, F. d'Alché-Buc, E. Fox, R. Garnett (eds.) Advances in Neural Information Processing Systems, vol. 32. Curran Associates, Inc. (2019)
27. Gordon, M., Duh, K., Andrews, N.: Compressing BERT: Studying the effects of weight pruning on transfer learning. In: Proceedings of the 5th Workshop on Representation Learning for NLP, pp. 143–155. Association for Computational Linguistics, Online (2020). https://doi.org/10.18653/v1/2020.repl4nlp-1.18. URL https://aclanthology.org/2020.repl4nlp-1.18
28. Shen, S., Dong, Z., Ye, J., Ma, L., Yao, Z., Gholami, A., Mahoney, M.W., Keutzer, K.: Q-BERT: hessian based ultra low precision quantization of BERT. In: AAAI, pp. 8815–8821. AAAI Press (2020)
29. Dettmers, T., Lewis, M., Belkada, Y., Zettlemoyer, L.: GPT-3.int8(): 8-bit matrix multiplication for transformers at scale. In: A.H. Oh, A. Agarwal, D. Belgrave, K. Cho (eds.) Advances in Neural Information Processing Systems (2022). URL https://openreview.net/forum?id=dXiGWqBoxaD
30. Xu, C., Zhou, W., Ge, T., Wei, F., Zhou, M.: BERT-of-theseus: Compressing BERT by progressive module replacing. In: Proceedings of the 2020 Conference on Empirical Methods in Natural Language Processing (EMNLP), pp. 7859–7869. Association for Computational Linguistics, Online (2020). https://doi.org/10.18653/v1/2020.emnlp-main.633. URL https://aclanthology.org/2020.emnlp-main.633
31. Graves, A.: Adaptive computation time for recurrent neural networks (2017)
32. Schwartz, R., Stanovsky, G., Swayamdipta, S., Dodge, J., Smith, N.A.: The right tool for the job: Matching model and instance complexities. In: Proceedings of the 58th Annual Meeting of the Association for Computational Linguistics, pp. 6640–6651. Association for Computational Linguistics, Online (2020). https://doi.org/10.18653/v1/2020.acl-main.593. URL https://aclanthology.org/2020.acl-main.593
33. Zhou, W., Xu, C., Ge, T., McAuley, J.J., Xu, K., Wei, F.: BERT loses patience: Fast and robust inference with early exit. In: NeurIPS (2020)

34. Leviathan, Y., Kalman, M., Matias, Y.: Fast inference from transformers via speculative decoding. arXiv preprint arXiv: 2211.17192 (2022)

35. Li, L., Lin, Y., Chen, D., Ren, S., Li, P., Zhou, J., Sun, X.: CascadeBERT: Accelerating inference of pre-trained language models via calibrated complete models cascade. In: Findings of the Association for Computational Linguistics: EMNLP 2021, pp. 475–486. Association for Computational Linguistics, Punta Cana, Dominican Republic (2021). https://doi.org/10.18653/v1/2021.findings-emnlp.43. URL https://aclanthology.org/2021.findings-emnlp.43

36. Varshney, N., Baral, C.: Model cascading: Towards jointly improving efficiency and accuracy of NLP systems. In: Proceedings of the 2022 Conference on Empirical Methods in Natural Language Processing, pp. 11007–11021. Association for Computational Linguistics, Abu Dhabi, United Arab Emirates (2022). URL https://aclanthology.org/2022.emnlp-main.756

37. Chowdhery, A., Narang, S., Devlin, J., Bosma, M., Mishra, G., Roberts, A., Barham, P., Chung, H.W., Sutton, C., Gehrmann, S., et al.: Palm: Scaling language modeling with pathways. arXiv preprint arXiv:2204.02311 (2022)

38. Google: Palm 2 (2023). URL https://ai.google/discover/palm2

39. Xu, C., Zhou, W., Ge, T., Xu, K., McAuley, J., Wei, F.: Beyond preserved accuracy: Evaluating loyalty and robustness of BERT compression. In: Proceedings of the 2021 Conference on Empirical Methods in Natural Language Processing, pp. 10653–10659. Association for Computational Linguistics, Online and Punta Cana, Dominican Republic (2021). https://doi.org/10.18653/v1/2021.emnlp-main.832. URL https://aclanthology.org/2021.emnlp-main.832

40. OpenAI: GPT-4 technical report. PREPRINT (2023)

41. Wu, Y., Rabe, M.N., Hutchins, D., Szegedy, C.: Memorizing transformers. In: International Conference on Learning Representations

42. Bulatov, A., Kuratov, Y., Burtsev, M.: Recurrent memory transformer. Advances in Neural Information Processing Systems **35**, 11079–11091 (2022)

43. Bulatov, A., Kuratov, Y., Burtsev, M.S.: Scaling transformer to 1m tokens and beyond with rmt. arXiv preprint arXiv:2304.11062 (2023)

44. Liang, X., Wang, B., Huang, H., Wu, S., Wu, P., Lu, L., Ma, Z., Li, Z.: Unleashing infinite-length input capacity for large-scale language models with self-controlled memory system. arXiv preprint arXiv: Arxiv-2304.13343 (2023)

45. Chen, S., Wong, S., Chen, L., Tian, Y.: Extending context window of large language models via positional interpolation. arXiv preprint arXiv: 2306.15595 (2023)

46. Li, A., Gong, B., Yang, B., Shan, B., Liu, C., Zhu, C., Zhang, C., Guo, C., Chen, D., Li, D., Jiao, E., Li, G., Zhang, G., Sun, H., Dong, H., Zhu, J., Zhuang, J., Song, J., Zhu, J., Han, J., Li, J., Xie, J., Xu, J., Yan, J., Zhang, K., Xiao, K., Kang, K., Han, L., Wang, L., Yu, L., Feng, L., Zheng, L., Chai, L., Xing, L., Ju, M., Chi, M., Zhang, M., Huang, P., Niu, P., Li, P., Zhao, P., Yang, Q., Xu, Q., Wang, Q., Wang, Q., Li, Q., Leng, R., Shi, S., Yu, S., Li, S., Zhu, S., Huang, T., Liang, T., Sun, W., Sun, W., Cheng, W., Li, W., Song, X., Su, X., Han, X., Zhang, X., Hou, X., Min, X., Zou, X., Shen, X., Gong, Y., Zhu, Y., Zhou, Y., Zhong, Y., Hu, Y., Fan, Y., Yu, Y., Yang, Y., Li, Y., Huang, Y., Li, Y., Huang, Y., Xu, Y., Mao, Y., Li, Z., Li, Z., Tao, Z., Ying, Z., Cong, Z., Qin, Z., Fan, Z., Yu, Z., Jiang, Z., Wu, Z.: Minimax-01: Scaling foundation models with lightning attention. arXiv preprint arXiv: 2501.08313 (2025)

47. Xiao, G., Tian, Y., Chen, B., Han, S., Lewis, M.: Efficient streaming language models with attention sinks. arXiv preprint arXiv: 2309.17453 (2023)

48. Liu, J., Zhu, D., Bai, Z., He, Y., Liao, H., Que, H., Wang, Z., Zhang, C., Zhang, G., Zhang, J., Zhang, Y., Chen, Z., Guo, H., Li, S., Liu, Z., Shan, Y., Song, Y., Tian, J., Wu, W., Zhou, Z., Zhu, R., Feng, J., Gao, Y., He, S., Li, Z., Liu, T., Meng, F., Su, W., Tan, Y., Wang, Z., Yang, J., Ye, W., Zheng, B., Zhou, W., Huang, W., Li, S., Zhang, Z.: A comprehensive survey on long context language modeling. arXiv preprint arXiv: 2503.17407 (2025)

49. Bai, Y., Zhang, J., Lv, X., Zheng, L., Zhu, S., Hou, L., Dong, Y., Tang, J., Li, J.: Longwriter: Unleashing 10,000+ word generation from long context LLMs. arXiv preprint arXiv:2408.07055 (2024)

50. Zhou, W., Jiang, Y.E., Cui, P., Wang, T., Xiao, Z., Hou, Y., Cotterell, R., Sachan, M.: Recurrentgpt: Interactive generation of (arbitrarily) long text. arXiv preprint arXiv: 2305.13304 (2023)

51. Que, H., Duan, F., He, L., Mou, Y., Zhou, W., Liu, J., Rong, W., Wang, Z.M., Yang, J., Zhang, G., Peng, J., Zhang, Z., Zhang, S., Chen, K.: Hellobench: Evaluating long text generation capabilities of large language models. arXiv preprint arXiv: 2409.16191 (2024)

52. Brown, T., Mann, B., Ryder, N., Subbiah, M., Kaplan, J.D., Dhariwal, P., Neelakantan, A., Shyam, P., Sastry, G., Askell, A., et al.: Language models are few-shot learners. Advances in neural information processing systems **33**, 1877–1901 (2020)

53. Touvron, H., Lavril, T., Izacard, G., Martinet, X., Lachaux, M.A., Lacroix, T., Rozière, B., Goyal, N., Hambro, E., Azhar, F., et al.: Llama: Open and efficient foundation language models. arXiv preprint arXiv:2302.13971 (2023)

54. Biderman, S., Schoelkopf, H., Anthony, Q., Bradley, H., O'Brien, K., Hallahan, E., Khan, M.A., Purohit, S., Prashanth, U.S., Raff, E., et al.: Pythia: A suite for analyzing large language models across training and scaling. arXiv preprint arXiv:2304.01373 (2023)

55. Du, Z., Qian, Y., Liu, X., Ding, M., Qiu, J., Yang, Z., Tang, J.: Glm: General language model pretraining with autoregressive blank infilling. In: Proceedings of the 60th Annual Meeting of the Association for Computational Linguistics (Volume 1: Long Papers), pp. 320–335 (2022)

56. Niu, W., Kong, Z., Yuan, G., Jiang, W., Guan, J., Ding, C., Zhao, P., Liu, S., Ren, B., Wang, Y.: Real-time execution of large-scale language models on mobile. arXiv preprint arXiv: Arxiv-2009.06823 (2020)

57. Zheng, Y., Chen, Y., Qian, B., Shi, X., Shu, Y., Chen, J.: A review on edge large language models: Design, execution, and applications. arXiv preprint arXiv: 2410.11845 (2024)

58. Bills, S., Cammarata, N., Mossing, D., Tillman, H., Gao, L., Goh, G., Sutskever, I., Leike, J., Wu, J., Saunders, W.: Language models can explain neurons in language models. https://openaipublic. blob.core.windows.net/neuron-explainer/paper/index.html (2023)

59. Radford, A., Wu, J., Child, R., Luan, D., Amodei, D., Sutskever, I.: Language models are unsupervised multitask learners (2019)

60. Meng, K., Sen Sharma, A., Andonian, A., Belinkov, Y., Bau, D.: Mass editing memory in a transformer. arXiv preprint arXiv:2210.07229 (2022)

61. Kaplan, J., McCandlish, S., Henighan, T., Brown, T.B., Chess, B., Child, R., Gray, S., Radford, A., Wu, J., Amodei, D.: Scaling laws for neural language models. arXiv preprint arXiv:2001.08361 (2020)

62. Tay, Y., Dehghani, M., Abnar, S., Chung, H.W., Fedus, W., Rao, J., Narang, S., Tran, V.Q., Yogatama, D., Metzler, D.: Scaling laws vs model architectures: How does inductive bias influence scaling? arXiv preprint arXiv: Arxiv-2207.10551 (2022)

63. Hu, S., Tu, Y., Han, X., He, C., Cui, G., Long, X., Zheng, Z., Fang, Y., Huang, Y., Zhao, W., Zhang, X., Thai, Z.L., Zhang, K., Wang, C., Yao, Y., Zhao, C., Zhou, J., Cai, J., Zhai, Z., Ding, N., Jia, C., Zeng, G., Li, D., Liu, Z., Sun, M.: Minicpm: Unveiling the potential of small language models with scalable training strategies. arXiv preprint arXiv: 2404.06395 (2024)

64. Wei, J., Tay, Y., Bommasani, R., Raffel, C., Zoph, B., Borgeaud, S., Yogatama, D., Bosma, M., Zhou, D., Metzler, D., Chi, E.H., Hashimoto, T., Vinyals, O., Liang, P., Dean, J., Fedus, W.: Emergent abilities of large language models. arXiv preprint arXiv: Arxiv-2206.07682 (2022)

65. Huang, Y., Hu, S., Han, X., Liu, Z., Sun, M.: Unified view of grokking, double descent and emergent abilities: A perspective from circuits competition. arXiv preprint arXiv: 2402.15175 (2024)

66. Wang, B., Yue, X., Su, Y., Sun, H.: Grokked transformers are implicit reasoners: A mechanistic journey to the edge of generalization. arXiv preprint arXiv: 2405.15071 (2024)

67. Nakkiran, P., Kaplun, G., Bansal, Y., Yang, T., Barak, B., Sutskever, I.: Deep double descent: Where bigger models and more data hurt. Journal of Statistical Mechanics: Theory and Experiment **2021**(12), 124003 (2021)
68. Gu, X., Pang, T., Du, C., Liu, Q., Zhang, F., Du, C., Wang, Y., Lin, M.: When attention sink emerges in language models: An empirical view. arXiv preprint arXiv: 2410.10781 (2024)
69. Kosinski, M.: Theory of mind may have spontaneously emerged in large language models. arXiv preprint arXiv: Arxiv-2302.02083 (2023)
70. Ferrando, J., Sarti, G., Bisazza, A., Costa-jussà, M.R.: A primer on the inner workings of transformer-based language models. arXiv preprint arXiv: 2405.00208 (2024)
71. LeCun, Y.: A path towards autonomous machine intelligence (2022). URL https://openreview. net/pdf?id=BZ5a1r-kVsf
72. Li, Y., Lin, C., Guerin, F.: CM-gen: A neural framework for Chinese metaphor generation with explicit context modelling. In: Proceedings of the 29th International Conference on Computational Linguistics, pp. 6468–6479. International Committee on Computational Linguistics, Gyeongju, Republic of Korea (2022). URL https://aclanthology.org/2022.coling-1.563
73. Li, Y., Wang, S., Lin, C., Guerin, F., Barrault, L.: FrameBERT: Conceptual metaphor detection with frame embedding learning. In: Proceedings of the 17th Conference of the European Chapter of the Association for Computational Linguistics, pp. 1558–1563. Association for Computational Linguistics, Dubrovnik, Croatia (2023). URL https://aclanthology.org/2023.eacl-main.114
74. Tang, C., Lin, C., Huang, H., Guerin, F., Zhang, Z.: EtriCA: Event-triggered context-aware story generation augmented by cross attention. In: Findings of the Association for Computational Linguistics: EMNLP 2022, pp. 5504–5518. Association for Computational Linguistics, Abu Dhabi, United Arab Emirates (2022). URL https://aclanthology.org/2022.findings-emnlp.403
75. Tang, C., Zhang, Z., Loakman, T., Lin, C., Guerin, F.: NGEP: A graph-based event planning framework for story generation. In: Proceedings of the 2nd Conference of the Asia-Pacific Chapter of the Association for Computational Linguistics and the 12th International Joint Conference on Natural Language Processing (Volume 2: Short Papers), pp. 186–193. Association for Computational Linguistics, Online only (2022). URL https://aclanthology.org/2022.aacl-short. 24
76. Bengio, Y.: The consciousness prior. ARXIV.ORG (2017)
77. Lee, M., Srivastava, M., Hardy, A., Thickstun, J., Durmus, E., Paranjape, A., Gerard-Ursin, I., Li, X.L., Ladhak, F., Rong, F., Wang, R.E., Kwon, M., Park, J.S., Cao, H., Lee, T., Bommasani, R., Bernstein, M.S., Liang, P.: Evaluating human-language model interaction. CoRR **abs/2212.09746** (2022). https://doi.org/10.48550/arXiv.2212.09746. URL https://doi.org/10. 48550/arXiv.2212.09746
78. Qian, R., Dong, X., Zhang, P., Zang, Y., Ding, S., Lin, D., Wang, J.: Streaming long video understanding with large language models. Advances in Neural Information Processing Systems **37**, 119336–119360 (2024)
79. Zeng, A., Du, Z., Liu, M., Wang, K., Jiang, S., Zhao, L., Dong, Y., Tang, J.: Glm-4-voice: Towards intelligent and human-like end-to-end spoken chatbot. arXiv preprint arXiv: 2412.02612 (2024)
80. OpenAI: hello-gpt-4o (2024). URL https://openai.com/index/hello-gpt-4o/
81. Yu, G.I., Jeong, J.S., Kim, G.W., Kim, S., Chun, B.G.: Orca: A distributed serving system for Transformer-Based generative models. In: 16th USENIX Symposium on Operating Systems Design and Implementation (OSDI 22), pp. 521–538. USENIX Association, Carlsbad, CA (2022). URL https://www.usenix.org/conference/osdi22/presentation/yu

Conclusion

14

Zekun Wang, Ge Zhang, Chenghua Lin and Jie Fu

In this book, we have offered a comprehensive exploration of Interactive Natural Language Processing, a burgeoning paradigm that situates language models as interactive agents within a diverse array of contexts. We have proposed a unified definition and framework for Interactive NLP (iNLP), followed by a systematic classification that deconstructs its integral components such as interactive objects, interfaces, and methods. Furthermore, we have elucidated the varied evaluation methodologies used in the field, showcased its numerous applications, discussed its ethical and safety issues, and pondered upon future research directions. By putting a spotlight on iNLP's ability to interact with humans, knowledge bases, models, tools, and environments, we have underscored the paradigm's potential for enhancing alignment, personalizing responses, enriching representations, avoiding hallucinations, decomposing complex tasks, and grounding language in reality, etc. Ultimately, this survey presents a wide-angle view of the current state and future potential of iNLP, serving as an essential reference point for researchers, eager to dive into this rapidly evolving field.

Z. Wang
Beihang University, Haidian District, Beijing, China
e-mail: zenmoore@buaa.edu.cn

G. Zhang
University of Michigan, Haidian District, Beijing, China
e-mail: gezhang@umich.edu

C. Lin
Department of Computer Science, University of Manchester, Manchester, UK
e-mail: chenghua.lin@manchester.ac.uk

J. Fu (✉)
Hong Kong University of Science and Technology, Hong Kong, China
e-mail: jiefu@ust.hk

© The Author(s), under exclusive license to Springer Nature Switzerland AG 2026   227
Z. Wang et al. (eds.), *Interactive Natural Language Processing*, Synthesis Lectures on Human Language Technologies, https://doi.org/10.1007/978-3-032-06264-2_14

Zeitfracht Medien GmbH
Ferdinand-Jühlke-Straße 7
99095 Erfurt, Deutschland
produktsicherheit@kolibri360.de